Contemporary Debates in Philosophy of Science

Contemporary Debates in Philosophy

In teaching and research, philosophy makes progress through argumentation and debate. *Contemporary Debates in Philosophy* presents a forum for students and their teachers to follow and participate in the debates that animate philosophy today in the Western world. Each volume presents pairs of opposing viewpoints on contested themes and topics in the central subfields of philosophy. Each volume is edited and introduced by an expert in the field, and also includes an index, bibliography, and suggestions for further reading. The opposing essays, commissioned especially for the volumes in the series, are thorough but accessible presentations of opposing points of view.

1. Contemporary Debates in Philosophy of Religion *edited by Michael L. Peterson and Raymond J. VanArragon*
2. Contemporary Debates in Philosophy of Science *edited by Christopher Hitchcock*

Forthcoming *Contemporary Debates* are in:

Aesthetics *edited by Matthew Kieran*
Applied Ethics *edited by Andrew Cohen and Christopher Heath Wellman*
Cognitive Science *edited by Robert Stainton*
Epistemology *edited by Matthias Steup and David Sosa*
Metaphysics *edited by Ted Sider, Dean Zimmerman, and John Hawthorne*
Moral Theory *edited by James Dreier*
Philosophy of Mind *edited by Brian McLaughlin and Jonathan Cohen*
Social Philosophy *edited by Laurence Thomas*

CONTEMPORARY DEBATES IN PHILOSOPHY OF SCIENCE

Edited by

Christopher Hitchcock

350 Main Street, Malden, MA 02148-5020, USA
108 Cowley Road, Oxford OX4 1JF, UK
550 Swanston Street, Carlton, Victoria 3053, Australia

First published 2004 by Blackwell Publishing Ltd

Library of Congress Cataloging-in-Publication Data

Contemporary debates in philosophy of science / edited by Christopher Hitchcock.
     p.   cm. – (Contemporary debates in philosophy ; 2)
Includes bibliographical references and index.
   ISBN 1-4051-0151-2 (alk. paper) – ISBN 1-4051-0152-0 (pbk. : alk. paper)
1. Science–Philosophy.   I. Hitchcock, Christopher, 1964–   II. Series.

   Q175.C6917 2004
   501–dc22

                                                                        2003016800

A catalogue record for this title is available from the British Library.

Set in 10/12½ pt Rotis serif
by SNP Best-set Typesetter Ltd., Hong Kong
Printed and bound in the United Kingdom
by MPG Books Ltd, Bodmin, Cornwall

For further information on
Blackwell Publishing, visit our website:
http://www.blackwellpublishing.com

Contents

Contents

Notes on Contributors

All of the information below was accurate at the time of publication. Readers should beware, however, that philosophers have sometimes been known to write new books, develop new interests, and even change locations.

James Robert Brown is in the Department of Philosophy at the University of Toronto. His interests include a wide range of topics, such as: thought experiments, which resulted in the book, *The Laboratory of the Mind: Thought Experiments in the Natural Sciences* (Routledge, 1991); scientific realism, as found in *Smoke and Mirrors: How Science Reflects Reality* (Routledge, 1994); the role of visual reasoning in mathematics, which is taken up in *Philosophy of Mathematics: An Introduction to the World of Proofs and Pictures* (Routledge, 1999); and the social relations of science, as presented in *Who Rules in Science? An Opinionated Guide to the Wars* (Harvard University Press, 2001). He is currently working on another book on the philosophy of mathematics, *What's Wrong With Mathematical Naturalism?*

Craig Callender works in philosophy of physics, philosophy of time, and the philosophy and metaphysics of science. He has worked on the measurement problem in quantum mechanics, especially Bohmian mechanics, philosophical issues confronting quantum gravity, the intersection of science and philosophy of time, the philosophy of spacetime and Humean supervenience. He has published extensively in academic journals and is author of an introductory book, *Introducing Time* (Icon, 2001), and editor of *Time, Reality & Experience* (Cambridge University Press, 2002) and, with Nick Huggett, *Physics Meets Philosophy at the Planck Length* (Cambridge University Press, 2001).

Peter Carruthers is Professor and Chair of the Philosophy Department at the University of Maryland. His primary research interests for most of the past dozen years have been in the philosophy of psychology. He has worked especially on theories of consciousness and on the role of natural language in human cognition. He is the author of eight books and editor of four, including *Phenomenal Consciousness: A Naturalistic Theory* (Cambridge University Press, 2000) and *The Philosophy of Psychology* (co-authored with George Botterill; Cambridge University Press, 1999). In a previous incarnation he trained as a Wittgensteinian and published a couple of monographs on Wittgensteins's *Tractatus*.

Fiona Cowie is Associate Professor of Philosophy at California Institute of Technology. Her research interests include innate ideas, learning and development, language and concept acquisition, the evolution of cognition, and the implications for psychology of neuroscience and genetics. Her first book, *What's Within? Nativism Reconsidered* (Oxford University Press, 1999) concerned innate ideas and language acquisition. She is currently writing a book about modularity and evolution with James Woodward; their *Naturalizing Human Nature* will be forthcoming with Oxford University Press.

Phil Dowe is Senior Lecturer in Philosophy at the University of Queensland, Australia. His interests in the metaphysics of science include causation, time, identity, and chance. He is currently writing a book on the philosophy of time travel.

Clark Glymour is Alumni University Professor of Philosophy at Carnegie Mellon University and Senior Research Scientist at the Institute for Human and Machine Cognition, University of West Florida.

Peter Godfrey-Smith received his B.A. from the University of Sydney in Australia, and his Ph.D. in philosophy from the University of California, San Diego. He is presently Associate Professor of Philosophy at Stanford University, although he is about to take up new posts at the Australian National University and Harvard University. He is the author of the books *Complexity and the Function of Mind in Nature* (Cambridge University Press, 1996) and *Theory and Reality* (The University of Chicago Press, 2003), along with various articles in the philosophy of biology and philosophy of mind.

Christopher Hitchcock is Professor of Philosophy at the California Institute of Technology. When he is not busy trying to get contributors to submit their papers on time, he is actually able to pursue his research, which primarily focuses on the theory of causation and explanation. He also works on confirmation and decision theory, and claims to have solved some beguiling paradoxes in probability. You may find Hitchcock's papers scattered throughout the leading journals in philosophy and philosophy of science.

Kevin T. Kelly is Professor of Philosophy at Carnegie Mellon University. His research interests include epistemology, philosophy of science, formal learning theory, and computability. He is the author of *The Logic of Reliable Inquiry* (Oxford University Press, 1996) and of articles on such topics as the problem of induction, Ockham's razor, infinite epistemic regresses, the reliability properties of normative theories of belief revision, and the problem of testing theories with uncomputable predictions. His current work involves a general derivation of Ockham's razor from truth-finding efficiency in statistical problems.

Harold Kincaid received his Ph.D. from Indiana University and is Professor of Philosophy at the University of Alabama at Birmingham. He is the author of *Philosophical Foundations of the Social Sciences: Analyzing Controversies in Social Research* (Cambridge University Press, 1996) and *Individualism and the Unity of Science: Essays on Reduction, Explanation and the Special Sciences* (Rowman and Littlefield, 1996). His research interests include the philosophy of the social sciences, especially economics, and general issues surrounding explanation and unification in science.

André Kukla is a professor in the Departments of Psychology and Philosophy at the University of Toronto. His main areas of research are the philosophy of science and cognitive science. He is the author of *Studies in Scientific Realism* (Oxford University Press, 1998), *Social Constructivism and the Philosophy of Science* (Routledge, 2000), and *Methods of Theoretical Psychology* (The MIT Press, 2001). He is currently writing a book on ineffable knowledge.

Jarrett Leplin is Professor of Philosophy at the University of North Carolina, Greensboro. He specializes in the philosophy of science and epistemology. He is the author of *A Novel Defense of Scientific Realism* (Oxford University Press, 1997), editor of *Scientific Realism* (University of California Press, 1984) and *The Creation of Ideas in Physics* (Kluwer, 1995), and author of some 50 papers on topics in the philosophy of science, such as scientific methods, scientific progress, explanation, theory change, realism, and confirmation. Further research interests include epistemic paradoxes and theories of justification.

Patrick Maher is Professor of Philosophy at the University of Illinois at Urbana-Champaign. He works on issues relating to probability and induction. His book *Betting on Theories* was published by Cambridge University Press in 1993.

John D. Norton is Professor and Chair of the Department of History and Philosophy of Science at the University of Pittsburgh. He works in history and philosophy of modern physics, especially general relativity, and his interest in Einstein, champion thought experimenter, led him to the philosophy of thought experiments. He also works in general philosophy of science and confirmation theory, and is a co-founder of http://philsci-archive.pitt.edu, the philosophy of science preprint server.

Huw Price is an Australian Research Council Federation Fellow at the University of Sydney, where he is Director of the Centre for Time and holds a Personal Chair in Natural Metaphysics in the Department of Philosophy. He is also Professor of Logic and Metaphysics at the University of Edinburgh. He is a Fellow of the Australian Academy of the Humanities, and a Past President of the Australasian Association of Philosophy. His publications include *Facts and the Function of Truth* (Blackwell, 1988),

Time's Arrow and Archimedes' Point (Oxford University Press, 1996), and a range of articles in journals such as *The Journal of Philosophy, Mind, The British Journal for the Philosophy of Science,* and *Nature.*

John T. Roberts received his B.S. in physics from the Georgia Institute of Technology in 1992 and his Ph.D. in philosophy from the University of Pittsburgh in 1999. He is currently Assistant Professor of Philosophy at the University of North Carolina at Chapel Hill. His research focuses on laws of nature and on general issues in the philosophy of physics. His publications include "Lewis, Carroll, and seeing through the looking glass," *Australasian Journal of Philosophy,* 76 (1998), and "Undermining undermined: Why Humean supervenience never needed to be debugged (even if it's a necessary truth)," *Philosophy of Science,* 68 (supplement) (2001). He goes to Cajun and Contra dances whenever his cat Tina allows him to do so.

Sahotra Sarkar is Professor of Integrative Biology and Philosophy at the University of Texas at Austin. His main research interests are in theoretical biology and the history and philosophy of science. Before coming to Texas he taught at McGill University and was a Fellow of the Institute for Advanced Study in Berlin. He is the author of *Genetics and Reductionism* (Cambridge University Press, 1998) and editor of *The Philosophy and History of Molecular Biology: New Perspectives* (Kluwer, 1996).

Jonathan Schaffer teaches philosophy at University of Massachusetts – Amherst. He wrote his dissertation on causation, and swore never to work on the topic again. Since then, he has written mainly on ontology and epistemology. But now he has lapsed, and started to work on causation again. Pitiful, really. Anyway, here are some of Schaffer's other papers: "Causes as probability-raisers of processes," *Journal of Philosophy* (2001); "Is there a fundamental level?" *Noûs* (forthcoming); and "Skepticism, contextualism, and discrimination," *Philosophy and Phenomenological Research* (forthcoming).

Joel Walmsley received his B.A. in psychology and philosophy from St. Anne's college, Oxford, and is currently a Ph.D. candidate in the department of philosophy at the University of Toronto, where he is a Junior Fellow of Massey College. His current research interests include the dynamical hypothesis in cognitive science, emergence, and the history of philosophical psychology.

James Woodward tries to teach philosophy at the California Institute of Technology. He is the author of *Making Things Happen: A Theory of Causal Explanation.* (Oxford University Press, 2003). He and Fiona Cowie are at work on a book on the evolution of cognition, tentatively titled *Naturalizing Human Nature,* also forthcoming from Oxford University Press.

Preface

Teachers: Has this ever happened to you? You are teaching a philosophy class that covers a number of different topics. You want to spend a week or so on, let's say, scientific realism. Naturally, you want to provide the students with some accessible readings on the subject. Moreover, you want to choose readings from both realists and anti-realists, partly so that the students will see both sides of the issue, and partly to give the students a glimpse of philosophers engaging in debate with one another. What you end up with, however, are readings that end up talking past one another: no two authors agree on what scientific realism is, so the realists are defending views that the anti-realists are not attacking. The students come away confused, and without any sense of the constructive value of debate.

Students: Has this ever happened to you? You are taking a philosophy class that covers a number of different topics. One of the topics covered is, let's say, scientific realism. There are assigned readings on the topic: they are challenging, but with some effort on your part and some help from your instructor, you get a pretty good sense of what each author is saying. The problem is that these authors seem to be talking about different things. Your professor says that they are debating with one another, but you just don't see it in the texts.

Researchers: Has this ever happened to you? You want to get a sense of what is currently going on in some area of philosophy – let's say, the philosophy of science. So, naturally enough, you look for a good anthology of papers on the subject. The anthology you pick up is divided into a number of sections, each devoted to a specific topic – scientific realism, for example. While each article makes some interesting points, you can't quite see how they fit together. No two articles seem to be addressing quite the same question, so it's just not clear what people are arguing about.

Whichever category you fall under, I bet this has happened to you. If so, the Blackwell *Contemporary Debates in Philosophy* series is just the thing for you. This volume is one of the first in this new series; others will have a similar format. The contributors to this volume address eight specific yes–no questions in the philosophy of science. For each question, there are two chapters: one arguing that the answer is "yes," the other that the answer is "no." These are original contributions, each one specifically written to defend an answer to a specific question. There is no room (or at any rate, as little room as there can possibly be) for misunderstanding or evading the crucial issue.

A number of different and conflicting desiderata went into the selection of the questions and the authors. There were, however, two nonnegotiable criteria. (Well, three if you count willingness to contribute!) The first was that each author had to be a recognized expert in the field about which she or he was writing. Some of the authors have long and distinguished careers; others are considered to be up-and-coming young stars; all are unimpeachable authorities. The second nonnegotiable criterion was that the paired authors agree on the exact wording of the question about which they were to disagree. Beyond this, I balanced a number different criteria: the authors should not merely be experts, but also clear and engaging writers; the authors should be ones who have brought new and interesting ideas to the problem areas; the topics should not only be philosophically significant, but of intrinsic interest to the nonspecialist reader; the topics, collectively, should provide broad coverage of the philosophy of science.

This last desideratum proved particularly difficult, since the volume was to contain essays on only eight topics, and each topic had to be constrained to a single question. In many cases, I was able to identify questions that could do double duty in order to provide broader coverage. The contributions of John Roberts and Harold Kincaid (chapters 7 and 8), for example, cover issues about the nature of laws, as well as important issues in the philosophy of the social sciences; the contributions by Peter Carruthers (chapter 15) and James Woodward and Fiona Cowie (chapter 16) cover topics in both the philosophy of biology and the philosophy of psychology; and so on. I made a conscious decision to focus the topics in core theoretical issues in the philosophy of science: confirmation, causation, the philosophies of specific sciences, and so on. There are many issues concerning the broader role of science in our society – the ethics of scientific research, the social value of science, the role of social forces within science, the status of women and minorities in science, and so on – that are fully deserving of the name "philosophy of science," yet receive no attention in this anthology. This decision reflects the fact that these issues are already highly visible to nonexperts, as well as the fact that there already exist many fine anthologies on these subjects. Indeed, at the time of this writing, Blackwell is slated to publish *Contemporary Debates in Applied Ethics*, which will include a couple of paired essays centered on questions that fall broadly within the philosophy of science. By the time you read this, there will no doubt be others under way. Teachers who are interested in using this volume as the basis of a course in the philosophy of science are encouraged to check other volumes in this series for pairs of essays that may be used to supplement those in this volume. The introduction to this volume locates the 16 contributions within the broader context of the philosophy of science. It also provides a number of suggestions for readers who want to pursue topics that are not covered by the chapters in this volume.

For each question, I presented the two responses in the order in which it made more sense to read them. Thus, the first chapter of each part typically spends more time laying out the basic issues, while the second chapter typically spends more time criticizing the sorts of positions and arguments described in the first. In each case, however, both chapters do some of each of these things. In most cases, it turned out that it made sense to put the affirmative answer first. The exception is the pair of essays on laws in the social sciences (chapters 7 and 8). John Roberts' chapter pressing the case for "no" presents a more traditional account of laws, and Harold Kincaid's companion piece spends some time responding to the sorts of objections raised by Roberts (as well as others). Readers are, of course, free to read the chapters in any order they choose.

A great many people helped with the preparation of this book. First and foremost, this volume would not exist without the work of the contributors – this is really their book and not mine. Alan Hájek and James Woodward offered a number of helpful suggestions for the introduction. A great many philosophical colleagues offered advice and suggestions for topics and authors in the early stages of this project. At the risk of leaving someone out, they are Frank Arntzenius, Gordon Belot, Jeremy Butterfield, the late James Cushing, Arthur Fine, Richard Grandy, Alan Hájek, Carl Hoefer, Nicholas Huggett, Paul Humphreys, Philip Kitcher, Marc Lange, Helen Longino, David Malament, Laura Ruetsche, Merrilee Salmon, Brian Skyrms, Elliott Sober, Paul Teller, Bas van Fraassen, and an anonymous reviewer for Blackwell.

Jeff Dean, Geoff Palmer, and Nirit Simon at Blackwell Publishing offered guidance throughout the process, and Barbara Estrada put in many long hours preparing the contributions. Finally, the love, patience, and indulgence of Ann Lindline enabled me to maintain my sanity even as deadlines loomed.

To all of the above: *thank you*!

Christopher Hitchcock
California Institute of Technology

Introduction: What is the Philosophy of Science?

Christopher Hitchcock

What is the philosophy of science? It is the application of philosophical methods to philosophical problems as they arise in the context of the sciences. That's not a particularly helpful answer as it stands, but at least it allows us to break our original question into parts: What are the methods of philosophy? What are philosophical problems? How do these problems arise within different scientific fields?

0.1 Philosophical Methods

The first question is the most difficult. In the first half of the twentieth century, a prominent school of thought (particularly associated with the Austrian philosopher Ludwig Wittgenstein) held that the philosopher's task was to clarify the meanings of words. The great problems of philosophy, it was thought, were mere confusions resulting from a failure to understand the meanings of the words used to frame those problems. Few philosophers today would subscribe to such an extreme view; nonetheless, the clarification of meanings is still an important part of the philosopher's repertoire. Particularly important is the ability to draw distinctions between different things that a term or phrase might mean, so as to assess more accurately claims involving those terms or phrases. The chapters on genetics by Sahotra Sarkar and Peter Godfrey-Smith (chapters 13 and 14), for example, involve careful analysis of the various things that one might mean by "information."

Perhaps even more fundamentally, philosophy involves the analysis of arguments, often aided by the formal methods and conceptual resources of symbolic logic (and other areas, such as probability theory). Philosophers, when defending a position, will construct arguments in support of that position. In addition, they will examine arguments that have been proposed by opponents. For each such argument, they may ask: What is the structure of the argument? Is it logically valid? If not, would it be valid if one were to add certain specific premises? Does it employ inferential methods other than those of deductive logic? What are the premises of the argument? Are the premises true? – and so on. Moreover, philosophers will try to anticipate objections to their own arguments, and defend their arguments against these objections before they are even raised. Almost every philosophy paper employs these methods to some extent or other; the two chapters on unobservable entities, by Jarrett Leplin and by André Kukla and Joel Walmsley (chapters 5 and 6), provide particularly clear examples – both chapters examine, criticize, and propose a variety of arguments on both sides of the debate.

Nonetheless, it is almost impossible to isolate any uniquely philosophical methods. In the philosophy of science, especially, there is no clear line where the philosophy ends and the science begins. While few (but still some!) philosophers actually conduct experiments, many philosophers will freely make use of empirical findings to support their positions. Consider chapters 15 and 16, by Peter Carruthers and by James Woodward and Fiona Cowie, for example. These chapters tackle the question "Is the mind a system of modules shaped by natural selection?" This involves traditional philosophical issues, such as the relationship between the mind and the brain; it involves careful analysis of the concept "module"; but it also requires the consideration of empirical results in psychology, as well as theoretical issues in evolutionary biology. Like empirical scientists, philosophers sometimes construct mathematical models of the "phenomena" that they seek to understand. In his chapter on scientific confirmation (chapter 3), Patrick Maher uses probability theory to construct a mathematical relation that, Maher argues, captures important features of the relation between scientific theory and empirical evidence. In general, then, it appears that philosophers are willing to employ almost any tools that can shed light on philosophical problems.

0.2 Philosophical Problems

It is hard to say what makes a problem "philosophical." There are, nonetheless, certain collections of problems that, over the past two and a half millennia, have come to be seen as paradigmatically philosophical problems. Three central areas of concern are *ethics*, *epistemology*, and *metaphysics*. This is by no means an exhaustive list – a fuller list would have to include aesthetics (the study of art and beauty), logic, social and political philosophy, the philosophies of language, mind, and religion (not to mention the philosophy of science itself), and the history of philosophy. Nonetheless, the core areas of ethics, epistemology, and metaphysics intersect with all these branches of philosophy; understood broadly, these three areas cover much of the field of philosophy.

Ethics deals with issues of right and wrong – both the morality of specific types of behavior and also more fundamental issues concerning the ultimate sources of moral value. Epistemology deals with the nature of knowledge and belief: What is knowledge, and how is it distinguished from mere belief? What are the sources of knowledge? What constitutes *justified* belief? Metaphysics is the most difficult to characterize; roughly, it involves the examination of concepts that play a fundamental role in other areas of philosophy, and in other disciplines. For example, metaphysical issues fundamental to ethics involve concepts such as the freedom of the will, and the nature of personal identity.

0.2.1 Ethical issues in science

Ethical issues can arise in a number of ways within the scientific context. Most obviously, technical innovation can create new possibilities whose moral status is in need of evaluation. For example, only recently has it become possible to clone large mammals such as sheep. It may soon be technologically possible to clone human beings (at the time of this writing, there are unsubstantiated reports that this has already happened). Many people react in horror at the thought of human cloning; similar reactions met other forms of technologically aided reproduction, such as artificial insemination and *in vitro* fertilization. Just what, if anything, is wrong with creating a genetic copy of a human being? Does this outweigh the possible benefits of cloning as a form of reproductive technology, especially for individuals or couples who have no other option? Obviously, ethical theorists such as Aristotle, Kant, and Mill were not able to anticipate these sorts of issues when developing their moral theories.

Another set of issues arises in connection with the treatment of experimental subjects. Presumably, the sub-atomic particles that are forced to follow very constrictive paths only to be annihilated in a super-collider are not *harmed* in any morally relevant sense. Experiments involving human beings, or even nonhuman animals, are more problematic. For human subjects, a consensus has emerged (although surprisingly recently) that *informed consent* is essential: experimentation upon human subjects is permissible only when the subjects have voluntarily given their consent after being informed of the potential risks and benefits involved. By their very nature, however, experimental treatments are such that the potential risks and benefits are not fully known in advance. Moreover, the notion of consent is much more complex than it appears. Various forms of coercion may affect a person's decision to participate in an experiment. In medicine, there is often a power asymmetry between patient and doctor, and a patient may feel that she has to participate in order to receive the best treatment. In psychology, it is a common practice for professors to require students to participate in experiments to receive course credit. In the case of animal subjects, informed consent is, of course, impossible. The key issues here involve the moral status of animals. Mammals, at least, are quite capable of suffering physical pain as well as some forms of psychological distress. How is this suffering to be weighed against the potential benefits of experimentation for human beings?

Recently, there has been considerable concern about the status of women and minorities in the sciences. There can be little doubt that the scientific profession has discriminated against women as well as members of racial and religious minorities in a number of ways. Perhaps most obviously, there have been considerable barriers preventing women and minorities from pursuing scientific careers (taking an extreme form, for example, in the expulsion of Jews from scientific posts in Nazi Germany). Some have argued that the exclusion of such alternative voices has harmed science by narrowing its vision.

To provide just one more example of ethical issues concerning science, let us remind ourselves that scientific research costs money. The funding that is necessary to support scientific research comes from a finite pool, and a decision to fund one research project is inevitably a decision to withhold funds from other projects, both within and outside of science. How are these decisions made? How can we evaluate the financial value of pure research as balanced against health care, education, defense, and other needs? How can we evaluate the financial value of one research project against another quite different one? Should these decisions be made by scientists alone, or should lay citizens participate in these decisions? If the latter, what form would lay participation take?

While these and related ethical issues in science are not taken up by the contributions in this volume, some of them will be covered in future volumes of the Blackwell *Contemporary Debates* series.

0.2.2 Epistemological issues in science

Science is in the business of producing knowledge, so it is not particularly surprising that epistemological problems arise in the scientific context. One of the most fundamental questions concerns the ultimate *source* of knowledge. *Empiricism* holds that all our knowledge of the world derives from sense experience. If you want to know what the world is like, you have to go out and look (or listen, feel, smell, or taste). It is a little tricky to say just what is meant by knowledge of *the world*, but there is an intended contrast with, for example, knowledge of mathematics or logic. Empiricism is most closely associated with three British philosophers of the seventeenth and eighteenth centuries: John Locke, George Berkeley, and David Hume. Locke, especially, held that experience is the ultimate source of all our *ideas*. More modern versions of empiricism hold that experience alone can provide the *justification* for our beliefs about the world: we may perhaps be able to formulate hypotheses without the benefit of sensory input, but only observation can tell us which hypotheses are correct. This form of empiricism is widely espoused by philosophers today.

Empiricism is most often contrasted with rationalism. This view, associated most strongly with the seventeenth-century philosophers René Descartes, Gottfried Leibniz, and Benedict de Spinoza, holds that human reason is the ultimate source of knowledge. Descartes, in particular, held that all knowledge should be constructed on the model of mathematics, deducing conclusions rigorously from basic premises whose truth could not be doubted (such as "I think, therefore I am"). Another alternative to empiricism can be traced to the work of the ancient Greek philosopher Plato (and hence is referred to as "Platonism"). Plato believed that an appropriately trained philosopher could acquire the ability to "see" past the appearances into the reality that lay behind those appearances. The word "see" here is metaphorical, and refers to a special kind of insight, rather than literal vision. This view has recently been revived in an interesting way by the philosopher James Robert Brown. Brown has argued that *thought experiments* can provide us with new knowledge of the world, even though by definition those "experiments" do not involve any new observations. Rather, thought experiments provide us with a kind of direct insight into the nature of things. Brown presents his views in chapter 1. In rebuttal, John Norton argues that thought experiments can be understood in empiricist terms, and calls for a more thorough analysis of the epistemology of thought experiments. The subject of thought experiments, then, provides one small arena in which the larger battles between empiricism, rationalism, and Platonism may be played out.

Despite these disputes, no one would deny that observational evidence plays a prominent (although perhaps not exclusive) role in supporting and undermining scientific hypotheses. How does this happen? In formal logic, there are explicit rules that tell us whether or not a certain conclusion follows from a particular set of premises. These rules are demonstrably truth-preserving: they guarantee that a logically valid inference from true premises will always yield a true conclusion. Let us put aside worries raised by Descartes and others about the reliability of our senses, and assume that the beliefs we form on the basis of direct observation are correct. Are there rules of inference, akin to the rules of deductive logic, that would will take us from these observational premises to theoretical conclusions with no risk of error? In general, this is not possible. Any interesting scientific hypothesis has implications whose truth cannot be established by direct observation. This may be because the hypothesis has implications for what goes on at distant locations, or in the future, or at scales too small to be seen by the human eye, or for any number of other reasons. There is thus little hope that we will be able to simply deduce the truth of scientific hypotheses and theories from observations in the way that conclusions can be deduced from their premises in logic. This gloomy conclusion is supported by the history of science, which tells us that even the best-confirmed theories (such as Newtonian gravitational theory) can be undermined by further evidence. Thus, while fields such as mathematics and logic trade in certainties, scientific hypotheses always remain at least partly conjectural.

In light of this situation, some philosophers have attempted to apply the concepts of probability to scientific theories and hypotheses. While it may be impossible to establish a scientific hypothesis with certainty, a hypothesis may be rendered more or less probable in light of evidence. Evidence that increases the probability of a theory is said to support or *confirm* that theory, while evidence that lowers the probability of a theory is said to undermine or *disconfirm* it. This way of thinking about the relationship between theory and evidence was pioneered by the eighteenth-century English clergyman Thomas Bayes, and further developed by the great French physicist Pierre Simon de Laplace. The probabilistic approach became very popular in the twentieth century, being championed in different ways by the economist John Maynard Keynes, the English *wunderkind* Frank Ramsey (who died at the age of 26), the Italian statistician Bruno de Finetti, the Austrian (and later American) philosopher Rudolf Carnap, and a host of later writers. One version (or perhaps a collection of interrelated versions) of this approach now goes by the name of "Bayesianism" (after the Reverend Thomas Bayes). The Bayesian position is sketched (and criticized) by Kevin Kelly and Clark Glymour in chapter 4. Patrick Maher, in his contribution to this volume (chapter 3), provides us with a different way of understanding confirmation in probabilistic terms.

A different line of response is most prominently associated with the Austrian (and later British) philosopher Karl Popper, who was knighted for his efforts (one of the perks of being British). This approach denies that it is appropriate to talk about the confirmation of theories by evidence, at least if this to be understood in terms of epistemic justification. The process whereby scientists subject their theories to empirical test is not one in which they seek to justify belief in that theory. The scientific method, rather, is one of formulating hypotheses, subjecting them to empirical test, and winnowing out (or at least modifying) those hypotheses that don't fit the results. It is possible that this process will eventually lead us to the truth, or at least to partial truth, but at no point will the empirical data collected to that point provide reason to believe in any of those hypotheses that remain. Kevin Kelly and Clark Glymour, the authors of chapter 4, present their own account of scientific inquiry that shares Popper's skepticism about the idea that data can partially support a general conclusion.

As we noted above, one of the reasons why there is a gap between observational evidence and scientific theory is that the latter often makes claims about entities that are *unobservable*. Are scientific claims about unobservable entities *especially* problematic? Questions of this sort are central to millennia-old debates between *realism* and *anti-realism*. Such debates arise in many different areas of philosophy – we will here be concerned with *scientific* realism and anti-realism, as opposed to, say, moral realism and anti-realism – and they can take on a number of different forms. Sometimes the debates are metaphysical in nature: George Berkeley held that nothing can exist without being perceived (although God perceives many things that are not perceived by humans). Sometimes the debates are semantic: members of the Vienna Circle, a school of philosophy centered in Vienna in the 1920s and 1930s, held that statements apparently referring to unobservable entities were to be reconstrued in terms of the empirical consequences of those statements. We will here be concerned with an epistemic form of the realism/anti-realism debate. In chapter 5, Jarrett Leplin argues that observational evidence can (at least sometimes) provide us with grounds for believing in unobservable entities, and in at least some of the assertions that scientific theories make about those entities. More specifically, Leplin argues that theories are particularly worthy of our belief when they successfully make novel predictions. André Kukla and Joel Walmsley (chapter 6) challenge this argument.

0.2.3 Metaphysical issues in science

Three of the most important concepts that appear throughout the sciences are those of law, causation, and explanation. Let us begin with law.

Almost every branch of science has basic principles referred to as "laws." In physics there are Snell's law, the Boyle–Charles law, the zeroth, first, and second laws of thermodynamics, Newton's laws of motion and gravitation, and so on. In addition, there are several "equations" that are essentially of the same character: Maxwell's equations in electromagnetic theory, Schrödinger's equation in quantum mechanics, and Einstein's field equations in the general theory of relativity. In biology, we have Mendel's laws and the Hardy–Weinberg law; in economics, Gresham's law and the law of supply and demand. The list could easily go on. In general, science seeks not only to discover what particular events take place where and when, but also to reveal the basic principles according to which these events unfold.

Just what makes something a law? According to one account championed by many empiricist writers, a law is a *regularity*. That is, a law is a pattern of the form "Whenever condition A is satisfied, condition B will be satisfied as well." There may be, however, any number of regularities of this form that are not laws – all senators for California in 2002 were women, but that is hardly a scientific law. Various proposals have been offered for discriminating true laws from such "accidental generalizations" – for example, laws must be fully general, and not make specific reference to any particular individuals, places or times – but none have become widely accepted. Another problem is that many of the "laws" in science are not universal regularities. There are, for example, certain types of genes (segregation-distorters) that do not obey the genetic "law" of random segregation. Such nonuniversal "laws" are sometimes called *ceteris paribus* laws, laws that hold true other things being equal. The thought is that there exists some specifiable set of conditions, as yet unknown to us, under which the regularity never fails. If these conditions are then built in to the condition A in the formulation "Whenever condition A is satisfied, condition B will be satisfied as well," then perfect regularity will be restored. John Roberts, in his contribution to this volume (chapter 7), champions the view that laws are regularities while arguing that *ceteris paribus* laws are no laws at all. As a consequence, he holds that all of the so-called "laws" of the social sciences, such as the law of supply and demand, are not true laws.

A very different approach to understanding laws, associated most prominently with the Australian philosopher David Armstrong, takes laws to comprise relationships of "necessitation" that hold between properties, rather than individual entities. James Robert Brown briefly discusses this account in his chapter on thought experiments (chapter 1). Harold Kincaid (chapter 8) argues that at least some laws serve to pick out certain kinds of causal tendency.

The concept of causation is closely related to that of law. According to one view, once widely held, one event A causes another event B just in case B follows A as a matter of law. Obviously, this account of causation will inherit the problems described above concerning the understanding of laws. Moreover, this account of causation will not be very illuminating if laws, in turn, must be understood in terms of causation. Even putting these worries aside, a number of problems remain. Consider, for example, the explosion of the Challenger space shuttle in 1986. One of the causes of this unfortunate event was the freezing of the rubber O-ring used to prevent the leaking of fuel. Are there laws that guarantee that whenever an O-ring freezes (and various other conditions also hold) then a space shuttle will explode? We certainly have found no such laws, and yet we nonetheless believe that the freezing of the O-ring did cause the explosion. Thus we are able to provide evidence for the truth of causal claims, even when that same evidence provides no support for an underlying law. Or suppose that I lick an ice cream cone on a sunny day, after which photons bounce off the cone at a velocity of (roughly) 300,000 kilometers per second. It certainly follows from the laws of physics that anytime I lick an ice cream cone on a sunny day, photons will bounce off the cone at just that speed. However, my licking the ice cream had nothing to do with this – it would have happened regardless of whether I licked the cone, or whether I foolishly watched it melt without ever licking it. So lawful succession appears to be neither necessary nor sufficient for causation.

In response to these problems, a number of alternative approaches to causation have been developed. Both Phil Dowe (chapter 9) and Jonathan Schaffer (chapter 10) canvass some of these alternatives. Dowe himself thinks that A causes B when they are connected by a *causal process* – a certain kind of physical process that is defined in terms of conservation laws. Jonathan Schaffer, in his chapter, argues that many causes are not so connected to their effects.

The third interrelated concept is that of explanation. At the beginning of the twentieth century, the French physicist Pierre Duhem claimed that physics (and, by extension, science more generally) cannot and should not explain anything. The purpose of physics was to provide a simple and economical system for describing the facts of the physical world. Explanation, by contrast, belonged to the domain of religion, or perhaps philosophy. The scientists of an earlier era, such as Sir Isaac Newton, would not have felt the need to keep science distinct from religion and philosophy; but by 1900 or so, this was seen to be essential to genuine progress in science. This banishment of explanation from science seems to rest on a confusion, however. If we ask "Why did the space shuttle Challenger explode?", we might mean something like "Why do such horrible things happen to such brave and noble individuals?" That is certainly a question for religion or philosophy, rather than science. But we might instead mean "What were the events leading up to the explosion, and the scientific principles connecting those events with the explosion?" It seems entirely appropriate that science should attempt to answer that sort of question.

Many approaches to understanding scientific explanation parallel approaches to causation. The German–American philosopher Carl Hempel, who has done more than anyone to bring the concept of explanation onto center stage in the philosophy of science, held that to explain why some event occurred is to show that it *had* to occur, in light of earlier events and the laws of nature. This is closely related to the "lawful succession" account of causation described above, and it inherits many of the same problems. Wesley Salmon, an American philosopher whose career spanned the second half of the twentieth century, was a leading critic of Hempel's approach, and argued for a more explicit account of explanation in terms of causation, to be understood in terms of causal processes. Salmon's view of explanation is thus closely related to (and indeed an ancestor of) Dowe's account of causation. (Hence it is potentially vulnerable to the sorts of objection raised in Schaffer's chapter 10.) A third approach, developed in greatest detail by Philip Kitcher, identifies explanation with *unification*. For example, Newton's gravitational theory can be applied to such diverse phenomena as planetary orbits, the tides, falling bodies on earth, pendula, and so on. In so doing, it shows that these seemingly disparate phenomena are really just aspects of the same phenomenon: gravitation. It is the ability of gravitational theory to unify phenomena in this way that makes it explanatory. While none of the chapters in this volume deals specifically with the problem of analyzing the concept of explanation, the subject of scientific explanation is discussed in a number of them, especially chapters 5, 6, 7, 8, 10, and 11.

0.3 The Sciences

In addition to the problems described above, which arise within science quite generally, there are a number of problems that arise within the context of specific scientific disciplines.

0.3.1 Mathematics

It isn't entirely clear whether mathematics should be regarded as a science at all. On the one hand, mathematics is certainly not an *empirical science*: Mathematicians do not conduct experiments and mathematical knowledge is not gained through observation of the natural world. On the other hand, mathematics is undoubtedly the most precise and rigorous of all disciplines. Moreover, in some areas of science, such as theoretical physics, it is often hard to tell where the mathematics ends and the science begins. A mathematician specializing in differential geometry and a theoretical physicist studying gravitation may well be working on the same sorts of problems (albeit with different notational conventions). Scientists in many disciplines solve equations and prove theorems, often at a very abstract level.

The most fundamental questions in the philosophy of mathematics ask what the *subject matter* of mathematics is, and how we acquire knowledge about it. Let's take a very simple case: arithmetic is about *numbers*. Just what are numbers? They are not "things" in the physical world, like planets, cells, or brains. Nor do we find out about them by observing their behavior. (Of course it may aid our understanding of arithmetic to play with blocks or marbles – if you put two marbles in an empty bag, and then another three, there will be five marbles in the bag. But it would be very odd indeed to consider this an empirical test of the hypothesis that $2 + 3 = 5$.) Of course, the standard method for acquiring knowledge in mathematics is proof: *theorems* are deduced from mathematical *axioms*. What a proof shows then is that the theorem is true *if* the axioms are true. But how do we know whether the axioms are true? We cannot derive them from further axioms, on pain of infinite regress. We cannot assess their truth by observation. One approach to this problem is to claim that mathematical axioms are not true or false in any absolute sense, but only serve to define certain kinds of mathematical system. For example, on this view Euclid's postulates are not assertions about any physical things but, rather, serve to define the abstract notion of a Euclidean geometry. Theorems that are derived from these axioms can only be said to be true in Euclidean geometry; in non-Euclidean geometries, these theorems may well turn out to be false. A mathematical system may be used to model a particular physical system, and it is an empirical matter whether or not the model fits, but this is an independent matter from that of whether the mathematics itself is true.

A different approach is that of mathematical Platonism, often associated with the Austrian mathematician Kurt Gödel of incompleteness theorem fame. (Like many of the great German and Austrian philosophers and mathematicians of the 1930s, he emigrated to America. He never became an American citizen, however, since he believed that there was a logical inconsistency in the American constitution.) According to Platonism, mathematical entities are in some sense *real*: there is an abstract realm in which numbers, sets, scalene triangles, and nondifferentiable functions all live. (This realm is sometimes referred to metaphorically as "Plato's heaven.") We are able to acquire knowledge of this realm by means of a kind of mathematical insight. Mathematical proof then becomes a tool for expanding our knowledge beyond the elementary basis of mathematical propositions that we can "see" to be true. Although this collection has no chapters specifically devoted to the philosophy of mathematics, James Robert Brown defends a Platonist view of mathematics in chapter 1.

0.3.2 Physics

Many philosophers of science have viewed physics as *the* science *par excellence*. It is certainly true that physics, and astronomy in particular, was the first empirical science to be rendered in a mathematically precise form. Even in the ancient world, it was possible to make very accurate predictions about the locations of the planets and stars. In the seventeenth century, Newton was able to formulate physical laws of unparalleled scope, unrivaled in any other branch of science for almost 200 years (Darwin's theory of evolution by natural selection and Mendeleev's periodic table perhaps being the only close competitors by the year 1900). It would not have been unreasonable, then, for philosophers to predict that all genuine branches of science would ultimately come to look like physics: a few simple laws of vast scope and power. Thus a full understanding of science could be gained simply by understanding the nature of physics.

In the twenty-first century, we have come to learn better. Chemistry and biology have certainly advanced to the stage of scientific maturity, and they look nothing like the model of a scientific system built upon a few simple laws. In fact, much of physics does not even look like this. Nonetheless, physics continues to pose a number or fascinating puzzles of a philosophical nature. The two most fundamental physical theories, both introduced in the early twentieth century, are *quantum mechanics* and *general relativity*. Newtonian physics provides an extremely accurate account of slow, medium-sized objects. It breaks down, however, at the level of very small (or more precisely, very *low energy*) objects such as sub-atomic particles; at the level of objects traveling at near-light velocity; and at the level of very massive objects such as stars. Quantum mechanics describes the behavior of very small objects, special relativity describes the behavior of very fast objects, and general relativity (which includes special relativity) describes very massive objects. All of these theories agree almost exactly with Newtonian mechanics for slow, medium-sized objects. As yet, however, there is no known way of incorporating quantum mechanics and general relativity into one unified theory.

Within quantum mechanics, the most substantial conceptual puzzle concerns the nature of *measurement*. According to the mathematical theory of quantum mechanics, which is extraordinarily accurate in its predictions, there are two different rules describing the behavior of physical systems. The first rule, Schrödinger's equation, describes a continuous and deterministic transition of states. This rule applies to a system unless it is being measured. When a system is measured, a new rule, named after Max Born, takes effect. Born's rule describes a transition that is discontinuous and indeterministic. When a system is measured, it is said to *collapse* into a new state, and the theory provides us only with probabilities for its collapse into one state rather than another. But how does a system "know" that it is being measured? Why can't we treat the original system, together with the measurement apparatus – whatever physical system is used to perform the measurement – as one big system that obeys Schrödinger's equation? And just what is a measurement anyway? It can't just be any old physical interaction, or else any multiple-particle system would be collapsing all the time. The Nobel laureate physicist Eugene Wigner even believed that *human consciousness* is the special ingredient that brings about collapse. Others have maintained that collapse is just an illusion. In the context of quantum mechanics, then, the concept of *measurement* is a particularly perplexing one.

General relativity raises a host of interesting questions about the nature of space and time. Between 1714 and 1716, Samuel Clarke, a close follower of Sir Isaac Newton, participated in a detailed correspondence with Gottfried Leibniz. (It is believed that Newton himself may have had a hand in drafting Clarke's letters; Clarke's strategy of writing a final reply after Leibniz's death in 1716 certainly smacked of Newton's vindictiveness.) They debated many different issues, including the nature of space and time. According to Newton's theory, *acceleration* has particular sorts of causes and effects. This seems to imply that there is a distinction between true accelerations and merely apparent ones. If I jump out of an airplane (with a parachute I hope!), I will accelerate toward the ground at a little under ten meters per second per second. But from my perspective, it may well seem that it is the ground that is accelerating up to me at that rate! In fact, however, only one of us (me) is being subject to a force sufficient to produce that acceleration. Newton (and hence Clarke) thought that this required the existence of an *absolute* space: one's true motion was one's change in location in absolute space, regardless of what other objects may be doing. Leibniz, by contrast, held that the only true motions were the motions of objects relative to one another. Absolute space was nothing more than a mathematical abstraction used to model the various relative motions. Einstein's special and general theories of relativity add new dimensions to this old debate. On the one hand, general relativity shows that one can formulate the laws of physics relative to *any* frame of reference: it does not matter which objects we think of as moving, and which we think of as being at rest. This would seem to undermine Newton's central reason for believing in an absolute space and time. On the other hand, in the framework of general relativity, matter (or more specifically, energy) interacts with spacetime. (Since, in relativity theory, space and time are intimately bound together, we refer to "spacetime" rather than to space and time separately.) The distribution of mass-energy affects the structure of spacetime, and the structure of spacetime determines how objects will move relative to one another. So in this framework, space and time seem able to causally interact with matter, which certainly suggests that they bear some kind of physical reality.

General relativity also introduces some interesting new physical possibilities, such as the collapse of massive stars into black holes. Perhaps most intriguingly, general relativity seems to be consistent with the existence of "closed causal curves," which would seem to admit the possibility of some kind of time travel. Such a possibility obviously presents serious challenges to our normal understanding of the nature of time. One important spin-off of the general theory of relativity is contemporary cosmology, including the well-confirmed "big bang" hypothesis. Of course, any theory that deals with issues such as the origins and eventual fate of the universe brings in its wake a host of philosophical questions.

One further area of physics that raises interesting philosophical problems is thermodynamics, developed in the first half of the nineteenth century by Clausius, Carnot, Kelvin, and others. This work was given new physical underpinnings by work in statistical mechanics in the second half of the nineteenth century, especially by Maxwell and Boltzmann. Of the three basic laws of thermodynamics, the second is by far the most philosophically interesting. It says that the entropy of a physical system can increase, but never decrease. Entropy, very informally, is the amount of "disorder" in a system – a more thorough explanation is given in chapter 11 by Huw Price. For example, if milk is poured into coffee, it will very quickly mix with the coffee until the milk is uniformly spread throughout the coffee. Once mixed, however, the milk will never spontaneously separate and jump back out into the pitcher. There are many asymmetries in time: time seems to move toward the future; we remember the past, but not the future; we seem to have some control over the future, but not the past; we would prefer to have our unpleasant experiences in the past, and our pleasant ones in the future; and so on. The second law of thermodynamics presents the promise of an explanation of these phenomena, or at the very least a physical underpinning for the idea that there is a fundamental difference between the past and the future. Unfortunately, the later work of Maxwell and Boltzman raised a number of problems for the second law. First, they showed that it is not, in fact, impossible for a system to decrease in entropy; it is only very unlikely (in some sense of unlikely – see Craig Callender's chapter 12) that this will happen. More fundamentally, the behavior of thermodynamic systems is ultimately determined by Newton's laws, which are completely symmetric with respect to the future and the past. There remains a mystery, then, as to how the asymmetry described by the second law can come about.

The two chapters by Price and Callender, chapters 11 and 12, deal with issues at the intersection of cosmology and thermodynamics. Price argues that the recent discovery that the universe was very "smooth" shortly after the big bang helps to explain why entropy increases in time. Price and Callender disagree, however, about whether the initial smoothness itself is something that is in need of explanation.

0.3.3 Biology

Most philosophers of biology have focused their attention on the theory of evolution by natural selection. According to this theory, there is variation within any species: individuals within that species do not all have the same characteristics. Some characteristics will give their bearers advantages in their competition with conspecifics for food, mates, and other resources. Individuals with advantageous characteristics, will, on average, produce more offspring than their rivals. Many of these characteristics can be inherited from one generation to the next. Those characteristics that have all three traits – variability, reproductive advantage, and heritability – will become more plentiful in subsequent generations. The gradual accumulation of such changes over time gives rise to the evolution of diverse life forms, many with highly complex adaptations to their specific environments.

One problem, the "units of selection" problem, concerns the "level" at which these processes occur. Strictly speaking, it is *genes* rather than *characteristics* that are passed on from one generation to another. Perhaps, then, we should also say that it is the genes, rather than individual organisms, that are competing with one another for the opportunity to reproduce themselves. In this picture, championed by the British biologist Richard Dawkins among others, individual organisms are just "vehicles" built by genes in order to aid those genes in reproducing themselves. Is this an appropriate way to describe what is going on? Just what is at stake in saying that it is the genes, rather than the individual organisms, that are being acted upon by natural selection?

Another issue concerns the extent to which evolution by natural selection can ground teleological claims, claims about the "purpose" or "function" of phenotypic traits (e.g., body morphology or behavioral predispositions). In the case of human artifacts, the function of an object is determined by the intentions of the designer. The function of a screwdriver is to turn screws, for that is what the screwdriver was designed for. I may use a screwdriver to jimmy open a door, or threaten a neighbor's dog, but these are not the *functions* of a screwdriver; they're not what the screwdriver is *for*. If (as so-called "intelligent design theorists" believe) organisms were created by an intelligent agent, then it would make perfect sense to talk about the purpose or function of a bird's feathers, a flower's petals, a snake's rattle, and so on: this would just be the use for which the designer intended the characteristic in question. If, however, an organism has evolved naturally, can we sensibly talk about the function of its various characteristics? Some think that we can. The function of a trait is that activity of the trait for which it was naturally selected. The heart is often used as an illustration. The heart does many things: it circulates the blood throughout the body, and it also makes rhythmic sounds. The second effect is beneficial, at least in humans – it allows various heart conditions to be easily diagnosed – but it is because of the first effect, and not the second, that organisms with hearts were able to reproduce themselves successfully in the past. Thus the circulation of blood, rather than the making of rhythmic noises, is the function of the heart.

A related issue concerns the viability of *adaptationism* as a research strategy in evolutionary biology. This strategy is an inference from the observation that a trait is capable of serving some purpose useful to the organism, to the historical claim that the trait has been naturally selected *because* it served that purpose. This strategy obviously has its limitations: Dr. Pangloss, a character in Voltaire's novel *Candide*, claimed that the purpose of the human nose was to hold up spectacles. The biologists Stephen Jay Gould and Richard Lewontin have argued that many of an organism's physical traits may arise as byproducts of developmental constraints on its basic body plan. The human chin is a standard example: there is just no way to get a larynx, an esophagus, a tongue capable of speech, and a jaw strong enough for chewing without getting a chin thrown in for free. (Or at any rate, this is impossible without a major overhaul in human architecture, an overhaul that just wouldn't be worth the effort from an evolutionary point of view). James Woodward and Fiona Cowie, in chapter 16, criticize the field of *evolutionary psychology* partly on the grounds that it is committed to an implausible form of adaptationism.

Recently, there has been increasing philosophical interest in other areas of biology, such as genetics and molecular biology. One issue concerns the relationship between these two fields. Within classical genetics, as originally formulated by Gregor Mendel in the nineteenth century, the gene is the basic unit of explanation. After the discovery of the structure of DNA by Watson and Crick in 1953, it has become possible to explore the internal structure of genes. This raises the question of whether classical genetics can survive as an autonomous branch of biology, or whether it "reduces" to molecular biology in the sense that all of its basic concepts and principles can be understood in terms of the concepts and principles of molecular biology. One problem concerns the difficulty of saying just exactly what it means for one branch of science to "reduce" to another. Issues about reduction arise within many other branches of science. Philosophy of chemistry, for example, is a very new field, and one of its most fundamental questions is whether chemistry reduces to physics.

Another issue concerns the extent to which genes determine the phenotypic features of an organism. All researchers agree that both genetic and environmental factors play some role here; and one or the other may play a stronger role in different traits. Nonetheless, some have argued that genes have a distinctive, if not decisive, role to play. This is sometimes expressed in metaphors such as "genetic coding" or "genetic information." The idea is that genes encode information for phenotypic traits in much the same way that strings of dots and dashes code for words in Morse code. (In genetics, the "words" would not be made up of dots and dashes, but of nucleotides.) In chapters 13 and 14, Sahotra Sarkar and Peter Godfrey-Smith debate the utility of this metaphor.

0.3.4 Psychology

The nature of the human mind has long been a concern of philosophers. René Descartes argued in the seventeenth century that the mind and the body are distinct. In fact, they are made of entirely different substances. This created a problem for Descartes, for he also held that physical matter could only act or be acted upon by contact with other physical matter. How, then is it possible for the physical world to affect the mind, as it does when we form beliefs about the world upon the basis of observation? How is it possible for the mind to affect the physical world, as it does when we form volitions that result in the motions of our bodies? This is the infamous *mind–body* problem. There is, within philosophy, a sub-field referred to as the philosophy of mind that deals with this and other problems about the nature of the mind. When the exploration of these issues makes contact with empirical psychology, we enter the realm of philosophy of psychology and hence, philosophy of science.

Just within the last quarter century, there have been extraordinary advances in neuroscience, due, in part, to technological advances that permit various forms of brain imaging. The empirical exploration of the neural activity that takes place as human subjects engage in various mental tasks would seem to have an obvious bearing on the mind–body problem. If there is an interface between the mind and the body, it is surely in the brain. (Descartes himself believed that it took place in the pineal gland.) It is unclear, however, whether the new understanding of the brain is giving us anything more than "more body" – more knowledge of the physical world – and whether it has got us any closer to making the jump across the mind–body gap. Nonetheless, neuroscience has taught us a great deal about the mind that is of considerable philosophical interest. For example, it has taught us about the role of the emotions in "rational" decision-making, and about the ways in which our conceptions of ourselves may be disrupted. The newly emerging field of *neurophilosophy* explores these connections between philosophy and neuroscience.

Earlier in the twentieth century, a new interdisciplinary field called *cognitive science* emerged out of the disciplines of psychology, philosophy, computer science, and linguistics. This field was driven, in part, by a general admiration for advances made in computer science. A computer is a physical entity made up of a vast network of transistors, made of doped silicon and embedded in chips, all housed in a container of metal and plastic, and requiring electricity to function. In theory, it may be possible to understand specific operations performed by a computer at this nuts and bolts level. But the operations of a computer can also be understood at a more abstract level, in terms of the *programs* that the computer executes. The guiding idea behind cognitive science was that the mind is to the brain as computer software is to hardware. (The brain is sometimes described as "wetware.") It should be possible, then, to understand the operations of the mind at a more abstract computational level. This way of thinking about the human mind is closely association with *functionalism*, the view that a particular mental state (such as the belief that it will rain today) is to be identified in terms of its role within the overall program.

Even more so than in other branches of science, it is often hard to tell where the philosophy of psychology ends, and psychology proper begins. This is, in part, because within academic psychology there is a very strong emphasis on experimentation and data-collecting. Much the same is true within neuroscience. While individual psychologists test and defend hypotheses about particular mental processes, there is a dearth of higher-level theory to unify and explain all of the various empirical findings. To a large extent, philosophers have filled the gap, effectively playing the role of theoretical psychologists.

One potential unifying theory within psychology is *evolutionary psychology*. According to this view, the mind consists of a large number of specialized modules. A module performs some one specific task, and does so in relative isolation from what is going on in the rest of the brain. The visual system seems to fit this description: it deals specifically with the interpretation of information gathered by the retina, and does so largely uninfluenced by data from, for example, other sensory systems. More controversially, a host of other tasks are said to be performed by modules. Some evolutionary psychologists maintain, for example, that we are equipped with a "cheater-detection" module, to identify those people who are exploiting the norms of social interaction in order to obtain an unfair advantage. These modules evolved in order to solve particular problems faced by our ancestors in an environment quite different from our own. This evolutionary perspective is thought to provide us with a useful explanatory framework for thinking about various mental traits. For instance, arachnophobia, while a nuisance now, may have been advantageous in an environment in which spider bites posed a genuine risk to life and limb. In chapter 15, Peter Carruthers defends this picture of the human mind, while James Woodward and Fiona Cowie (chapter 16) are highly skeptical of both the methods and the conclusions of evolutionary psychology.

0.3.5 The social sciences

"Social science" is a broad term that encompasses a number of different fields, especially sociology and economics, but also including parts of political science, anthropology, and linguistics. One central question concerns whether the social sciences are genuinely scientific. By and large, the social sciences do not make the sorts of precise predictions that can be cleanly checked against the results of observation. There are two principal reasons for this. The first is the sheer complexity of the systems studied by social scientists. Particle physics requires a tremendous amount of training and mathematical background, to be sure, but when one is attempting to explain and predict the behavior of single (perhaps even indivisible) particles, it is not so surprising that it is sometimes possible to do so with great precision. Even the simplest organic compounds are substantially more complex than that, a simple living organism considerably more complex than that, and a human brain vastly more complex still; social networks and institutions consisting of large numbers of human beings ... well, we have come a long, long way from sub-atomic particles. Social scientists study complex systems whose basic units literally have minds of their own.

A second, related reason why prediction is so difficult in the social sciences is that it is so difficult to find or create social systems where all but a handful of factors can be ignored. It is not surprising that astronomy was the first successful predictive science: the objects under study are, for all intents and purposes, affected only by gravity. To be sure, the earth is subject to light pressure from the sun, it has a magnetic field and interacts with charged particles floating around in space, and so on. But these further factors have such a tiny effect on the earth's motion that they can safely be ignored. In experimental physics, it is possible to isolate a small number of significant forces by carefully shielding the experimental system from unwanted sources of influence. By contrast, even when social scientists are able to identify a few of the most important factors that affect the evolution of an economy, social institution, or cultural practice, there are simply too many potentially disrupting forces to be anticipated. Natural disasters, foreign wars, epidemics, political crises, technological advances, and even personal idiosyncrasies can and typically do derail even the best understood social processes. Moreover, in the social sciences, practical and ethical considerations typically prohibit the sort of artificial shielding that takes place in experimental physics. These two factors combine to make precise predictions (as opposed to predictions about general trends) virtually impossible in most areas of social science. (Of course, some physical systems, such as the earth's climate, are also enormously complex, and we are not very good at making predictions about them either. And some areas of the social sciences – for example, microeconomics – do at least sometimes deal with systems that are on a manageable scale.)

A related challenge to the scientific status of the social sciences claims that science aims at the discovery of *laws*, and that there can be no genuine laws of social science. In the third quarter of the twentieth century, especially, it was believed that laws were essential for both explanation and confirmation by evidence. In their contributions to this volume (chapters 7 and 8), John Roberts and Harold Kincaid address the issue of whether there can be social-scientific laws. Kincaid claims that the social sciences can have laws that have much the same character as (at least some of) the laws of physics. Roberts argues that there are no laws of social science, and hence that the social sciences are fundamentally different from physics, although he does not go so far as to deny that the social sciences are scientific.

A different sort of challenge to the status of the social sciences emerges from the claim that the social sciences are largely involved in *interpretation*. Anthropologists, in particular, are often interested in the symbolism involved in certain social rituals, the hidden meaning that lies behind the nominal purpose of the activity. This involves interpreting a culture's practices, much as a psychoanalyst might interpret a dream in order to uncover its hidden significance. Kincaid addresses this challenge in chapter 8.

The field of economics raises a different sort of worry. That field's basic principles are to a large extent *a priori*. These principles lay down rules for how an individual or firm *ought* to behave, on the assumption that they are interested in maximizing their own well-being. (In the case of firms, well-being is effectively identified with financial profit; in the case of human beings, there is at least some lip-service paid to the idea that money isn't everything.) These rules are then used to determine how such individuals and firms will behave in the marketplace, the prices at which they will be willing to buy and sell goods, the financial risks they will be willing to undertake, and so on. Predictions derived from these principles are then applied to actual economic situations, despite the fact that they are ultimately predicated on purely *a priori* assumptions about the nature of economic agents. This approach raises questions about the epistemology of economics: in particular, it seems to clash with the empiricist doctrine that our knowledge of the world ultimately stems from our experience. Recently, however, there has been an increasing interest in behavioral and experimental economics, which attempt to gather empirical evidence concerning the economic behavior of actual agents; indeed, the 2002 Nobel Memorial Prize in economics was awarded to two pioneers in this area.

In addition to these issues concerning the scientific status of the social sciences, the fields of economics and political science overlap with and influence areas of ethics and social and political philosophy. For example, ethical questions about the most just way of distributing goods within a society cannot be completely divorced from economic questions about the effects of distributing goods in different ways. It may be, for example, that distributing goods on the basis of productivity (rather than simply dividing them equally, for example) can produce a system of incentives for productivity that benefits all, or at least most, members of that society.

0.4 Conclusion

While this brief overview has only scratched the surface, I hope that it provides the reader with some sense of the sorts of issues that comprise the philosophy of science. Moreover, it should help the reader better understand how the eight topics covered by the contributions to this volume fit into the larger picture.

Further reading

All of the chapters in this volume provide references to the most important works on the specific topics covered. Here, I will offer some suggestions for those interested in reading in more detail about some of the other topics covered in this introduction.

There are a number of good anthologies containing some of the most significant contributions to the philosophy of science; among the best are Balashov and Rosenberg (2002), Curd and Cover (1998), Klemke et al. (1998), and Schick (1999).

Westview Press publishes a series of excellent introductory texts, written by leaders in their respective fields. These include *Philosophies of Science: Feminist Theories* (Duran, 1997), *Philosophy of Biology* (Sober, 1999), *Philosophy of Physics* (Sklar, 1992), and *Philosophy of Social Science* (Rosenberg, 1995). Blackwell is the leading publisher of reference works in philosophy: the two most directly relevant are *The Blackwell Guide to the Philosophy of Science* (Machamer and Silberstein, 2002) and *A Companion to the Philosophy of Science* (Newton-Smith, 2000). As with any topic, there is a cornucopia of material on philosophy of science available on the world wide web, much of it unreliable. One outstanding resource is the *Stanford Encyclopedia of Philosophy*, at http://plato.stanford.edu (Zalta, 1995–). Still expanding, this online encyclopedia already has a very strong collection of entries in the philosophy of science, all of them written by recognized experts.

Cohen and Wellman (forthcoming) is another volume in the Blackwell *Contemporary Debates* series, dealing with issues in applied ethics. This volume will contain pairs of chapters on cloning and on animal rights; the paired essays will be in much the same format as the contributions to this volume. Munson (1999) is an excellent collection of some of the most important works in bioethics, with extensive introductions and discussions by the editor. This anthology is now in its sixth edition and is updated regularly. (The fourth edition (1991) is now available in mass market paperback form.) Blackwell has several reference works in bioethics, including Kuhse and Singer (2001), and Frey and Wellman (2003). Kitcher (2001) is an engaging discussion of the place of science in a democratic society, including an evaluation various strategies for the funding of scientific research.

Cahn (2002) is an excellent anthology of the works of the great philosophers. It contains, *inter alia*, excerpts from the works of Plato, Descartes, Spinoza, Leibniz, Locke, Berkeley, and Hume, including some of their most important works in epistemology. It is currently in its sixth edition, but earlier editions will have these works as well. Salmon (1990) provides an excellent overview of work on scientific explanation in the twentieth century.

Shapiro (2000) is a good overview of central issues in the philosophy of mathematics. Albert (1992) is an engagingly written book that explains the basic structure of quantum mechanics as painlessly as possible, and provides a critical appraisal of various attempts to solve the measurement problem. Sklar (1974), almost encyclopedic in scope, provides an excellent overview of problems concerning the nature of space and time. For those with a background in general relativity, Earman (1995) provides a rigorous discussion of philosophical issues in cosmology.

Sober (1994) contains an excellent selection of important papers in the philosophy of biology, including papers on the units of selection, functions, adaptionism and the reduction of genetics to molecular biology. Sterelny and Griffiths (1999) provides a good introduction to some of the major issues in the philosophy of biology.

Cahn (2002) includes many of the classic discussions of the mind–body problem, including Descartes's *Meditations*. Churchland (1996) is a highly readable discussion of philosophical issues involving neuroscience; Bechtel et al. (2001) collects a number of important papers on the subject. Clark (2001) is a nice introduction to philosophical issues in cognitive science, while Haugeland (1997) contains a number of classic papers on the subject. Blackwell also publishes *A Companion to Cognitive Science*

(Bechtel and Graham, 1998). For the philosophy of social science, Martin and McIntyre (1994) contains some of the most important works.

Bibliography

Albert, D. 1992: *Quantum Mechanics and Experience.* Cambridge, MA: Harvard University Press.

Balashov, Y. and Rosenberg, A. (eds.) 2002: *Philosophy of Science: Contemporary Readings.* London and New York: Routledge.

Bechtel, W. and Graham, G. (eds.) 1998: *A Companion to Cognitive Science.* Oxford: Blackwell.

Bechtel, W., Mandik, P., Mundale, J., and Stufflebeam, R. S. (eds.) 2001: *Philosophy and the Neurosciences: A Reader.* Oxford: Blackwell.

Cahn, S. (ed.) 2002: *Classics of Western Philosophy,* 6th edn. Indianapolis: Hackett.

Churchland, P. 1996: *The Engine of Reason, the Seat of the Soul: A Philosophical Journey into the Brain.* Cambridge, MA: The MIT Press.

Clark, A. 2001: *Mindware: An Introduction to the Philosophy of Cognitive Science.* New York: Oxford University Press.

Cohen, A. and Wellman, C. (eds.) forthcoming: *Contemporary Debates in Applied Ethics.* Oxford: Oxford University Press.

Curd, M. and Cover, J. A. (eds.) 1998: *Philosophy of Science: The Central Issues.* London and New York: Norton.

Duran, J. 1997: *Philosophies of Science: Feminist Theories.* Boulder, CO: Westview Press.

Earman, J. 1995: *Bangs, Crunches, Whimpers, and Shrieks.* Oxford: Oxford University Press.

Frey, R. and Wellman, C. (eds.) 2003: *A Companion to Applied Ethics.* Oxford: Blackwell.

Haugeland, J. 1997: *Mind Design II: Philosophy, Psychology, and Artificial Intelligence.* Cambridge, MA: The MIT Press.

Kitcher, P. 2001: *Science, Truth, and Democracy.* Oxford: Oxford University Press.

Klemke, E. D., Hollinger, R., Rudge, D. W., and Kline, D. (eds.) 1998: *Introductory Readings in the Philosophy of Science.* New York: Prometheus.

Kuhse, H. and Singer, P. (eds.) 2001: *A Companion to Bioethics.* Oxford: Blackwell.

Machamer, P. and Silberstein, M. (eds.) 2002: *Blackwell Guide to the Philosophy of Science.* Oxford: Blackwell.

Martin, M. and McIntyre, L. (eds.) 1994: *Readings in the Philosophy of Social Science.* Cambridge, MA: The MIT Press.

Munson, R. (ed.) 1999: *Intervention and Reflection,* 6th edn. New York: Wadsworth.

Newton-Smith, W. (ed.) 2000: *A Companion to the Philosophy of Science.* Oxford: Blackwell.

Rosenberg, A. 1995: *Philosophy of Social Science.* Boulder, CO: Westview Press.

Salmon, W. 1990: *Four Decades of Scientific Explanation.* Minneapolis: University of Minnesota Press.

Shapiro, S. 2000: *Thinking about Mathematics.* Oxford: Oxford University Press.

Schick, T. (ed.) 1999: *Readings in the Philosophy of Science.* Mountain View, CA: Mayfield.

Sklar, L. 1992: *Philosophy of Physics.* Boulder, CO: Westview Press.

— 1974: *Space, Time, and Spacetime.* Los Angeles and Berkeley: University of California Press.

Sober, E. 1999: *Philosophy of Biology,* 2nd edn. Boulder, CO: Westview Press.

— (ed.) 1994: *Conceptual Issues in Evolutionary Biology,* 2nd edn. Cambridge, MA: The MIT Press.

Sterelny, K. and Griffiths, P. E. 1999: *Sex and Death: An Introduction to Philosophy of Biology.* Chicago: The University of Chicago Press.

Zalta, E. 1995–: *Stanford Encyclopedia of Philosophy.* http://plato.stanford.edu

DO THOUGHT EXPERIMENTS TRANSCEND EMPIRICISM?

In a nutshell, empiricism is the view that all of our knowledge of the world ultimately derives from sense experience. There is, arguably, no other philosophical principle of comparable scope and historic significance that is so widely held by philosophers today. James Robert Brown argues that *thought experiments* provide a counterexample to the principle of empiricism. Thought experiments, which are widely used throughout the sciences, seem to give us insight into the nature of the world; yet at least some thought experiments do not rely, even implicitly, upon knowledge gained through observation. In defending this view, Brown criticizes a position held by John Norton, to the effect that thought experiments are colorful presentations of ordinary arguments from empirical premises. In his response to Brown, Norton challenges Brown's critique of empiricism, and calls for a more thorough investigation into the epistemology of thought experiments. In addition to its implications for broader debates about empiricism, the subject of the role of thought experiments within science is of interest in its own right.

Why Thought Experiments
Transcend Empiricism

James Robert Brown

1.1 Galileo's Rationalism

A recent issue of *Physics World* lists the ten "most beautiful experiments" of all time, the result of a poll of contemporary physicists (Crease, 2002; see also Rogers, 2002). "Beauty" was left undefined, but judging from the list of winners it seems to involve a combination of simplicity in design and execution with deep and far reaching consequences. The examples include the double-slit experiment with single electrons, Millikan's oil-drop experiment to measure the charge on an electron, Newton's decomposition of sunlight through a prism, and so on.[1] They are wonderful examples.

Galileo's experiment on falling bodies is ranked number two, but there are reasons for thinking that this might not be appropriate. Galileo is supposed to have climbed the Leaning Tower of Pisa and dropped objects of different weights. His result, that all bodies fall at the same rate regardless of their weight, is taken to refute the prevailing Aristotelian view, a theory that said that heavy objects fall faster than light ones and that their rates of fall vary in proportion to their weights. There are several problems with proposing that this is one of the most beautiful experiments ever. For one thing, dropping things from a tower isn't really that innovative. As novel ideas go, it's not in the same league as the double-slit idea. Moreover, it is unlikely that Galileo actually performed such an experiment. More striking yet, when a variety of different objects are dropped, the do *not* all fall at the same rate. In particular, cannon balls will hit the ground before musket balls when released at the same time. True,

---

1 According to the poll, the top ten are: (1) Young's double-slit experiment applied to the interference of single electrons; (2) Galileo's experiment on falling bodies; (3) Millikan's oil-drop experiment; (4) Newton's decomposition of sunlight with a prism; (5) Young's light-interference experiment; (6) Cavendish's torsion-bar experiment; (7) Eratosthenes's measurement of the earth's circumference; (8) Galileo's experiments with rolling balls down inclined planes; (9) Rutherford's discovery of the nucleus; (10) Foucault's pendulum.

Figure 1.1 Galileo and the determination of rates of fall.

they do not fall in the way Aristotle claimed, but they do not fall as Galileo contended either.

At this point in a discussion of falling objects one often hears the interjection, "But in a vacuum . . . " This caveat must be quickly set aside – Galileo did not perform any experiment in a vacuum, and Aristotle had arguments against the very possibility. It is certainly true that all objects fall at the same rate in a vacuum, but this was not an experimental discovery made by Galileo.

Instead of performing a real experiment on falling bodies, Galileo did something much more clever and profound. He established his result by means of a thought experiment (*Discoursi*, p. 66f). It was, arguably, the most beautiful thought experiment ever devised. It would certainly get my vote as the "most beautiful" ever. It's brilliantly original and as simple as it is profound.

Aristotle and common sense hold that a heavy body will fall faster than a light one. (We can symbolize this as H > L). But consider figure 1.1, where a heavy cannon ball is attached to a light musket ball (H + L). It must fall faster than the cannon ball alone (H + L > H). Yet the compound object must also fall slower (H + L < H), since the lighter part will act as a drag on the heavier part. Now we have a contradiction (H + L > H & H + L < H). That's the end of Aristotle's theory. But we can go further. The right account of free fall is now perfectly obvious: they all move at the same speed (H = L = H + L).

Galileo once remarked that "Without experiment, I am sure that the effect will happen as I tell you, because it must happen that way" (*Diologo*, p. 145). What he had in mind in this surprising pronouncement is that often a thought experiment will yield the result. An actual experiment is not needed and may even be impossible. The great French historian of science, Alexandre Koyré, once remarked "Good physics is made *a priori*" (1968, p. 88), and he claimed for Galileo "the glory and the merit of having known how to dispense with [real] experiments" (1968, p. 75). This remarkable assessment of Galileo is right. Some thought experiments do transcend empirical sensory experience.

1.2 Some Examples

It's difficult to say precisely what a thought experiment is. However, it is not important. We know them when we see them, and that's enough to make talking about them possible. A few features are rather obvious. Thought experiments are carried out in the mind and they involve something akin to experience; that is, we typically "see" something happening in a thought experiment. Often, there is more than mere observation. As in a real experiment, there might be some calculating, some application of theory, and some guesswork and conjecture.

Thought experiments are often taken to be idealizations. This is sometimes true, but idealization is neither necessary nor sufficient. Thinking about how something might move along a frictionless plane needn't be a thought experiment at all, since it might involve nothing more than a calculation. And no idealizations are involved in Schrödinger's cat. Some people also want to include the claim that a genuine thought experiment *cannot* be performed in the actual world. I doubt this very much, since what seems to matter is whether we can get the result by thinking, not whether we can get it in some other way, as well. For now I'd prefer simply to leave these considerations out of the definition, allowing that maybe these features should or maybe they should not be part of the definition. These are things to be argued, debated, and, with luck, resolved at the end of inquiry, not fixed by stipulation at the outset. The best way to get a grip on what thought experiments are is to simply look at lots of examples. For the sake of illustration, I'll briefly give a few.

One of the most beautiful early examples is from Lucretius's *De Rerum Natura* (although it has an earlier history). It attempts to show that space is infinite. If there is a boundary to the universe, we can toss a spear at it. If the spear flies through, then it isn't a boundary after all. And if the spear bounces back, then there must be something beyond the supposed edge of space, a cosmic brick wall that stopped the spear, a wall that is itself in space. Either way, there is no edge of the universe; so, space is infinite.

Figure 1.2 Throwing a spear at the edge of space.

This example nicely illustrates many of the common features of thought experiments: We visualize some situation; we carry out an operation; we see what happens. Although we use empirical concepts, we often can't carry out an empirical test. It also illustrates their fallibility. In this case we've learned how to conceptualize space so

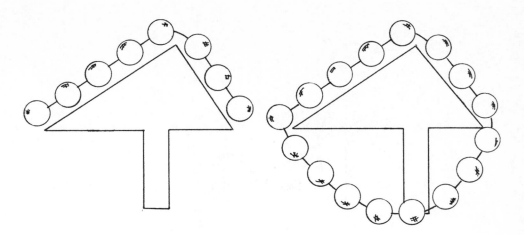

Figure 1.3 A chain draped over frictionless planes.

that it is both finite and unbounded. It also illustrates the fact that a thought experiment is a scheme that could be implemented in different ways. Here, a spear is thrown. John Norton (see chapter 2, this volume) gives a slightly different version, stemming from an older source. It's still reasonable to call them the same thought experiment.

Ernst Mach (who seems to have coined the much used expression *Gedankenexperiment*) developed an interesting empiricist view in his classic, *The Science of Mechanics*. He claims we have lots of "instinctive knowledge" picked up from experience. This needn't be articulated at all, but we become conscious of it when we consider certain situations, such as in one of his favorite examples, due to Simon Stevin. A chain is draped over a double frictionless plane (figure 1.2, left). How will it move? We could plausibly imagine it sliding one way or the other or staying fixed. But we don't know which. Add some links (as in figure 1.2, right) and the correct answer becomes obvious. It will remain fixed. So, the initial setup must have been in static equilibrium. Otherwise, we would have a perpetual motion machine. Our experience-based "instinctive knowledge," says Mach (1960, p. 34ff), tells us that this is impossible.

This beautiful example illustrates another feature of thought experiments: they often use background information, just as real experiments do – in this case, that there is no perpetual motion.

According to Maxwell's theory of electrodynamics, light is an oscillation in the electromagnetic field. The theory says that a *changing* electric field, E, gives rise to a magnetic field, M, and a *changing* magnetic field gives rise to an electric field. If I were to jiggle a charge, it would change the electric field around it, which in turn would create a magnetic field which in turn would create an electric field, and so on. This wave motion spreading through the electromagnetic field with velocity c is none other than light itself.

When he was only 16, Einstein wondered what it would be like to run so fast as to be able to catch up to the front of a beam of light. Perhaps it would be like running toward the shore from the end of a pier stretched out into the ocean, with a wave coming in alongside. There would be a hump in the water that would be stationary

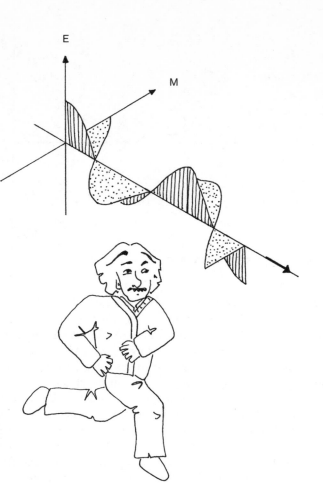

Figure 1.4 Einstein chases a light beam.

with respect to the runner. Would it be the same for a light wave? No, since change is essential; if either the electric or the magnetic field is static, it will not give rise to the other and hence there will be no electromagnetic wave. "If I pursue a beam of light with the velocity c (velocity of light in a vacuum)," said Einstein, "I should observe such a beam of light as a spatially oscillatory electromagnetic field at rest. However, there seems to be no such thing, whether on the basis of experience or according to Maxwell's equations" (1949, p. 53).

Newton's bucket is one of the most famous thought experiments ever. It's also perfectly doable as a real experiment, as I'm sure just about everyone knows from personal experience. The two-spheres example, which is also described in the famous *scholium* to definition 8 of the *Principia*, is not actually doable. So let's consider it. We image the universe to be completely empty except for two spheres connected by a cord. The spheres are of a material such that they neither attract nor repel one another. There is a tension in the cord joining them, but the spheres are not moving

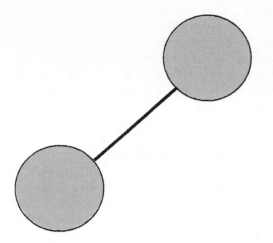

Figure 1.5 Newton's spheres in otherwise empty space.

toward one another under the force in the cord. Why not? Newton offers an explanation: they are rotating with respect to space itself; their inertial motion keeps them apart. And so, Newton concludes, absolute space must exist.

1.3 Different Uses of Thought Experiments

Even with only a few examples such as these, we can begin to draw a few conclusions. For one thing, thought experiments can mislead us, not unlike real experiments. The Lucretius example leads us to infinite space, but we have since learned how to conceive of an unbounded but still finite space. (A circle, for example, is a finite, unbounded one-dimensional space.) Some of these thought experiments couldn't be actually performed. Einstein can't run that fast, so catching up to a light beam is something done in thought alone. Needless to say, Newton's two-spheres thought experiment is impossible to perform, since we can't get rid of the entire universe to try it out. On the other hand, the thought experiments of Galileo and Stevin come close to being doable – not exactly, perhaps, but we can approximate them rather well. This, I think, shows that being able or not being able to perform a thought experiment is rather unimportant to understanding them. The so-called counterfactual nature of thought experiments is overstressed. And, as I mentioned earlier, so is idealization.

More interesting is the different roles that they play (for a taxonomy, see Brown, 1991, ch. 2). Some play a negative or refuting role. The Einstein example amounts to a *reductio ad absurdum* of Maxwell's electrodynamics, in conjunction with what were then common-sense assumptions about motion (i.e., Galilean relativity). Stevin, by contrast, teaches us something positive, something new about static equilibrium. Newton's spheres do not so much give us a new result but, rather, give us a remarkable phenomenon, something that needs to be explained. The thought experiment establishes a phenomenon; the explanation comes later. And the best explanation, according to Newton, is the existence of absolute space. This way of looking at it is confirmed, it seems to me, by the way in which Berkeley and Mach reacted to the thought experiment. They didn't deny that rotation with respect to absolute space is the best explanation for the tension in the cord. Instead, they denied there would be any tension in the first place, if the spheres are not moving together. Or the two spheres would move together, if there were any tension. That is, they didn't bother to challenge the explanation of the phenomena that Newton posited; they challenged the alleged phenomenon itself.

1.4 *A priori* Knowledge of Nature

The most interesting example is surely Galileo's. This seems to play a negative role – it refutes Aristotle by means of a *reductio ad absurdum* – then, in a positive vein, it establishes a new theory of motion. There are lots of wonderful thought experiments, but only a small number work in this way. Elsewhere I have called them Platonic (Brown, 1991). I think they are rather remarkable – they provide us with *a priori* knowledge of nature.

My reasons for calling this thought experiment *a priori* – it transcends experience – are rather simple and straightforward. First, although it is true that empirical knowledge is present in this example, there are no new empirical data being used when we move from Aristotle's to Galileo's theory of free fall. And, second, it is not a logical truth. After all, objects could fall on the basis of their color, say, with red things falling faster than blue things. If this were the case, then Galileo's thought experiment wouldn't be effective in refuting it. After all, sticking two red things together doesn't make one thing that is even redder.

This last point is worth a moment's reflection. Properties such as weight or mass are commonly called "extensive," while other properties such as color or heat are known as "intensive." Extensive properties can be added in a way that suggests that they have a structure similar to adding real numbers; the intensive properties don't. Thus, physically adding two bodies at 50 kg each yields one body at 100 kg. But physically adding two bodies at 50 degrees will not yield one body at 100 degrees. Galileo's thought experiment leads to a result that is very much more powerful than is initially apparent, even though the initial result is already very powerful. What Galileo has in effect shown (though it is not manifest), is that for any extensive property, any theory along the lines of Aristotle will succumb to the same problem that the particular case based on weight did. Suppose that bodies fall due to X (where X is an extensive property; for example, weight, height, and so on), and that their (supposed) rates of fall differ due to differing quantities of X. We could now follow the Galileo pattern: we could join two bodies with different amounts of X and ask how fast this new body will fall. The inevitable answer will be: "On the one hand, faster because . . . , and on the other hand, slower because . . . " We would have generated a contradiction.

As an exercise, readers are invited to try it for themselves with various extensive properties such as height, volume, and so on. They should also try Galileo's style of argument on the so-called "fifth force," which was postulated a few years ago as a slight modification to gravity. Objects would fall, according to this theory, at slightly different rates, due to their differing chemical composition. Would the Galileo style of thought experiment work in this case? Is the property *being made of aluminum* intensive or extensive?

In considering an arbitrary X, I have remained focused on X being the cause of *falling*. But I'm sure this could be generalized too, although I'm not sure to what extent and in what directions this generalization might go. In any case, the importance of Galileo's discovery goes well beyond the case of objects falling *due to their heaviness*. This alone is enough to say that we have *a priori* knowledge of nature. The additional power of the generalized Galileo pattern of thinking is rather clearly not derivable from empirical premises; it goes too far beyond anything we have experienced. Yet it is clear as day. This, it seems to me, is another case of *a priori* intuition at work.

Let me call the argument so far *the prima facie case for a priori knowledge of nature*. It is, I admit, not yet a completely decisive argument. Two things are needed to strengthen it. One is an explanation of how we are able to acquire this *a priori* knowledge via thought experiments. The second is to cast doubt on rival views.

1.5 Mathematical Intuitions and Laws of Nature

Even if one thought the Galileo example impressive, one might still resist the idea of *a priori* knowledge of nature, since it seems so contrary to the currently accepted principles of epistemology, where empiricism reigns supreme. How is it even possible to have experience that transcends knowledge? I can't address all concerns, but I will try to develop an analogy that will give an account of how thought experiments might work. I'll start with mathematical Platonism. Kurt Gödel famously remarked,

> Classes and concepts may, however, also be conceived as real objects . . . existing independently of our definitions and constructions. It seems to me that the assumption of such objects is quite as legitimate as the assumption of physical bodies and there is quite as much reason to believe in their existence. They are in the same sense necessary to obtain a satisfactory system of mathematics as physical bodies are necessary for a satisfactory theory of our sense perceptions . . . (1944, p. 456f)

> . . . despite their remoteness from sense experience, we do have something like a perception also of the objects of set theory, as is seen from the fact that the axioms force themselves upon us as being true. I don't see any reason why we should have any less confidence in this kind of perception, i.e., in mathematical intuition, than in sense perception, which induces us to build up physical theories and to expect that future sense perceptions will agree with them and, moreover, to believe that a question not decidable now has meaning and may be decided in the future. The set-theoretical paradoxes are hardly more troublesome for mathematics than deceptions of the senses are for physics . . . [N]ew mathematical intuitions leading to a decision of such problems as Cantor's continuum hypothesis are perfectly possible . . . (1947/1964, p. 484)

I take these passages to assert a number of important things, including the following: mathematical objects exist independently from us; we can perceive or intuit them;

our perceptions or intuitions are fallible (similar to our fallible sense perception of physical objects); we conjecture mathematical theories or adopt axioms on the basis of intuitions (as physical theories are conjectured on the basis of sense perception); these theories typically go well beyond the intuitions themselves, but are tested by them (just as physical theories go beyond empirical observations but are tested by them); and in the future we might have striking new intuitions that could lead to new axioms that would settle some of today's outstanding questions. These are the typical ingredients of Platonism. The only one I want to focus on is the perception of abstract entities, commonly called intuition, or seeing with the mind's eye. Gödel took intuitions to be the counterparts of ordinary sense perception. Just as we can see some physical objects – trees, dogs, rocks, the moon – so we can intuit some mathematical entities. And just as we can see that grass is green and the moon is full, so we can intuit that some mathematical propositions are true.

I cannot argue here for mathematical Platonism, since that would take us too far off course. I present it dogmatically, with only Gödel's authority to strengthen its plausibility (for more detail, see Brown, 1999). The key thing to take from this is that we can have a kind of perception, an intuition, of abstract entities. This, of course, is *a priori* in the sense that ordinary sense perception is not involved. Seeing with the mind's eye transcends experience.

In addition to mathematical Platonism, we need a second ingredient, this time involving laws of nature, in order to have the full analogy that I'm trying to develop.

What is a law of nature? What is it about a law that makes it a law? Let's take "Ravens are black" to be a law. It's not a very good example of a law, but it's simple and will easily illustrate the issue. There are several accounts of laws – the one favored most by modern empiricists is some version of Hume's account (for further discussion, see the contributions to this volume by John Roberts and Harold Kincaid, chapters 7 and 8). According to Hume, a law is just a regularity that we have noticed and come to expect in the future. We have seen a great many ravens and they have all been black; and we expect future ravens also to be black. "All events seem entirely loose and separate." says Hume: "One event follows another, but we never can observe any tie between them." (*Enquiry*, p. 74) ". . . after a repetition of similar instances, the mind is carried by habit, upon the appearance of one event, to expect its usual attendant, and to believe that it will exist" (*Enquiry*, p. 75). Causality and the laws of nature are each nothing more than regularities that are expected. To say that fire causes heat, or that it is a law of nature that fire is hot, is to say nothing more than that fire is constantly conjoined with heat. Hume defined cause as "an object, followed by another, and where all the objects similar to the first are followed by objects similar to the second" (*Enquiry*, p. 76). We can't see a "connection" between fire and heat such that if we knew of the one we could know that the other must also occur. All we know is that whenever in the past we have experienced one we have also experienced the other – hence the "regularity" or "constant conjunction" view of causality and laws of nature.

The appeal to empiricists is evident. All that exists are the regular events themselves; there are no mysterious connections between events – no metaphysics to cope with. The general form of a law is simply a universal statement. "It is a law that As are Bs" has the form $(\forall x)\ (Ax \supset Bx)$.

There are all sorts of problems with this view. For one thing, laws are subjective. If there were no conscious beings with expectations, there would be no laws of nature. This is very counterintuitive. Secondly, there is a problem with so-called accidental generalizations. Consider: "All the books in Bob's office are in English." Suppose this is true and, moreover, that in the entire history of the universe a non-English book never makes its way into Bob's office. When I pick any book from the shelf I fully expect it to be in English, and I am never wrong. This has the form of a law of nature, on Hume's view. But this too seems absurd.

Problems such as these and other difficulties have cast serious doubt on the hope that a simple empiricist account might work. A new, rival account of laws has been proposed (independently) by David Armstrong, Fred Dretske, and Michael Tooley. Each claims that laws of nature are relations among universals; that is, among abstract entities which exist independently of physical objects, independently of us, and outside of space and time. At least this is so in Tooley's version (which I prefer); it's a species of Platonism.

The "basic suggestion," according to Tooley, "is that the fact that universals stand in certain relationships may logically necessitate some corresponding generalization about particulars, and that when this is the case, the generalization in question expresses a law" (1977, p. 672).

Thus, a law is not a regularity; rather, it is a link between properties. When we have a law that Fs are Gs, we have the existence of universals, F-ness and G-ness, and a relation of necessitation between them (Armstrong symbolizes this as $N(F,G)$). A regularity between Fs and Gs is said to hold in virtue of the relation between the universals F and G. "[T]he phrase 'in virtue of universals F and G' is supposed to indicate," says Armstrong, that "what is involved is a real, irreducible, relation, a particular species of the necessitation relation, holding between the universals F and G . . . " (1983, p. 97).

The law entails the corresponding regularity, but is not entailed by it. Thus we have

$$N(F,G) \rightarrow (\forall x)(Fx \supset Gx)$$

and yet

$$(\forall x)(Fx \supset Gx) \nrightarrow N(F,G)$$

The relation N of nomic necessity is understood to be a primitive notion. It is a theoretical entity posited for explanatory reasons. N is also understood to be contingent. At first sight, this seems to be a contradiction. How can a relation of necessitation be contingent? The answer is simple: In this possible world, Fs are required to be Gs, but in other possible worlds Fs may be required to be something else. The law $N(F,G)$ is posited only for this world; in other worlds perhaps the law $N(F,G')$ holds.

Some of the advantages of a realist view of laws are immediately apparent. To start with, this account distinguishes – objectively – between genuine laws of nature and accidental generalizations. Moreover, laws are independent of us – they existed before we did and there is not a whiff of subjectivity or relativism about them. Thus, they can be used to *explain* and not merely summarize events. As with mathematical

Platonism, I cannot attempt to defend this account of laws of nature beyond what I have briefly mentioned. I can only hope to have said enough about it that readers can glimpse its advantage and see why it might be preferred over a Humean account.

The reason for this digression into mathematical Platonism and the realist account of laws of nature is to try to develop some idea of how it is that a thought experiment could transcend experience; that is, how it could yield *a priori* knowledge of nature. I'll now try to explain with a simple argument. According to Platonism, we can intuit some mathematical objects, and mathematical objects are abstract entities. Thus, we can (at least in principle) intuit abstract entities. According to the realist account of laws of nature, laws are also abstract entities. Thus, we might be able (at least in principle) to intuit laws of nature as well. There is one situation in which we seem to have a special access to the facts of nature, namely in thought experiments. Thus, it is possible that thought experiments (at least in some cases) allow us to intuit laws of nature. Intuitions, remember, are nonsensory perceptions of abstract entities. Because they do not involve the senses, they transcend experience and give us *a priori* knowledge of the laws of nature.

Let me head off one concern right away. *A priori* knowledge is not certain knowledge. Intuitions are open to mistakes, just as ordinary sense perceptions are. Mathematics often presents itself as a body of certain truths. This is not so. The history of mathematics is filled with false "theorems." Just think of Russell's paradox, for example. But certainty, as Gödel remarked in the passage quoted above, is not a part of contemporary Platonism. And certainty need not be part of any thought experiment.

1.6 Norton's Arguments and Empiricism

Let me review the argument so far. I described the Galileo thought experiment concerning falling objects, then gave what I called the *prima facie* case for *a priori* knowledge of nature. This was followed by an account of how such *a priori* knowledge could come about, drawing on mathematical Platonism and the realist view of laws of nature. Now I move to the final ingredient in my overall argument, which is to cast doubt on rivals. Ideally, I would refute them all. That's impossible. What I can do is look at one alternative, John Norton's, which I take to be my most serious rival, and try to undermine his account. Norton himself surveys a number of rival accounts of thought experiments in chapter 2 of this volume (see his section 2.5). His descriptions and criticisms of these various views are admittedly brief, but (aside from his criticism of my own view, of course) I'm inclined to agree with what he says about these other views.

In arguing that thought experiments do not transcend experience, Norton is making two claims that give him this conclusion:

1 (Argument thesis): A thought experiment is an argument. It may be disguised, but on reconstruction it begins with premises and the conclusion is derived by means of the rules of deductive logic or inductive inference. (These rules are

understood quite liberally.) Unlike a real experiment, no observation or observation-like process occurs.

2 (Empiricism): All knowledge stems from sensory experience.

These two assumptions lead to the following conclusion:

> A thought experiment is a good one (or "reliable," as Norton puts it) insofar as the premises are empirically justified and the conclusion follows by good rules of inference.

From this, he can now derive the crucial conclusion:

> Thought experiments do not transcend experience.

This conclusion does indeed follow from his two key assumptions, so if I am to reject his conclusion I must attack his premises. One way is to refute empiricism. The other way is to refute the thesis that thought experiments are really just arguments. Before going into this, a few remarks are in order.

A way of seeing the difference between Norton and me is to consider, first, real experiments. We would agree (as would most people) that a real experiment carries us from a perception (and some possible background propositions) to a proposition (a statement of the result). The so-called experimental result may be the culmination of a great deal of theorizing and calculating, but somewhere along the way the experimenter has had to look at something; for example, a thermometer, a streak in a cloud chamber, or a piece of litmus paper. I hold that a thought experiment has a similar structure. The only difference is that the perception is not a sense perception but, rather, is an intuition, an instance of seeing with the mind's eye. In other words, a thought experiment (or at least some of them) carries us from a *(nonsensory) perception* (and some possible background propositions) to a proposition, just as a real experiment does. Norton would deny this similarity and instead claim that a thought experiment carries us from a proposition (and some possible background propositions) to a proposition. For him, it is argument and inference all the way; there is no perception of any kind involved during the thought experiment. Let me quickly add a warning about a possible misunderstanding. In a thought experiment one sees *in the imagination* falling bodies, just as in a real thought experiment. Norton and I would both agree that one "sees" something in this sense. But I would claim – and he would deny – that one also "sees" something else, a law of nature, and that this is a non-sensory experience that is different from any inference that might be made in the thought experiment.

Norton has two important considerations in favor of his deflationary view. One is that his account involves only unproblematic ingredients; that is, empirical observations and empirically acceptable inference patterns. The naturalist-minded take a view such as mine, involving intuitions of abstract entities, to be highly problematic and they regard the idea of seeing with the mind's eye as hopeless. Secondly, Norton has so far managed to reconstruct every thought experiment that he has examined into the argument pattern that he champions. There seem to be no counterexamples. These two considerations make Norton's view very plausible. He goes so far as to claim that unless his view is defeated, it should be accepted as the "default" account. I think he's right about this. The current philosophical climate favors empiricism and opposes abstract entities. Since my view flies in the face of this, the burden of proof is on me.

A second aside, this time historical: the debate between Norton and myself is somewhat reminiscent of an older debate over how to understand Descartes's *cogito, ergo sum*. Is it an argument or an immediate intuition, or something else? The term "*ergo*" suggests an argument, with "I think" as the premise and "I exist" as the conclusion. As an argument, however, it's grossly invalid as it stands. The alternative that I'm inclined to favor is that "I exist" is an immediate intuition, evident on noticing that I happen to be thinking. I will not pursue this point, but I think the parallel with interpreting thought experiments is interesting and possibly instructive (for a discussion, see Hintikka, 1962).

In trying to make my case, let me raise a small point, first. Even if every thought experiment could be reconstructed in Norton's argument form, this would not guarantee that this is what thought experiments essentially are. Norton calls it the "elimination thesis." It says: "Any conclusion reached by a (successful) scientific thought experiment will also be demonstrable by a non-thought-experimental argument" (1991, p. 131). This claim – bold though it is – may be true without thought experiments actually being arguments, disguised or not. For instance, most of us would say that we make judgments of the relative size of other people based on our perception of their geometric shape. Suppose that Norton claimed this is not so, and that what we really do is count molecules. He then shows us that every time we judge that person A is bigger than person B, it turns out that on his (laborious) reconstruction, A, has more molecules than B. Even a success rate of 100 percent in reconstructing judgments of size does not refute the common-sense claim that we judge size by means of visual shape.

There is no denying that such reconstructions would be significant. But for Norton to make his case for arguments, mere reconstruction is not enough. He must also make the case that a thought experiment gives some sort of clue to the (hidden) argument; perhaps it's there in a sketchy or embryonic form and can be readily grasped. Norton acknowledges this point when he remarks that ". . . we should expect the schema of this logic not to be very complicated, so that they can be used tacitly . . . " (this volume, chapter 2, p. [13] – I'll return to this point below). If this were not the case, then there must be something else going on of great epistemic significance. This, of course, is what I think, even if a Norton-type reconstruction is possible in every case.

My strategy will be to find examples of thought experiments that fail to fit the Norton pattern of being an argument with empirically justified premises. It is the argument ingredient that I will focus upon, but the empiricism ingredient will also be challenged as a byproduct. To do this I will use a mathematical example, so the condition requiring the premises to be empirically justified will be set aside for now. My aim is to undermine Norton's claim that a thought experiment is an argument. I'll begin with an easy, but impressive, warm-up example.

Consider the following simple picture proof of a theorem from number theory.

THEOREM. $1 + 2 + 3 + \ldots + n = n^2/2 + n/2$.

Proof:

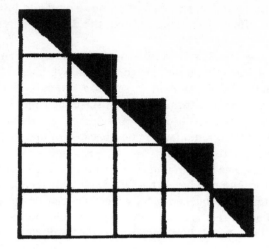

Figure 1.6 A picture proof.

This is a kind of thought experiment, since the evidence is visual. It's quite different than a normal mathematical proof, which is a verbal/symbolic entity. Norton would grant that the picture works as a proof, but only because there is a traditional proof that we can give of this theorem. That is, we can reconstruct such pictorial evidence with proper proofs, in this case a proof by mathematical induction. Such a proof would, of course, be an argument in Norton's sense.

It is certainly true that a traditional proof (by mathematical induction) exists for this theorem. But I think we can still ask: Does the picture proof suggest the traditional proof? When given a "proof sketch" we can fill in the details, but is something like that going on here? Speaking for myself, I don't "see" induction being suggested in the proof at all. Perhaps others do. For Norton to be right, the picture must be a crude form of induction (or a crude form of some other correct argument form); otherwise, he has not made his case. The mere fact that an argument *also* exists does not refute the claim that the thought experiment works evidentially in a nonargument way. Norton, as I mentioned above, recognizes this point. In describing his generalized logic (his allowable rules of inference in a argument), he says (I repeat the quote): ". . . we should expect the schema of this logic not to be very complicated, so that they can be used tacitly by those who have the knack of using thought experiments reliably."

An example such as this one is unlikely to be decisive. I would claim that the picture is conclusive evidence in its own right for the theorem. Norton would deny this, claiming that the real evidence is the proof by mathematical induction. I might then claim that the inductive proof is not "tacitly" contained in the picture. He might disagree. We're at a standoff. At this point, readers might try to decide for themselves which account is more plausible and persuasive.[2]

2 I have often presented this example to audiences of mathematicians and asked if any "see" induction in the picture. About half of any audience says "Yes," while the other half assert just as strenuously that they see no induction at all.

Because this example is not decisive, I'll try another, one that is more complex and certainly bound to be more controversial. It is, however, quite interesting in its own right, and worth the effort of working through the details, though it will not likely be decisive either.

Christopher Freiling (1986) produced a remarkable result that has gone largely unnoticed by philosophers. He refuted the continuum hypothesis. Did he *really* refute the continuum hypothesis? That's hard to say. The standards for success in such a venture are not normal mathematical standards, since the continuum hypothesis (CH) is independent of the rest of set theory. So any "proof" or "refutation" will be based on considerations outside current mathematics. Because of this, Freiling calls his argument "philosophical."

First of all, we need some background. The cardinality of the set of natural numbers is the first infinite cardinal number. If N = {0, 1, 2, . . . }, then $|N| = \aleph_0$. Any set this size or finite is called *countable* because its members can be paired with the counting numbers. But not all sets are countable. The infinite cardinal numbers increase without bound: $\aleph_0, \aleph_1, \aleph_2, \ldots$ It is known that the cardinality of the real numbers is uncountable; it is greater than that of the natural numbers, $|R| = 2^{\aleph_0} > \aleph_0$. The interesting question is which one of the cardinal numbers this is. Does $|R|$ equal $\aleph_1$ or $\aleph_{27}$, or which? The continuum hypothesis is the conjecture that $|R| = \aleph_1$. Is it true? It has been shown that CH is independent of the rest of set theory, which means that it cannot be proven or refuted on the basis of the existing standard axioms. In short, it cannot be proven or refuted in the normal sense of those terms.

The second thing to mention is that we shall take ZFC for granted. ZF is Zermelo–Frankel set theory, which is standard. The C refers to the axiom of choice, which we also assume. An important consequence of ZFC is the so-called *well ordering principle*. It says that any set can be well ordered; that is, can be ordered in such a way that every subset has a first element. The usual ordering, <, on the natural numbers is also a well ordering. Pick any subset, say {14, 6, 82}; it has a first element, namely 6. But the usual ordering on the real numbers is not a well ordering. The subset (0, 1) = {x: 0 < x < 1}, for instance, does not have a first element. Nevertheless, ZFC guarantees that the real numbers can be well ordered by some relation, <, even though no one has yet found such a well ordering. Now we can turn to CH.

Imagine throwing darts at the real line, specifically at the interval [0,1]. Two darts are thrown and they are independent of one another. The point is to select two random numbers. We assume ZFC and further assume that CH is true. Thus, the points on the line can be well ordered so that, for each $q \notin [0, 1]$, the set {$p \notin [0, 1]$: $p < q$} is countable. (Note that < is the well ordering relation, not the usual *less than*, <.) The well ordering is guaranteed by ZFC; the fact that the set is countable stems from the nature of any ordering of any set that has cardinality $\aleph_1$. To get a feel for this, imagine the set of natural numbers. It is infinite, but if you pick a number in the ordering, there will be only finitely many numbers earlier – and infinitely many numbers later. Similarly, pick a number in an well ordered set that is $\aleph_1$ in size and you will get a set of earlier members that is at most $\aleph_0$ in size, and possibly even finite. In any case, it will be a countable set.

We shall call the set of elements that are earlier than the point q in the well ordering S_q. Suppose that the first throw hits point q and the second hits p. Either $p < q$

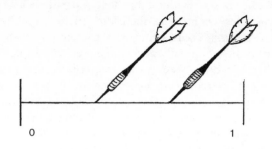

Figure 1.7 Darts picking out independent real numbers at random.

or vice versa; we'll assume the first. Thus, $p \in S_q$. Note that S_q is a *countable* subset of points on the line. Since the two throws were independent, we can say that the throw landing on q defines the set S_q "before" or "independently from" the throw that picks out p. The measure of any countable set is 0. So the probability of landing on a point in S_q is 0. (It is not important to understand the notion of *measure*, only the consequence for probability theory.) While logically possible, this sort of thing is almost never the case. Yet it will happen every time there is a pair of darts thrown at the real line. Consequently, we should abandon the initial assumption, CH, since it led to this absurdity. Thus, CH is refuted and so the number of points on the line is greater than $\aleph_1$.

Notice that if the cardinality of the continuum is $\aleph_2$ or greater, then there is no problem (at least as set out here), since the set of points S_q earlier in the well ordering need not be countable. Thus, it would not automatically lead to a zero probability of hitting a point in it.[3]

There is one aspect of this example on which I should elaborate. The darts give us a pair of real numbers picked at random. These are "real random variables" says Mumford (2000), whose version of the thought experiment I have followed. The concept of random variable at work here is not the mathematical concept found in measure theory (a defined concept inside set theory that will not yield ~CH). Moreover, the two real numbers are picked independently; either could be considered as having been chosen "first." This means that the example cannot be dismissed in the way we might dismiss someone who said of a license number, say, 915372, on a passing car: "Wow, there was only a one in a million chance of that!" We're only impressed if the number is fixed in advance. The independence and randomness of the darts guarantees the symmetry of the throws, so either could be considered the first throw that fixes the set of real numbers that are earlier in the well ordering.

An example such as this must be controversial. Only a minority of mathematicians has accepted it. I am going to assume that it works as a refutation of CH. But I realize that Norton could quite reasonably dismiss it. However, because the example is of the utmost intrinsic interest and is of the greatest importance to mathematics, and because it might prove decisive in the debate with Norton, I find it wholly irresistible.

3 Freiling (1986) actually goes on to show that there are infinitely many cardinal numbers between zero and the continuum.

Recall Norton's view: We start with established propositions and then, using deductive or inductive principles, we infer a conclusion. There is nothing in the process of getting to the conclusion that could count as a perception of any sort. Norton also requires the premises to be empirically acceptable, but we can ignore that here, since the example is mathematical. My view is that we perhaps start with some established propositions and rules of inference, but somewhere in the process we make some sort of empirical experience-transcending observation, we have an intuition, we see something with the mind's eye, and this is essential in reaching the conclusion. Argument alone from already established premises is not sufficient. Which pattern, Norton's or mine, does the CH example follow?

It cannot follow the Norton pattern. The conclusion cannot be derived from the initial assumptions, since CH is independent of the rest of set theory. That is, if we tried to formalize the Freiling example, we would violate existing principles of mathematics. So it cannot be a deductive argument from established premises. Is it perhaps inductive? This seems most unlikely too. Of course, we might be suspicious of the reasoning involved, but if it is correct, it feels far too tight to be called an inductive argument. If it's neither deductive nor inductive, then it's not an argument at all.

Of course, there are sub-arguments within it. In that respect, it is similar to a real experiment, which often includes calculation and theory application. But, as I stressed above, there is at least one point in an experiment at which an observation is made, perhaps as trivial as reading a thermometer or checking the color of some litmus paper. I say, analogously, that there is at least one point in the CH thought experiment at which an intuition occurs. Where? I don't know, but I will conjecture. I suspect that it has to do with the perception (with the mind's eye) of randomness and independence of the darts. I'm not at all confident of this, but ignorance does not upset the case for intuitions. We may be in the same situation as the famous chicken-sexers who can indeed distinguish male from female day-old chicks, and they do this through sense perception, but they have no idea what it is that they actually see. We could be having an appropriate intuition without knowing what it is or when it happened.

1.7 The Moral of the Story

If the analysis of the dart-throwing thought experiment is right, then the moral is fairly straightforward. Mathematical thought experiments are sometimes not arguments and they sometimes involve mathematical intuitions. It is now a small step by analogy to a similar moral for thought experiments in the natural sciences. If we have established that experience-transcending intuitions can happen in the case of a mathematical thought experiment, then they should be able to happen when the laws of nature are the issue.

Recall the Galileo case. There, I argued that we have good reason to think that intuitions, not sensory experience, are at work. Then I tried to explain how all of this might work, tying it to mathematical Platonism and a realist account of the laws of nature. Finally, we saw the dart-throwing example, which tends to reinforce mathematical Platonism, but in a way that is highly analogous to scientific thought experiments. These different strands seem to dovetail very nicely in support of my general contention that thought experiments are sometimes not arguments from empirically justified premises and they sometimes involve nonsensory intuitions of laws of nature. In other words, some thought experiments transcend experience.

Acknowledgments

I am very grateful to John Norton for numerous discussions over many years. We adopted our very different views at the same time and quite independently of one another, and when almost no one else was the least interested in the topic. That was in 1986. Since then, he has made me re-think and modify my views over and over again. I count myself very lucky to have him as a friendly rival. I am also very grateful to my daughter, Elizabeth, for preparing the diagrams.

Bibliography

Armstrong, D. 1983: *What is a Law of Nature?* Cambridge: Cambridge University Press.

Arthur, R. 1999: On thought experiments as *a priori* science. *International Studies in the Philosophy of Science*, 13, 215–29.

Benacerraf, P. and Putnam, H. (eds.) 1984: *Philosophy of Mathematics*, 2nd edn. Cambridge, UK: Cambridge University Press.

Brown, J. R. 1986: Thought experiments since the scientific revolution. *International Studies in the Philosophy of Science*, 1, 1–15.

— 1991: *Laboratory of the Mind: Thought Experiments in the Natural Sciences*. London: Routledge.

— 1993a: Seeing the laws of nature [author's response to Norton, 1993]. *Metascience* (new series), 3, 38–40.

— 1993b: Why empiricism won't work. In Hull et al., op. cit., pp. 271–9.

— 1999: *Philosophy of Mathematics: An Introduction to the World of Proofs and Pictures*. London: Routledge.

— 2002: Peeking into Plato's heaven. Manuscript prepared for Philosophy of Science Association Biennial Meeting, Milwaukee, Wisconsin.

Crease, R. 2002: The most beautiful experiment. *Physics World*, September 2002.

Dretske, F. 1977: Laws of nature. *Philosophy of Science*, 44, 248–68.

Einstein, A. 1949: Autobiographical notes. In P. A. Schilpp (ed.), *Albert Einstein: Philosopher–Scientist*. La Salle, IL: Open Court, 1–95.

Freiling, C. 1986: Axioms of symmetry: throwing darts at the real number line. *Journal of Symbolic Logic*, 51(1), 190–200.

Galileo Galilei 1967: (*Dialogo*), *Dialogue Concerning the Two Chief World Systems*, trans. S. Drake. Berkeley, CA: University of California Press (original work published 1632).

— 1974: (*Discoursi*), *Two New Sciences*, trans. S. Drake. Madison, WI: University of Wisconsin Press (original work published 1638).

Gendler, T. S. 1998: Galileo and the indispensability of scientific thought experiment. *The British Journal for the Philosophy of Science*, 49, 397–424.

— 2002: Thought experiments rethought – and reperceived. Manuscript prepared for Philosophy of Science Association Biennial Meeting, Milwaukee, Wisconsin.

Genz, H. 1999: *Gedanken-experimente*, Weinheim: Wiley–VCH.

Gödel, K. 1944: Russell's mathematical logic. Reprinted in Benacerraf and Putnam, op.cit.

— 1947/1964: What is Cantor's continuum problem? Reprinted in Benacerraf and Putnam, op. cit.

Gooding, D. 1993: What is experimental about thought experiments? In Hull et al., op. cit., pp. 280–90.

Hacking, I. 1993: Do thought experiments have a life of their own? In Hull et al., op. cit., pp. 302–8.

Häggqvist, S. 1996: *Thought Experiments in Philosophy*. Stockholm: Almqvist & Wiksell International.

Hintikka, J. 1962: *Cogito, ergo sum*: inference or performance? *Philosophical Review*, 71, 3–32.

Horowitz, T. and Massey, G. (eds.) 1991: *Thought Experiments in Science and Philosophy*. Savage, MD: Rowman and Littlefield.

Hull, D., Forbes, M., and Okruhlik, K. (eds.), *PSA 1992*, vol. 2. East Lansing, MI: Philosophy of Science Association.

Hume, D. 1975: *(Enquiry), Enquiry Concerning Human Understanding*. Oxford: Oxford University Press.

Humphreys, P. 1994: Seven theses on thought experiments. In J. Earman *et al.* (eds.), *Philosophical Problems of the Internal and External World*. Pittsburgh: University of Pittsburgh Press.

Koyré, A. 1968: Galileo's treatise *De Motu Gravium*: the use and abuse of imaginary experiment. In *Metaphysics and Measurement*. London: Chapman and Hall, 44–88.

Kuhn, T. S. 1964: A function for thought experiments. In *L'Aventure de la Science, Mélanges Alexandre Koyré*, vol. 2. Paris: Hermann, 307–34. Reprinted in *The Essential Tension: Selected Studies in Scientific Tradition and Change*. Chicago: University of Chicago Press, 1977, 240–65.

Mach, E. 1906: Über Gedankenexperimente. In *Erkenntnis und Irrtum. Skizzen zur Psychologie der Forschung*, 2nd edn. Leipzig: Verlag von Johann Ambrosius, 183–200. Translated from the 5th edn. as: On thought experiments. In *Knowledge and Error: Sketches on the Psychology of Enquiry*, trans. T. J. McCormack. Dordrecht: D. Reidel, 1976, 134–47.

— 1960: *The Science of Mechanics*, trans. J. McCormack, 6th edn. La Salle, IL: Open Court.

Maddy, P. 1997: *Naturalism in Mathematics*. Oxford and New York: Oxford University Press.

McAllister, J. 1996: The evidential significance of thought experiments in science. *Studies in History and Philosophy of Science*, 27, 233–50.

— 2002: Thought experiments and the belief in phenomena. Manuscript prepared for Philosophy of Science Association Biennial Meeting, Milwaukee, Wisconsin.

Miscevic, N. 1992: Mental models and thought experiments. *International Studies in the Philosophy of Science*, 6, 215–26.

Mumford, D. 2000: Dawning of the age of stochasticity. In V. Arnold et al. (eds.), *Mathematics: Frontiers and Perspectives*. Washington, DC: American Mathematical Society, 197–218.

Newton, I. 1999: *(Principia), Mathematical Principles of Natural Philosophy*, trans. I. B. Cohen. Berkeley, CA: University of California Press (original work published 1687).

Nersessian, N. 1993: In the theoretician's laboratory: thought experimenting as mental modeling. In Hull et al., op. cit., pp. 291–301.

Norton, J. D. 1991: Thought experiments in Einstein's work. In Horowitz and Massey, op. cit., pp. 129–48.

— 1993: Seeing the laws of nature [review of Brown, 1991]. *Metascience*, 3 (new series), 33–8.

— 1996: Are thought experiments just what you always thought? *Canadian Journal of Philosophy*, 26, 333–66.

— 2002: On thought experiments: Is there more to the argument? Manuscript prepared for 2002 Philosophy of Science Association Biennial Meeting, Milwaukee, Wisconsin.

Rogers, P. 2002: Editorial: the double slit experiment. *Physics World*, September 2002.

Schabas, M. 2002: Hume on thought experiments. Unpublished manuscript, presented at the Canadian Society for History and Philosophy of Science, Toronto, June.

Sorensen, R. 1992: *Thought Experiments*. Oxford: Oxford University Press.

Tooley, M. 1977: The nature of laws. *Canadian Journal of Philosophy*, 7, 667–98.

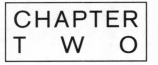

CHAPTER TWO

Why Thought Experiments do not Transcend Empiricism

John D. Norton

2.1 Introduction

The epistemological problem of thought experiments   The essential element in experimentation is the natural world. We learn about the natural world by watching what it does in some contrived circumstance. Just imagining what the world might do if we were to manipulate it in this way or that would seem futile, since it omits this essential element. Yet the literature of science frequently leads us to just such imaginary experiments, conducted purely in the mind, and with considerable apparent profit. These are "thought experiments." We imagine a physicist trapped in a box in remote space, that the box is accelerated by some outside agent, and, from tracing what we imagine the physicist would see in the box, we arrive at one of the fundamental physical principles that Einstein used to construct his general theory of relativity. If this can be taken at face value, thought experiments perform epistemic magic. They allow us to use pure thought to find out about the world. Or at least this is dubious magic for an empiricist who believes that we can only find out about the world from our experience of the world.

Can thought experiments perform this magic? My concern in this chapter is restricted to this one question, which we can label the epistemological problem of thought experiments in the sciences:

> Thought experiments are supposed to give us knowledge of the natural world. From where does this knowledge come?

I shall also restrict myself to thought experiments as they are used in the sciences, although I expect that my analysis and conclusions can be used in other contexts. My concern is *not* directly with the many other interesting facets of thought experiments: their effective exploitation of mental models and imagery, their power as rhetorical devices; their entanglement with prior conceptual systems; their similarity to real experiments; and so on. More precisely, I am concerned with these facets only insofar as they bear directly on the epistemological problem.

This chapter   My goal in this chapter is to state and defend an account of thought experiments as ordinary argumentation that is disguised in a vivid pictorial or narrative form. This account will allow me to show that empiricism has nothing to fear from thought experiments. They perform no epistemic magic. Insofar as they tell us about the world, I shall urge that thought experiments draw upon what we already know of it, either explicitly or tacitly; they then transform that knowledge by disguised argumentation. They can do nothing more epistemically than can argumentation. I will defend my account of thought experiments in section 2.3 by urging that the epistemic reach of thought experiments turns out to coincide with that of argumentation, and that this coincidence is best explained by the simple view that thought experiments just are arguments. Thought experiments can err – a fact to be displayed by the thought experiment – anti thought experiment pairs of section 2.2. Nonetheless thought experiments can be used reliably and, I will urge in section 2.4, this is only possible if they are governed by some very generalized logic. I will suggest on evolutionary considerations that their logics are most likely the familiar logics of induction and deduction, recovering the view that thought experiment is argumentation. Finally, in section 2.5 I will defend this argument-based epistemology of thought experiments against competing accounts. I will suggest that these other accounts can offer a viable epistemology only insofar as they already incorporate the notion that thought experimentation is governed by a logic, possibly of very generalized form.

2.2 Thought Experiment – Anti Thought Experiment Pairs

A test for any epistemology of thought experiments   How are we to know that we have a viable epistemology of thought experiments? I propose a simple test. It is presented here as a gentle warm-up exercise that the reader is asked to bear in mind as the chapter unfolds and various epistemologies are visited.

We can have cases in which one thought experiment supports a result and another thought experiment supports the negation of the same result. These I will call "thought experiment – anti thought experiment pairs." An epistemology of thought experiments must give us some account of why at least one of these fails. It is not enough for us to know by other means that one or other fails. We must be able to explain what went wrong in the failed thought experiment itself. Consider the analogous situation with real experimentation. We may be convinced that the result reported by some experiment is incorrect; it may contradict firmly held theory, for example. If we are to retain confidence in experimentation, we must – at least in principle – be able to explain how the experiment could produce a spurious result.[1]

1   So, when D. C. Miller repeated the famous Michelson–Morley experiment in 1921 and reported evidence of the motion of the earth through the ether, Einstein suggested that the result could be due to tiny thermal gradients in the equipment. See Pais (1982, pp. 113–14).

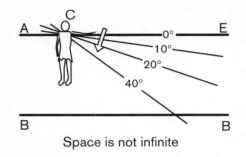

Space is not infinite   Space is not finite

Figure 2.1 Is the world spatially finite?

In the following, I present three thought experiment – anti thought experiment pairs that are unified by the theme of rotation.[2]

2.2.1 Is the world spatially finite?

Aristotle argued, in *On the Heavens*, Book 1, Ch. 5, 272^{a}8–21, that the universe cannot be infinite, since an infinite universe cannot rotate uniformly, as he believed our universe does. In such an infinite universe, he imagined a line ACE, infinite in the direction of E, rotating with the world about the center C and asked when it would cut another infinite line BB. We can make his analysis more thought experimental by imagining that ACE is the ray indicated by a pointing hand that turns with the universe (in the clockwise direction in figure 2.1). At the 0 degrees position shown, ACE is parallel to BB. Prior to attaining that position, ACE does not cut BB; afterwards, it does. But when does it *first* cut BB? It is not at the 40 degrees position, since it has already cut BB at the 20 degrees position; and it was not at the 20 degrees position, since it had already cut BB at the 10 degrees position; and so on, indefinitely. No position greater than 0 degrees is the first, but ACE has not cut BB at 0 degrees. So ACE never cuts BB, which is impossible. (It is interesting that the thought experiment does not seem to depend on the rotation of the universe; all it requires is rotation of a pointer.)

The corresponding anti thought experiment is ancient, most famous, and due to the Pythagorean Archytas. If the universe is finite and I go to the edge: "... could I stretch my hand or my stick outside, or not? That I should not stretch it out would be absurd, but if I do stretch it out, what is outside will be either body or place ..." (Simplicius, *Phys.* 467, 26–32, as quoted in Sorabji, 1988, p. 125).

---

2   A fourth pair that retains the theme of rotation is the thought experiment of Newton's bucket that favors absolute space. Mach's anti thought experiment imagines the bucket walls to be made several leagues thick; it is usually interpreted as blocking Newton's claim. See Norton (1996, pp. 347–9) and Mach (1893, p. 284). See also Norton (forthcoming) for another thought experiment – anti thought experiment pair and for criticism of alternative epistemologies.

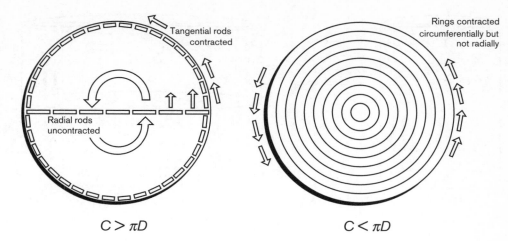

$C > \pi D$ $C < \pi D$

Figure 2.2 What is the geometry of space for a rotating observer?

2.2.2 What is the geometry of space for a rotating observer?

At a decisive moment in the course of his discovery of the general theory of relativity, sometime in 1912, Einstein realized that the geometry of space for an accelerated observer may be non-Euclidean. He showed this for the case of a uniformly rotating observer in special relativity by means of a thought experiment concerning a rigidly rotating disk (see Stachel, 1980). Einstein imagined that the geometry of the surface of the disk is investigated by the usual method of laying down measuring rods. If the disk's diameter is D, what will we measure for its circumference C? Will it be the Euclidean $C = \pi D$? The lengths of measuring rods laid radially are not affected by the Lorentz contraction of special relativity, since their motion is perpendicular to their length. But rods laid tangentially along the circumference move in the direction of their length and will be contracted. Thus *more* rods will be needed to cover the circumference than according to Euclidean expectations. That is, we will measure a non-Euclidean $C > \pi D$ (see figure 2.2).

While Einstein's thought experiment gives a non-Euclidean geometry with $C > \pi D$, an anti thought experiment gives the opposite result of a geometry with $C < \pi D$. The alternative was proposed, for example, by Joseph Petzold in a letter to Einstein dated July 26, 1919 (see Stachel, 1980, p. 52). It is, in effect, that the rotating disk be conceived as concentric, nestled, rotating rings. The rings are uncontracted radially, so the diameter of the disk is unaffected. But the rings are contracted in the circumferential direction, the direction of motion due to the rotation, so their length is *less* that the corresponding Euclidean length. That is, the lengths on the disk conform to $C < \pi D$.

Another anti thought experiment, investigated by Ehrenfest in 1910 and Varicak in 1911, gives the Euclidean result $C = \pi D$. The positions of distance markers on the rotating disk are transferred at some instant to superimposed but nonrotating tracing paper and the geometric figures on the disk reconstructed. The result, Varicak urged,

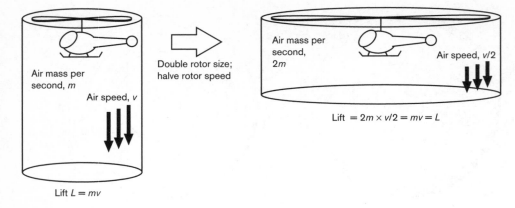

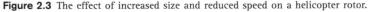

Figure 2.3 The effect of increased size and reduced speed on a helicopter rotor.

would be recovery of Euclidean figures, because the surface of nonrotating tracing paper conforms to Euclidean geometry (see Klein et al., 1993, pp. 478–80).

2.2.3 What is the lift of an infinite rotor at rest?

Imagine a helicopter rotor. When it rotates, it generates lift as a reaction force that results from the momentum imparted to the current of air that it directs downward. If the rotor moves a mass m of air in one second at speed v, then the lift L generated is just mv. What would happen if we were to double the radius of the rotor? To answer, let us assume that it is part of the design of rotors of varying size that the speed of the air currents they generate is proportional to the rotational speed of the rotor. (This can be achieved by flattening the rotor blades more, further from the center.) Since the area swept by the rotor has increased by a factor of $2^2 = 4$, if we leave the rotational speed of the rotor fixed, in one second the rotor will move a mass $4m$ of air at speed v. So the lift will have increased by a factor of 4, to $4mv$. To keep the lift constant at $L = mv$ we should now also reduce the rotational speed of the rotor by a factor of 2. That halves the speed of the air to $v/2$ and also halves the mass moved from $4m$ to $2m$. The lift is now $(2m)(v/2) = mv = L$, which is the original lift.[3]

In short, the lift stays constant at L as we double the rotor size and halve its speed. Repeat this process endlessly in thought, indefinitely doubling the rotor size and halving the rotational speed. In the limit of infinitely many doublings, we have a rotor of infinite size that is not rotating but still generates the original lift L.

The obvious anti thought experiment yields no lift for an infinitely large rotor at rest. A finitely sized rotor that does not turn generates no lift. This is true if we double its size. In the limit of infinitely many doublings, we have an infinitely large rotor that does not rotate and generates no lift.

---

3 The (minimum) power required to sustain the lift is just the kinetic energy of the air moved per unit time; that is, $P = mv^2/2$. So in this process the power is halved to $(2m)(v/2)^2/2 = mv^2/(2 \times 2) = P/2$. Thus, in the limit of the infinite rotor, no power is needed to sustain the lift L. "That must be how angels work. Wide wing spans." – Jeremy Butterfield.

The challenge It is hard to resist the puzzle of determining which (if either) of the members of a pair gives the correct result and what is wrong with the other one. That sort of exercise is part of the fun of thought experiments. But it is not my principal concern here. My concern is to ask how different epistemologies diagnose the existence of the competing pairs; how they explain why one succeeds and the other fails; and how the epistemologies can do this while still preserving the reliability of thought experiments as instruments of inquiry.

2.3 Thought Experiments are Arguments

Why arguments? My account of thought experiments is based on the presumption that pure thought cannot conjure up knowledge, aside, perhaps, from logical truths. All pure thought can do is transform what we already know. This is the case with thought experiments: they can only transform existing knowledge. If thought experiments are to produce knowledge, then we must require that the transformations that they effect preserve whatever truth is in our existing knowledge; or that there is at least a strong likelihood of its preservation. The only way I know of effecting this transformation is through argumentation; the first case is deductive and the second inductive.

Thus I arrive at the core thesis of my account:

(1) Thought experiments are arguments.

which forms the basis of my earlier account of thought experiments (Norton, 1991, 1996).[4] To put it another way, if thought experiments are capable of producing knowledge, it is only because they are disguised, picturesque arguments. That does not assure us that all thought experiments do produce knowledge. They can fail to in just the same way that arguments can fail; that is, either may proceed from false premises or employ fallacious reasoning.

How experience enters a thought experiment Thought experiments need not produce knowledge of the natural world. There are, for example, thought experiments in pure mathematics (for examples, see Brown, 1993b, pp. 275–6) and these, I have argued, are merely picturesque arguments (see Norton, 1996, pp. 351–3). However, the thought experiments that interest me here are those of the natural sciences that do yield contingent knowledge of the natural world. According to empiricism, they can only do so if knowledge of the natural world is supplied to the thought experiment; that is, if this

---

4 In my original account (Norton, 1991), I required that:

Thought experiments are arguments which:
(i) posit hypothetical or counterfactual states of affairs, and
(ii) invoke particulars irrelevant to the generality of the conclusion.

where (i) and (ii) are conditions necessary for an argument to be a thought experiment, but not sufficient. The analysis of Norton (1991) was intended in part to investigate the ramifications of the existence of these necessary conditions.

knowledge comprises a portion of the premises upon which the argument proceeds. It may enter as explicitly held knowledge of the world. We assert on the authority of an empirical theory, special relativity, that a moving rod shrinks in the direction of its motion. Or it may enter as tacit knowledge. We just know that the space of our experience never runs out; we have never seen a boundary in space beyond which we could not pass, unless there was already something past the boundary to obstruct us.

I do not seek here to argue for empiricism; the debate between empiricism and other epistemologies is as ancient as philosophy itself and not likely to be advanced fundamentally here. However, empiricism is overwhelmingly the predominant epistemology in philosophy of science, so that an account that accommodates thought experiments to empiricism in a simple and straightforward manner ought to be accepted as the default, as opposed to some more extravagant account. I claim this default status for the view advocated here.

Two forms of the thesis   The thesis that thought experiments are arguments requires some elucidation. Is the claim merely that thought experiments can do no more than argumentation when it comes to justifying claims? Or is it, in addition, that the actual execution of a thought experiment is just the execution of an argument? Following Norton (1996, p. 354), I intend the stronger version and urge both:

(1a)   *(Context of justification)*[5] All thought experiments can be reconstructed as arguments based on tacit or explicit assumptions. Belief in the outcome–conclusion of the thought experiment is justified only insofar as the reconstructed argument can justify the conclusion.

(1b)   *(Context of discovery)* The actual conduct of a thought experiment consists of the execution of an argument, although this may not be obvious, since the argument may appear only in abbreviated form and with suppressed premises.

Justifying (1a)   As indicated above, the first thesis (1a) derives from the assumption that pure thought cannot conjure up new knowledge. There is a second and more practical justification. As far as I know, all thought experiments can in fact be reconstructed as arguments, and I have little hope of finding one that cannot. Indeed, this expectation supplies a quite stringent test of thesis (1a). It can be defeated merely by finding a thought experiment that cannot be reconstructed as an argument. Norton (1991, 1996) contain many examples of reconstruction of typical thought experiments from various different areas of the physical sciences, including those have been offered as opaque to such reconstruction. The ease of their reconstruction suggests that a counterexample will not be found. The reconstructions are generally rather straightforward and often differ little from the original narrative of the thought experiment. Einstein's rotating disk thought experiment is a typical example. It can be reconstructed in summary as follows:

(D1)   In Euclidean geometry, the measured circumference of a disk is π times its diameter. (Premise)

5   Given condition (ii) above, that thought experiment arguments invoke particulars irrelevant to the generality of the conclusion, thesis (1a) entails that thought experiments may be eliminated from our discourse without loss of demonstrative power, although the actual arguments that replace them may well be harder to follow. This is the "elimination thesis" (Norton, 1991, p. 131).

(D2) The geometry of a nonrotating disk is Euclidean. (Premise)

(D3) The motion of a radial element on a rotating disk is perpendicular to its length, so that (according to special relativity) the length is unaltered. (Premise)

(D4) The motion of a circumferential element on a rotating disk is along its length, so that (according to special relativity) the length is contracted. (Premise)

(D5) Therefore the measured circumference of a rotating disk is more than π times the measured diameter. (From D2, D3, D4)

(D6) Therefore the geometry of a rotating disk is not Euclidean. (From D1, D5)

Justifying (1b)   The situation with the second thesis (1b) is not so straightforward. It is both a thesis in the philosophy of thought experiments and also a thesis in empirical psychology. Perhaps prudence should instruct us to assert only (1a) and remain agnostic on (1b), awaiting the verdict of empirical work in psychology. Indeed, (1a) with agnosticism on (1b) already amounts to a strongly empiricist restriction on what thought experiments can teach us. However, it seems to me that this contracted account is unnecessarily timid. There are several indications that favor (1b).

In spite of their exotic reputation, thought experiments convince us by quite prosaic means. They come to us as words on paper. We read them and, as we do, we trace through the steps to complete the thought experiment. They convince us without exotic experiences of biblical moment or rapturous states of mind. At this level of description, thought experimenting does not differ from the reading of the broader literature in persuasive writing. A long tradition in informal logic maintains that this activity is merely argumentation and that most of us have some natural facility in it. The text prompts us to carry out arguments tacitly and it is urged that reconstructing the arguments explicitly is a powerful diagnostic tool. I merely propose in (1b) that matters are no different in thought experimenting. Parsimony suggests that we make this simplest of accounts our default assumption.

Thesis (1a) supplies a stronger reason for accepting (1b). Whatever the activity of thought experimenting may be, if we accept (1a), we believe that the reach of thought experimenting coincides exactly with the reach of argumentation. If thought experimentation opens up some other channel to knowledge, how curious that it should impersonate argumentation so perfectly! How are we to explain this coincidence, if not by the simple assumption that thought experimenting merely is disguised argumentation? Analogously, we would accord no special powers to a clairvoyant whose prognostications coincided precisely with what could be read from our high school graduation year book. We would strongly suspect a quite prosaic source for the clairvoyant's knowledge.

Thought experiment – anti thought experiment pairs   This account of the nature of thought experiments can readily accommodate the existence of these pairs. We can have two arguments whose conclusions contradict. It then follows that at least one of the arguments is not sound; it has a false premise or a fallacious inference. The diagnosis is the same for a pair of thought experiments that produce contradictory outcomes. The argument of at least one of them has a false premise or fallacious step, and we resolve the problem by finding it. Thus the existence of these pairs presents no special obstacle to the reliability of thought experiments. If they fail, they do so for an identifiable reason, although finding the false premise or fallacy may not be easy. Thought experiments have the same transparency and reliability as ordinary argumentation.

2.4 The Reliability Thesis

There is a further justification for the epistemology of thought experiments advocated here and it is independent of empiricism. It is summarized as follows:

> (2) *Reliability thesis*. If thought experiments can be used reliably epistemically, then they must be arguments (construed very broadly) that justify their outcomes or are reconstructible as such arguments.

The thesis will be explained and justified below. To preclude confusion, I stress here that this thesis invokes a notion of argumentation that is far more general than the one usually invoked in logic texts, so the claim is weaker than it may first appear. In addition, however, I will suggest below that the efforts of logicians to codify new inferential practices ensures that the presently familiar logics of deduction and induction will in practice suffice as the logic of present, reliable thought experiments.

Reliability   Thought experiments are commonly taken to be more than just generators of interesting hypotheses. Thought experimentation is also taken to be a reliable mode of inquiry. I take this to mean that we can have good reason to believe the outcome of at least some thought experiments; that is, there is a way of using thought experiments so that we do have grounds for believing their outcomes.

We would feel licensed to believe in the reliability of thought experiments if their conclusions were inerrant, or at least almost always so. Take an oracle as an analogy. We might not know how the oracle works, but we would have good grounds for believing its reliability if it has a very strong history of successful predictions. The complication with thought experiments is that we have no such history. Thought experiments have proven far too malleable. Proponents of virtually all scientific theories, from the profound to the profoundly false, have had little trouble calling up thought experiments that speak in their favor. The existence of thought experiment – anti thought experiment pairs displays the problem vividly. The situation is more to akin to a plethora of oracles wildly generating predictions indiscriminately and willing to foresee whatever touches our fancy. So why do at least some of us believe that thought experiments can be used reliably? Is there a mark of reliability for the trustworthy thought experiments? Such a mark is possible, but it will require a small detour to find it.

The most general notion of a logic   It is easy to think of logic as some fixed domain of inquiry, so that when we seek to reconstruct a thought experiment as an argument, we must rely on a fixed codification of logic that is already in the logic literature. This view underestimates the ingenuity of logicians and the fertility of logic. In recent centuries, the history of logic is a history of growth. Deductive logic has grown from the simple syllogistic logic of Aristotle through to many variant forms of predicate logic and symbolic logic. Inductive logic has grown from the much maligned enumerative induction to a bewildering abundance that spans from elaborations of enumerative induction to accounts that draw on the resources of the mathematical theory of probability. Sometimes the growth is driven merely by the recognition that this or that extension of an existing logic is possible. On other occasions it is driven by the recognition that there are uncodified argument forms in use. Over the past one or two centuries, as science has become considerably more complicated, so have the inductive maneuvers of scientists. This has been a stimulus for the growth of the inductive logic and confirmation theory that seeks to systematize their methods of inference.

How far can this extension of logic go? We have gone too far if we say that we have a new logic, but all we do is to supply a list of valid inferences, without any apparent connections between the inferences on the list. For the extension to count as a logic, there must be some systematic, identifiable feature of the allowed inferences so that we can distinguish the valid from the invalid. In the new logic, a valid argument will have two parts: the identifiable feature and the aspects peculiar to the particular inference at hand. This is just the familiar distinction between the form and content of an argument; that is, the distinction between a schema and the propositions, terms, and the like inserted into its slots. To count as a logic in this most general sense, the specification of the admissible forms must admit some systematization; for practical reasons, we would hope that they are tractable and communicable.

As far as I can see, this systematizable distinction of form and content is all we need to say that we have a logic in the most general sense. One might be tempted to impose further restrictions. But naturally arising restrictions seem to me to be too restrictive. We might demand that the content must be finite sentences formed from a finite alphabet of symbols, as in traditional symbolic logics. But that would exclude Bayesian confirmation theory, currently the most popular form of inductive logic, for its content may include real-valued functions of continua (probability densities) that represent belief states. Or, in keeping with the notion that logics are truth-preserving and not truth-creating, we might demand that the arguments be supplied with premises that are independently known to be true. But that would contradict a standard practice in deductive logic of using tautologies as premises, where tautologies are sentences that the logic itself assures us are true in the sense of being assertable. Or we might want to insist that the argument forms licensed not be too indiscriminate; they should not end up licensing contradictions, for example. But specifying what counts as "too indiscriminate" might be difficult. Indeed, standard inductive argumentation can end up licensing contradictions. (Consider enumerative induction on the white swans of Europe and then on the black swans of Australia.) So, rather than denying the honorific term of "logic" to such indiscriminate systems, we should think of the indiscriminate logics merely as less useful and eschew them, much as we might ignore a logic, at the other extreme, that is so discriminating as to license nothing.

The mark We now seek the mark that identifies successful thought experiments, that is, those that succeed in justifying their outcomes. Without it, we have no way of determining whether some new thought experiment will succeed in justifying its result; and no way to check that a claim of successful justification is properly made.

The mark cannot be something external to the thought experiment; that is, something about the person who authors the thought experiment or about the context in which it is proposed. A thought experiment is quite portable and moves wherever its written account goes. Independently of its history, we read the account of the experiment; recreate it in our minds; and decide its success or failure. So the mark must lie within the thought experiment itself.

What can this internal mark be? It cannot reside in the brute fact that this *one* thought experiment succeeds; or in the brute fact of the success of some disparate collection of thought experiments. Exactly because they are *brute* facts, they would have to be supplied externally – a separate certificate of truth that must be carried with the thought experiments to assure the reader of their success. The internal mark must be some identifiable property of a successful thought experiment shared with others, or some identifiable relationship between the successful thought experiments. The mark cannot embrace everything in the thought experiment. Some elements can be freely changed. Einstein's celebrated elevator may be wood or steel or brass. So a successful thought experiment has a structural property shared by other successful thought experiments and freely variable content. But that demand is just that thought experiments be governed by the very general notion of a logic introduced above, by schemas into which we can insert freely variable content. The mark is just that the thought experiment either uses an argument form licensed by a logic or can be reconstructed as one.

The mark designates which thought experiments succeed in justifying their results. So we should not expect the associated logic to be a wildly arbitrary codification of admissible arguments. If we are to recognize the logic as delimiting the successful thought experiments, there must be something in the logic that evidently confers the power of a thought experiment to justify its conclusion. For example, deductive logics are characterized by their preservation of truth and inductive logics by the preservation of its likelihood, so that a thought experiment using these logics will have a justified outcome if it proceeds from true premises. In addition, we should expect the schemas of this logic not to be very complicated, so that they can be used tacitly by those who have the knack of using thought experiments reliably.

In sum, we expect thought experiments to be governed by a simple logic that licenses the ability of a thought experiment to justify its outcome.

Will the familiar logics suffice? An evolutionary argument   The reliability of thought experiments leads us to conclude that thought experimentation is governed by a generalized logic. However, it does not prescribe the nature of the logic beyond the expectations that it underwrite justification and be sufficiently tractable for use. In principle, the logic may be of a very exotic type. We shall see below that some accounts portray thought experiments as manipulating mental models. Perhaps they are accompanied by their own exotic logic. That eventuality would be quite in accord with my view of thought experiments as arguments. Indeed, it would be a nice extension of it.

However, I think there are some reasons to believe that no new, exotic logic is called for. In outlining the general notion of logic above, I recalled the evolutionary character of the logic literature in recent times. New inferential practices create new niches and new logics evolve to fill them. Now, the activity of thought experimenting in science was identified and discussed prominently a century ago by Mach (1906) and thought experiments have been used in science actively for many centuries more. So logicians and philosophers interested in science have had ample opportunity to identify any new logic that may be introduced by thought experimentation in science. So my presumption is that any such logic has *already* been identified, insofar as it would be of use in the generation and justification of scientific results. I do not expect thought experiments to require logics not already in the standard repertoire. This is, of course, not a decisive argument. Perhaps the logicians have just been lazy or blind. It does suggest, however, that it will prove difficult to extract a new logic from thought experiments of relevance to their scientific outcomes – else it would already have been done!

The case against the likelihood of a new logic is strengthened by our failure to identify a thought experiment in science that cannot be reconstructed as an argument using the familiar corpus of deductive and inductive logics. My own view is that thought experiments justify by means already employed more generally in science, which makes it even more likely that their implicit logic has already been investigated and codified.

Independence from empiricism We have inferred from the reliability of thought experiments to the outcome that they are arguments or reconstructible as arguments. This inference does not require the presumption of empiricism. Assume for a moment that thought experiments do somehow tap into a nonempirical source of knowledge. Since they can err but, we believe, can still be used reliably, the above analysis can be repeated to recover the same outcome.

2.5 Alternative Accounts of Thought Experiments

Challenges So far, I have tried to show that the notion of thought experiments as picturesque, disguised arguments allows us to develop a simple empiricist epistemology for thought experiments in the natural sciences. As I indicated in the introduction, my concern here is narrowly with the epistemic problem of thought experiments. I have no illusions that portraying thought experiments as picturesque arguments says everything that can be said about them. I do claim, however, that it does supply a complete epistemology in the sense that all there is to learn about a thought experiment's epistemic power can be recovered from considering it as an argument.

There are other accounts of thought experiments and I will try to list the more important ones below. Some clearly contradict the view developed here. Others may be compatible with it, typically assimilated as a refinement of the argument view. My concern in this section is to defend my epistemology of thought experiments. So I will take issue with these accounts only insofar as they contradict that epistemology.

A generic defense In formulating responses to these alternatives, it has become apparent to me that these responses are drawn from a short list of generic responses, which is suggested by my view and which seems flexible enough to accommodate all challenges. Insofar as the alternatives differ from my view, they proffer some extra factor – let us call it "factor X" – that thought experiments are supposed to manifest but arguments cannot. It is then concluded that:

- thought experiments cannot be arguments, for arguments lack this factor X; or
- the factor X confers some additional epistemic power on thought experiments, so that the account of thought experiments as picturesque arguments cannot offer a complete epistemology.

If thought experiments are picturesque arguments and this view supports a viable epistemology, then the objection must fail. It may fail in one of four ways:

(3a) *Denial.* Thought experiments do not manifest the supposed factor X; or
(3b) *Incorporation.* Arguments can also manifest the factor X; or
(3c) *Epistemic Irrelevance.* The factor X is irrelevant to the epistemic power of a thought experiments; or
(3d) *Unreliability.* A thought experiment cannot employ the epistemic channel proposed by factor X reliably. (Thus, if factor X is essential to thought experiments, they are unreliable epistemically.)

From this list, either responses (3a) or (3b) must succeed to defeat the objection that the factor X shows that thought experiments are not arguments. Any of (3a)–(3d) must succeed to defeat the objection that factor X shows a deficiency in my epistemology of thought experiments. It is quite natural to conjoin (3b) and (3c) for a factor X that can be manifested by particular arguments, while at the same time adding that factor is irrelevant to the epistemic power of the argument. Or we may well want to deny that factor X exists (3a), while also urging that even if it were somehow essential to thought experiments, it would defeat their reliability (3d).

2.5.1 Platonism

Brown (1991, 1993a,b, and this volume) has advanced a radical epistemology of thought experiments. He maintains that the laws of nature reside in a Platonic world and that the right sort of thought experiments allows us to grasp those laws directly. While his epistemology differs in the extreme from mine, I am very sympathetic to one aspect of his project. If you decide to eschew the simple empiricist view that I advocate, then no half measures can suffice. You are committed to explaining how we can gain knowledge of the world without relevant experience of the world. Only a quite radical, alternative epistemology will suffice. Brown has not flinched from advocating such an epistemology. However, as I explained in Norton (1993, 1996), I do not believe that Brown's alternative succeeds.

My criticism, augmented with ideas developed above, follows. Brown's factor X is that thought experiments allow us to grasp directly the Platonic world of laws. Several of the responses (3) are applicable.

(3a)   Denial. I do not believe that there is a world inhabited by Platonic laws. Since the debate over such worlds extends well beyond our concerns here, I will restrict

myself to noting that nothing in the phenomena of thought experiments requires such Platonic worlds and the associated Platonic perception. I have tried to show elsewhere (Norton, 1996) that Brown's favorite examples of Platonic thoughts experiments can be accommodated quite comfortably in my view. And, as I suggested above, if an austere empiricist epistemology of thought experiments succeeds, it should be accepted as the default.

Let us set aside doubts about the Platonic worlds. Even if we accept the existence of such worlds, it is a serious problem for Brown's account that we have no systematic understanding of how Platonic perception works and when it works. The difficulties arising from this opacity can be expressed in two ways:

(3b) Incorporation. Might it simply be that argumentation is *the* way of accessing the Platonic world? Then Brown's account of thought experiments would be annexed by the argument view and his account would persist only as a commitment to a superfluous ontology.

(3d) Unreliability. When a thought experiment depends upon Platonic perception, if the perception is more than mere argumentation, we have no way of knowing internal to the thought experiment that the perception was successful and that misperception did not occur. The epistemology of Platonic perceptions provides no means of adjudicating the competing claims of thought experiment – anti thought experiment pairs.

Brown (1991, pp. 65–6) tries to deflect these concerns over the opacity of Platonic perception by drawing an analogy with ordinary perceptions. The latter are accepted even though the full process from vision to belief formation is still poorly understood. I also accept Brown's (1993a) rejoinder to me that ordinary perceptions were credible well before we had the elaborate modern accounts of perception, such as vision as the reception of photons in the retina, and so on. The crucial disanalogy, however, between the two forms of perception pertains to reliability. With ordinary perception, even rudimentary experiences quickly give us abundant indicators of when ordinary perceptions will succeed or fail. Sight fails if we cover our eyes, but only depth perception suffers if we cover just one eye; sight, sound, and smell are enhanced by proximity and weakened by distance; sight is compromised by weak light, but smell is enhanced by favorable breezes – and so on, in innumerable variations. With Platonic perception, we have nothing whatever comparable to tell us when we perceive or misperceive.

The "disproof" of the continuum hypothesis Brown has conceived an ingenious candidate for a Platonic thought experiment in his chapter in this volume (see chapter 1). While I do not agree that the example succeeds, we do agree that the crucial phase is the establishment of what Brown (forthcoming) calls the Freiling Symmetry Axiom (FSA). It asserts that for every function f that maps real numbers onto countable sets of real numbers, we can always find a pair of reals x,y such that y is not in $f(x)$ and x is not in $f(y)$. It turns out to be equivalent to the negation of the continuum hypothesis.

Brown attributes a recognition of the truth of FSA to Platonic perception. Insofar as it works at all, I just find it to be the result of prosaic argumentation of an informal kind – just the sort of thing I say arises commonly in thought experiments. The recognition depends on seeing that there is a zero probability of picking a number at random from a measure zero set. We infer that result from reasoning by analogy with dart throws. On a real dartboard, there is only a small probability of hitting the thin wires. The probability drops to zero for infinitely thin wires, the analog of a measure zero set.

Brown also urges that there is no precise, formal mathematical argument that corresponds to this thought experiment. I would not be troubled if he were right, since I have always urged that thought experiments may be informal arguments. However, Brown has misstated the relevant mathematics. FSA cannot be derived as a theorem of Zermelo–Frankel set theory with the axiom of choice, where we understand such a theorem to be a result that can be derived without additional premises. However, it certainly could be derived if suitable, additional premises were allowed. Given the liberal amount of vague additional material introduced in the discussion of dart throws, I don't see how to rule out that such premises are not already at hand.

Finally, although it is again immaterial to the issues that separate us, I believe that our dart-throwing intuitions do not allow us to arrive at FSA after all, whether by argument or Platonic intuition, and the entire disproof is mistaken (see Norton, forthcoming).

2.5.2 Constructivism

Kuhn (1964) and Gendler (1998) have described how thought experiments can serve the function of revealing problems in and allowing reform of a scientist's system of concepts. As Kuhn (1964, p. 252) suggests, these thought experiments teach scientists about their mental apparatus instead; so they avoid the problem of teaching us about the world without drawing on new information from the world. For example, Kuhn recalls thought experiments due to Galileo that force Aristotelians to distinguish the concepts of average and instantaneous speed. Gendler also analyses a celebrated thought experiment of Galileo's that forces Aristotelians to see the incompatibility of assuming that heavier masses fall faster, but conjoined masses fall at an intermediate speed. The escape is the recognition that all bodies fall alike.

This constructivist view is interesting and important. From the epistemological point of view, I have two reactions. First, whatever its merits may be, the view cannot supply a complete epistemology of thought experiments in science. There are many thought experiments in science that do not yield a reform of the scientist's conceptual systems. Thought experiments may merely demonstrate results within a theory. (Mach (1893, p. 269) uses one to show that his definition of equality of mass must be transitive, else energy conservation will be violated.) This yields the response *(3a) Denial* insofar as not all thought experiments manifest this factor X.

Secondly, Gendler (1998, sections 3.1 and 3.2) has urged that mere argumentation cannot reconfigure conceptual schemes, the factor X of the objection. As far as I can tell, however, the constructivist reconfiguring is fully compatible with the view that thought experiments are arguments.[6] So my response is *(3b) Incorporation*. I base this on the well known power of argumentation to produce the reform of conceptual schemes. A celebrated and quite profound example is supplied by Russell's paradox of naïve set theory. A basic principle of naïve set theory is that any property defines a set – those entities that manifest the property. Russell revealed the untenability of this conception in a celebrated *reductio ad absurdum*. If the principle holds, then the set of all sets that do not have themselves as a member is a legitimate set. But it has contradictory properties. It follows quickly that this set must both be a member and not be a member of itself. The *reductio* is complete; the principle must be rejected; and our conceptual system is profoundly changed.

Now there are complications in this change. Any *reductio* argument ends in a contradiction. In principle, any of the premises of the argument – tacit or explicit – may be taken to have been refuted. In each case we arrive at a different consistent subset of beliefs. Analogously, for any given set of premises we may derive arbitrarily many conclusions; pure logic alone cannot tell us which we should derive. Just as the format of the premises may suggest that we draw one conclusion rather than another, so the rhetorical formulation of a thought experiment may direct us to one result rather than another. Contrary to Gendler, I do not see any special epistemic power in this fact. Is the proposal that the format of the thought experiment somehow directs us to the true consistent subset of beliefs in a way that transcends the reach of argument?[7] How does it accrue this power?

The alternative account offered by Gendler (1998, pp. 414–15) is that the reconfiguration is arrived at by "a sort of *constructive participation* on the part of the reader"; "the person conducting the experiment asks herself 'What would I say/judge/expect were I to encounter circumstances XYZ?' and then *finds out* the apparent answer" (her emphasis). My concern is the reliability of this procedure. Insofar as it involves anything more than argumentation, why should the results of such introspection be credible? Gendler mentions Mach's idea of stores of tacit experiential knowledge. Even if we tap into these, they must be converted into the final result. If the conversion is not through argumentation, then how is it truth-preserving? How can constructive participation allow us to adjudicate thought experiment – anti thought experiment pairs? In short, if the reconfiguration is not effected by argumentation, my response is *(3d) Unreliability*.

2.5.3 Visualization and simulation

Thought experiments seem to tap into an uncanny ability of the mind to simulate the real world. As we engage in the thought experiment, we watch an experiment unfold not with our real senses but with our mind's eye in the laboratory of pure thought.[8]

6 Boorsboom et al. (2002) similarly underestimate argumentation when they claim that a particular thought experiment in probability theory cannot be reconstructed as an argument, since it is not a derivation within a theory but creates a conceptual framework for a theory.

7 This seems to be the import of Gendler's (1998, section 2.4) reflection that the Galilean thought experimenter ignores many logically admissible escapes from the *reductio* contradiction.

8 Gooding (1992, p. 285) writes: "Visual perception is crucial because the ability to visualize is necessary to most if not all thought experimentation."

Should we look to this power to underwrite the epistemic prowess of thought experiments, our elusive factor X? I think not. This prowess cannot be sustained *solely* by the mind's power of visualization or, more generally, simulation. If that were all that counts, I could readily concoct a spurious thought experiment in which the conservation of energy is violated. In my laboratory of thought I connect the shaft of an (elaborately imagined) rapidly turning electric motor to an (elaborately imagined) generator; and then I direct the electric current generated through wiring back to the motor; and I note that in my mental simulation there is surplus electrical power, as revealed by the positive reading on a (shiny brass) wattmeter. Examples such as this show that any epistemic power attributed to the mind's ability to visualize or simulate must at best be subsidiary.[9] An account that tries to explicate this subsidiary epistemic power must face the problem that the mind can visualize or simulate quite spuriously; it must explain how such visualization can be epistemically potent in one case and not in another (such as in the case of thought experiment – anti thought experiment pairs) and how we are to adjudicate.

My view is that it is merely rhetorical window dressing that, for psychological reasons, may well ease acceptance of the result. In many cases, this superfluity is easy to see, since the elements visualized can be supplied in many ways that will not affect the outcome. When Galileo's Salviati imagined a large stone and a small one dropped from a tall tower, he could equally have imagined a cannon ball and a musket ball, or a brick and pebble. All that matters is that one is heavier than the other and neither experiences much air resistance. The epistemic power of the thought experiment comes from what is common to the many formulations – the argument. The variable details, visualized so powerfully, are epistemically neutral; changing them does not change the outcome. In sum, my response to much of the talk of visualization and simulation is to dismiss it as irrelevant epistemically – *(3c) Epistemic Irrelevance.*

2.5.4 Mental models

There is an exception to this last response. In a most promising approach, Nersessian (1993) and Palmieri (2003) call on the literature on mental modeling in cognitive science to explicate mental simulations in thought experiments. This literature accounts for the relevant cognition through the formation of mental models that guide our cognition. An extremely simple example – a further simplified version of Johnson-Laird (1989, p. 472) – is drawn from two assertions:

> The fork is to the left of the plate.
> The knife is to the right of the plate.

They allow formation of the mental model

Fork   Plate   Knife

9   Arthur (1999, p. 228) endorses a subsidiary epistemic power in visualization when he concludes "... I do not think thought experiments are simply reducible to arguments without epistemic loss ... Thought experiments go beyond arguments in providing an easily visualizable – or ... graspable – imaginative reconstruction of the phenomenon at issue."

This mental model in turn licenses further assertions, such as

The plate is to the right of the fork.
The fork is to the left of the knife.

If these mental models can somehow be grounded properly in experience, then why should they not produce knowledge of the world if they are used in thought experiments? For me, the real question is how they can do this *reliably*. Here's how. These models are built from templates into which we slot particulars. In the above example, the template is

Object 1   Object 2   Object 3

and we substitute fork/plate/knife for object 1/object 2/object 3. If this template correctly reflects the nature of space, then the resulting model can be used reliably. But now we see immediately that this reliability is purchased exactly by incorporating just the sort of generalized logic discussed above in the context of section 2.4 ("The Reliability Thesis"). The templates are just the schema of a generalized logic. Thus my response is *(3b) Incorporation.* Use of this mental model literature implements exactly the sort of generalized logic that I had in mind. Knowledge of the world enters the thought experiment in the factual templates that ground the physical scenarios imagined.[10]

In principle, mental models may function this way in thought experiments. However, the case is yet to be made. That mental modeling might be a good account of ordinary cognition does not entail that it is a good account of thought experiments, a highly contrived activity within science with a deceptively simple name. What tempers my optimism is that I know of no example of a thought experiment in science that depends essentially on such mental modeling; or at least all the good candidates I have seen are indistinguishable from argumentation concerning pictures and schemes, much as proofs in Euclidean geometry are just arguments about certain figures. And there are many cases in which the core of the thought experiment is an explicit mathematical derivation in a physical theory and that is *unambiguously* an argument. (An example is Bohr's version of Einstein's clock-in-the-box thought experiment: its essential content is the computation of a relativistic correction to the timing of a process – see Bishop (1999).)

What of the remaining cases in which the decision is unclear? Mental modeling and traditional argumentation can be very close and thus hard to distinguish. Cognitive theorists allow that the supposed mental models of thought experiments can be reconstructed as arguments. The reverse may also be possible: the explicit argumentation of thought experiments can be simulated by mental models. Which is it? When they are close, I favor argumentation, since thought experimentation is a far more highly refined activity than simple cognition about the placing of knives and

10   That templates in logic can also be contingent is actually quite familiar. It is a contingent fact that "If something is human, then it is mortal." That fact licenses inferences as well. From "Socrates is human." we are licensed by it to infer to "Socrates is mortal."

forks on a table. It must be unambiguously communicable in a few words, and its outcome objectively verifiable – conditions that suggest a need for something more secure, such as argumentation. Or perhaps the evolutionary processes described above in section 2.4 ("The Reliability Thesis") have already culled out from the generalized logic of mental models those logics that might explicitly serve science.

I leave the matter to those with expertise in cognitive science, since either way I believe that characterization of thought experimenting as argumentation survives, if the activity is to be reliable. However, the scope of the program is limited by the presence of explicit argumentation (as derivations within theories) at the heart of many thought experiments.

2.5.5 Experimentalism

Thought experiments are so named since they mimic real experiments, the ultimate epistemic channel to the world, at least in the empiricists' view. Might thought experiments attain some epistemic power from their mimicking of this ultimate ideal? Experimentalism answers that they do. This mimicry is the factor X. My principal response is *(3c) Epistemic Irrelevance*. The reason is simple and obvious. Mimicking an experiment is just not the same thing as doing an experiment; one doesn't learn about the world by the mere fact of feigning contact with it. Here, I set aside whether *(3b) Incorporation* is also an appropriate response, since I do not want to decide what it takes for a thought experiment to be experiment-like in the way that (purportedly) gains it epistemic powers. Certainly, if all that is required to be experiment-like is that the thought experiment describe an imaginary experiment and even trace its execution, then that can be done by an argument.

There are two general accounts of how thought experiments gain epistemic powers from their experiment-like character. Sorensen (1992, p. 205) defines a thought experiment as "... an experiment ... that purports to achieve its aim without the benefit of execution." They have the power "to justify beliefs in the same fashion as unexecuted experiments" (p. 213). And this last power derives from the fact that ordinary experiments persuade (justify?) in two ways: first, by "injection of fresh information about the world"; and, secondly, by "... armchair inquiry: by reminder, transformation, delegation, rearrangement, and cleansing" (p. 251). Thought experiments avail themselves of the second mode only. Insofar as the second is merely a synonym for argumentation, perhaps from tacit knowledge, then obviously Sorensen's view would accord, in the end, with the argument view. But Sorensen apparently does not accept that thought experiments are arguments.[11] Recalling the discussion of section 2.4 ("The

---

11 More precisely, he refuses to reply with "direct denial" (p. 214) and proceeds instead with a rather transparent evasion of the question. He offers a "parity thesis": "thought experiments are arguments if and only if experiments are arguments" and urges that, if we believe that thought experiments are arguments, we must take the burden of proving that real experiments are arguments. What makes the evasion curious is that there seems no strong reason to accept the parity thesis. It is not even clear what it asserts. Does "[real] experiments are arguments" mean that they are *entirely* arguments, a claim that is obviously false? Or does it mean that real experiments contain *some* argumentation, a claim that would be easily sustained if we allow that the notion of experiment includes even some minimal interpretation of the raw data read by the experimenter?

Reliability Thesis") above, this invites the response of *(3d) Unreliability*, unless Sorensen can offer another reliable, truth-preserving way of transforming or rearranging what we are reminded of.

The second account accords epistemic power to a thought experiment through their realizing some idealized limit of actual or possible experiments (for such an account and for examples, see Laymon, 1991). The immediate problem with this account is that it cannot provide a complete epistemology, since many thought experiments are not idealized limiting cases.[12] Insofar as not all thought experiments manifest this factor X, the response is *(3a) Denial.* In those cases in which the factor X is present, the obvious response is *(3c) Epistemic Irrelevance.* Unless we are simply inferring the results from the properties assumed of the ideal limit, why should our imagining of this limit have any epistemic power? I will not pursue this line since Laymon (1991) seems to agree, insofar as he analyses thought experiments as tacit argumentation.

2.5.6 Other views

The survey above omits a number of views.[13] Most prominent of these is that of Mach (1906). He accounts for thought experiments as the manipulation of instinctively gained raw experience by a few simple schemes, such as variations of the conditions that determine the result. For example, he considers (p. 138) the distance above the earth of a falling stone. If that distance is increased in thought to the height of the moon, we would still expect the stone to fall in some diminished degree, suggesting that the moon, which is composed of many stones, also falls toward the earth. The difficulty with Mach's view is that it is readily assimilated to nearly all viewpoints. I see his raw experience as supplying premises for the arguments that implement the manipulations. Nersessian (1993, p. 292) sees much in common between Mach's and her view. Gendler (1998, p. 415) calls on Mach for help in one stage of her account. Sorensen (1991) finds an evolutionary epistemology in Mach. So I am not sure how to categorize it.

Bishop (1999) has proposed a most ingenious demonstration of why thought experiments cannot be arguments. He reflects on Einstein's celebrated clock-in-the-box thought experiment, which was conducted in a classical spacetime. Bohr replicated it in a relativistic spacetime and recovered a different outcome. It is the one thought experiment, Bishop urges, but it must be reconstructed as two arguments; so thought exper-

12   For example, a thought experiment quickly establishes that the time reversibility of physical law is not directly expressed in the phenomena. The phenomena manifest a decided unidirectionality in time. To see this, we need only imagine that we locate a familiar process in a device capable of reversing its time order. If the device is large enough to host a banquet, we would find elegantly dressed diners regurgitating the content of their stomachs, chewing it back to pristine morsels and modestly conveying them back to their plates with their forks – a process compatible with the physical laws but otherwise never seen. The thought experiment does not employ a continuous approach to some ideal limit, such as the gradual elimination of friction. Indeed, the thought experiment is more effective the more we avoid idealization; that is, the more realistic we make the processes subject to time reversal.

13   See also Kühne (2001), which includes an account of Oersted's views on thought experimentation. I also pass over McAllister's (1996) claim that thought experimentation is evidentially inert unless one accepts a Galilean doctrine of phenomena, since the view does not supply an alternative epistemology but explores the foundations of all epistemologies of thought experiments. For criticism of his view, see Norton (forthcoming).

iments cannot be arguments. In my view, the difficulty is that Einstein and Bohr *do* have two different, but similar, thought experiments; and they correspond to two different, but similar, arguments. We can convert the two thought experiments into one by ignoring the different spacetimes of each. The different spacetime settings are then responsible for the different outcomes. If that is admissible, then the same stratagem works for the arguments. Ignoring premises pertaining to the spacetime setting, the two arguments proceed from the same experimental premises. They arrive at different results only because of the differences in the premises pertaining to spacetime setting.

Finally, I also correct a persistent confusion concerning my view. Some (e.g., Gooding, 1993, p. 283; Hacking, 1993, p. 303) report that I demand that the argument in a thought experiment must be *deductive*; others suggest that the argument must be *symbolic* (or so I have seen reported in a manuscript version of a paper); and others (Boorsboom et al., 2002) that the arguments must be derivations within some definite theory. A brief review of what I have written will show that none of these restrictions are a part of my view, which allows inductive and informal argumentation and premises from outside any fixed theory.[14]

2.6 Conclusion

I have defended my view that thought experiments in science are merely picturesque arguments. Their epistemic reach can always be replicated by an argument and this is best explained by their merely being arguments. I have also urged that thought experiments can only be used reliably if they are governed by some sort of logic, even if of a very general kind, and proposed that the natural evolution of the literature in deductive and inductive logic would extract and codify the implicit logic of thought experiments. So thought experiments are arguments, but not because thought experimenters have sought to confine themselves to the modes in the existing literature on argumentation; it is because the literature on argumentation has adapted itself to thought experiments.

This argument view provides a natural home for an empiricist account of thought experiments. Insofar as a thought experiment provides novel information about the world, that information was introduced as experientially based premises in the arguments. The argument view may not be the only view that can support an empiricist epistemology. I have surveyed other accounts above, and at least constructivism, mental modeling, and experimentalism may support an empiricist epistemology. However, I have also urged that these accounts are merely variants of the argument view, insofar as they are viable, and that fact may already account for their hospitality to empiricism.[15]

14 Häggqvist (1996, pp. 89–91) criticizes me precisely because he finds me too lenient in admitting inductive inference into the treatment of thought experiments as arguments.

15 What is my resolution of the thought experiment – anti thought experiment pairs? In the case of the rotating disk, the anti thought experiment fails. A rigid ring cannot be set into rotation preserving rigidity; it would shatter exactly because of the Lorentz contraction. In the case of the infinite rotor, the assumption that is false is that the limit of infinitely many doublings (of the moving rotor) produces a physically admissible system. An air current is required for lift and that is absent in the limit. Sometimes limits can yield nonsense.

Acknowledgments

I am grateful to Greg Frost and Wendy Parker for helpful comments, and to Jim Brown for over 15 years of stimulating debate.

Bibliography

Arthur, R. 1999: On thought experiments as *a priori* science. *International Studies in the Philosophy of Science*, 13, 215–29.

Bishop, M. 1999: Why thought experiments are not arguments. *Philosophy of Science*, 66, 534–41.

Boorsboom, G., Mellenbergh, G., and van Heerden, J. 2002: Functional thought experiments. *Synthese*, 130, 379–87.

Brown, J. 1991: *The Laboratory of the Mind: Thought Experiments in the Natural Sciences*. London: Routledge.

— 1993a: Seeing the laws of nature [author's response to Norton, 1993]. *Metascience*, 3 (new series), 38–40.

— 1993b: Why empiricism won't work. In Hull et al., op. cit., pp. 271–9.

— forthcoming: Peeking into Plato's heaven. Manuscript prepared for Philosophy of Science Association Biennial Meeting, Milwaukee, Wisconsin. To be published in *Philosophy of Science*.

Gendler, T. S. 1998: Galileo and the indispensability of scientific thought experiments. *The British Journal for the Philosophy of Science*, 49, 397–424.

Gooding, D. 1993: What is *experimental* about thought experiments? In Hull et al., op. cit., pp. 280–90.

Hacking, I. 1993: Do thought experiments have a life of their own? Comments on James Brown, Nancy Nersessian and David Gooding. In Hull et al., op. cit., pp. 302–8.

Häggqvist, S. 1996: *Thought Experiments in Philosophy*. Stockholm: Almqvist & Wiksell International.

Horowitz, T. and Massey, G. J. (eds.) 1991: *Thought Experiments in Science and Philosophy*. Savage, MD: Rowman and Littlefield.

Hull, D., Forbes, M., and Okruhlik, K. (eds.) 1993: *PSA 1992: Proceedings of the 1992 Biennial Meeting of the Philosophy of Science Association*, vol. 2. East Lansing, MI: Philosophy of Science Association.

Johnson-Laird, P. N. 1989: Mental models. In M. Posner (ed.), *Foundations of Cognitive Science*. Cambridge, MA: The MIT Press, 469–500.

Klein, M., Kox, A. J., Renn, J., and Schulmann, R. 1993: Einstein on length contraction in the theory of relativity. In *The Collected Papers of Albert Einstein*, vol. 3: *The Swiss Years, 1909–1911*. Princeton, NJ: Princeton University Press, pp. 478–80.

Kuhn, T. S. 1964: A function for thought experiments. In *L'Aventure de la Science, Mélanges Alexandre Koyré*, vol. 2. Paris: Hermann, 307–34. Reprinted in *The Essential Tension: Selected Studies in Scientific Tradition and Change*. Chicago: University of Chicago Press, 1977, 240–65.

Kühne, U. 2001: Die Methode der Gedankenexperimente: Untersuchung zur Rationalität naturwissenschaftler Theorienform. Manuscript.

Laymon, R. 1991: Thought experiments by Stevin, Mach and Gouy: thought experiments as ideal limits and as semantic domains. In Horowitz and Massey, op. cit., pp. 167–91.

Mach, E. 1893: *The Science of Mechanics: A Critical and Historical Account of its Development*, 6th edn., trans. T. J. McCormack. La Salle, IL: Open Court, 1960.

—— 1906: Über Gedankenexperimente. In *Erkenntnis und Irrtum. Skizzen zur Psychologie der Forschung*, 2nd edn. Leipzig: Verlag von Johann Ambrosius, 183–200. Translated from the 5th edn. as: On thought experiments. In *Knowledge and Error: Sketches on the Psychology of Enquiry*, trans. T. J. McCormack. Dordrecht: D. Reidel, 1976, 134–47.

McAllister, J. 1996: The evidential significance of thought experiment in science. *Studies in History and Philosophy of Science*, 27, 233–50.

Nersessian, N. 1993: In the theoretician's laboratory: thought experimenting as mental modeling. In Hull et al., op. cit., pp. 291–301.

Norton, J. D. 1991: Thought experiments in Einstein's work. In Horowitz and Massey, op. cit., pp. 129–48.

—— 1993: Seeing the laws of nature [review of Brown, 1991]. *Metascience*, 3 (new series), 33–8.

—— 1996: Are thought experiments just what you thought? *Canadian Journal of Philosophy*, 26, 333–66.

—— forthcoming: On thought experiments: Is there more to the argument? Manuscript prepared for 2002 Philosophy of Science Association Biennial Meeting, Milwaukee, Wisconsin. To be published in *Philosophy of Science*.

Palmieri, P. 2003: Mental models in Galileo's early mathematization of nature. *Studies in History and Philosophy of Science, Part A*, 34(2), 229–64.

Pais, A. 1982: *Subtle is the Lord. . . . The Science and Life of Albert Einstein*. Oxford: The Clarendon Press.

Sorabji, R. 1988: *Matter, Space and Motion: Theories in Antiquity and their Sequel*. London: Duckworth.

Sorensen, R. 1991: Thought experiments and the epistemology of laws. *Canadian Journal of Philosophy*, 22, 15–44.

—— 1992: *Thought Experiments*. Oxford: Oxford University Press.

Stachel, J. 1980: Einstein and the rigidly rotating disk. In A. Held (ed.), *General Relativity and Gravitation. One Hundred Years after the Birth of Albert Einstein*, vol. 1. New York: Plenum Press, 1–15. Reprinted in D. Howard and J. Stachel (eds.) 1989: *Einstein and the History of General Relativity: Einstein Studies Vol. 1*. Boston: Birkhäuser, 48–62.

DOES PROBABILITY CAPTURE THE LOGIC OF SCIENTIFIC CONFIRMATION OR JUSTIFICATION?

Scientific hypotheses and theories, unlike theorems of mathematics and logic, can never be established conclusively. Most philosophers believe, however, that theories can be *confirmed* by empirical evidence. In other terminology, empirical evidence can (at least partially) *justify* our belief in scientific theories. Many of these philosophers – as well as researchers in other fields, such as statistics and economics – have found it natural to understand the relation between theory and evidence in terms of probability: evidence confirms a theory when it renders that theory more probable, and disconfirms a theory when it renders the theory less probable. One popular probabilistic approach is *Bayesianism*, which is outlined at the beginning of the chapter by Kevin Kelly and Clark Glymour. According to many versions of Bayesianism, probability represents an individual's subjective beliefs. Patrick Maher details a different way of understanding confirmation in terms of probability, according to which the confirmation relation is more closely analogous to the entailment relation of logic. This allows him to avoid many of the difficulties brought on by the subjective element of traditional Bayesianism. Maher demonstrates that his account captures many of the intuitive features of confirmation. Kelly and Glymour are critical of Bayesianism and other probabilistic approaches to confirmation, arguing that they fail to connect with the aims of science. Science seeks truth (in some broad sense), and philosophy of science ought to be concerned with the evaluation of strategies for getting us there. Such strategies may be *justified* by demonstrating their reliability and efficiency. Bayesians and other fellow travelers ask: To what extent is belief in this theory justified by this evidence? For Kelly and Glymour, if a method is efficient or inefficient at finding the truth, further talk of "justification" adds nothing. Their positive proposal draws upon the resources of *formal learning theory* rather than probability. Together, these chapters illustrate how formal tools can fruitfully be brought to bear on philosophical issues.

CHAPTER THREE

Probability Captures the Logic of Scientific Confirmation

Patrick Maher

3.1 Introduction

"Confirmation" is a word in ordinary language and, like many such words, its meaning is vague and perhaps also ambiguous. So if we think of a logic of confirmation as a specification of precise rules that govern the word "confirmation" in ordinary language, there can be no logic of confirmation. There can, however, be a different kind of logic of confirmation. This can be developed using the philosophical methodology known as *explication* (Carnap, 1950, ch. 1). In this methodology, we are given an unclear concept of ordinary language (called the *explicandum*) and our task is to find a precise concept (called the *explicatum*) that is similar to the explicandum and is theoretically fruitful and simple. Since the choice of an explicatum involves several desiderata, which different people may interpret and weight differently, there is not one "right" explication; different people may choose different explicata without either having made a mistake. Nevertheless, we can cite reasons that motivate us to choose one explicatum over another.

In this chapter I will define a predicate "C" which is intended to be an explicatum for confirmation. I will establish a variety of results about "C" dealing with verified consequences, reasoning by analogy, universal generalizations, Nicod's condition, the ravens paradox, and projectability. We will find that these results correspond well with intuitive judgments about confirmation, thus showing that our explicatum has the desired properties of being similar to its explicandum and theoretically fruitful. In this way, we will develop parts of a logic of confirmation. The predicate "C" will be defined in terms of probability and in that sense we will conclude that probability captures the logic of scientific confirmation.

3.2 Explication of Justified Degree of Belief

I will begin by explicating the concept of the degree of belief in a hypothesis H that is justified by evidence E. A little more fully, the explicandum is the degree of belief in H that we would be justified in having if E was our total evidence. We have some opinions about this but they are usually vague, and different people sometimes have different opinions.

In order to explicate this concept, let us begin by choosing a formalized language, like those studied in symbolic logic; we will call this language L. We will use the letters "D," "E," and "H," with or without subscripts, to denote sentences of L. Let us also stipulate that L contains the usual truth-functional connectives "~" for negation, "∨" for disjunction, "." for conjunction, and "⊃" for material implication.

Next, we define a two-place function p which takes sentences of L as arguments and has real numbers as its values. The definition of p will specify, for each ordered pair of sentences H and E in L, a real number that is denoted $p(H|E)$. This number $p(H|E)$ is intended to be our explicatum for the degree of belief in H that is justified by evidence E; we will therefore choose a definition of p using the desiderata for such an explicatum, namely:

1 The values of p should agree with the judgments about justified degrees of belief that we have. For example, if we judge that E justifies a higher degree of belief in H_1 than in H_2 then we will want to define p in such a way that $p(H_1|E) > p(H_2|E)$. In this way, we ensure that the explicatum is similar to the explicandum.
2 The function p should be theoretically fruitful, which means that its values satisfy many general principles.
3 The function p is as simple as possible.

I assume that this function p will satisfy the mathematical laws of conditional probability. We will express these laws using the following axioms. Here D, E, E', H, and H' are any sentences of L and "=" between sentences means that the sentences are logically equivalent.[1]

AXIOM 1. $p(H|E) \geq 0$.

AXIOM 2. $p(E|E) = 1$.

AXIOM 3. $p(H|E) + p(\sim H|E) = 1$, *provided that E is consistent.*

---

1 The following formulation of the axioms of probability is like that of von Wright (1957, p. 93) in several respects. In particular, I follow von Wright in taking $p(H|E)$ to be defined even when E is inconsistent. However, I differ from von Wright about which axiom is not required to hold for inconsistent E; the result is that on my axiomatization, but not von Wright's, if E is inconsistent then $p(H|E) = 1$ for all H (Proposition 2, section 3.12.1). Since E entails H if E is inconsistent, the value $p(H|E) = 1$ accords with the conception of probability as a generalization of deductive logic.

AXIOM 4. $p(E.H|D) = p(E|D)p(H|E.D)$.

AXIOM 5. *If* $H' = H$ *and* $E' = E$, *then* $p(H'|E') = p(H|E)$.

3.3 Explication of Confirmation

Now we will explicate the concept that is expressed by the word "confirms" in statements such as the following. (These are from news reports on the web.)

> New evidence confirms last year's indication that one type of neutrino emerging from the Sun's core does switch to another type en route to the earth.

> New evidence confirms rapid global warming, say scientists.

> Tree-ring evidence confirms Alaskan Inuit account of climate disaster.

If we were to examine these examples in detail, we would find that the judgment that some evidence E confirms some hypothesis H makes use of many other, previously known, pieces of evidence. Let us call this other evidence the *background evidence*. Then our explicandum may be expressed more precisely as the concept of E confirming H given background evidence D.

In looking for an explicatum for this concept, we will be guided by the idea that E confirms H given D iff (if and only if) the degree of belief in H that is justified by E and D together is higher than that justified by D alone. The corresponding statement in terms of our explicatum p is $p(H|E.D) > p(H|D)$. So let us adopt the following definition.

DEFINITION 1. $C(H, E, D)$ iff $p(H|E.D) > P(H|D)$.

Thus $C(H, E, D)$ will be our explicatum for the ordinary language concept that E confirms H given background evidence D.

In ordinary language we sometimes say that some evidence E_1 confirms a hypothesis H more than some other evidence E_2 does. Plausibly, such a statement is true iff the degree of belief in H that is justified by E_1 and the background evidence is higher than that justified by E_2 and the background evidence. The corresponding statement in terms of our explicatum p is $p(H|E_1.D) > p(H|E_2.D)$. So let us adopt the following definition:

DEFINITION 2. $M(H, E_1, E_2, D)$ iff $p(H|E_1.D) > p(H|E_2.D)$.

Thus $M(H, E_1, E_2, D)$ will be our explicatum for the ordinary language concept that E_1 confirms H more than E_2 does, given D.

In the following sections we will use these explications to derive some of the logic of scientific confirmation. Specifically, we will state and prove theorems about our explicata $C(H, E, D)$ and $M(H, E_1, E_2, D)$. These results will be seen to correspond closely to intuitive opinions about the ordinary language concept of confirmation, thus verifying that our explicata are similar to their explicanda.

3.4 Verified Consequences

Scientists often assume that if E is a logical consequence of H, then verifying that E is true will confirm H. For example, Galileo's argument that falling bodies are uniformly accelerated consisted in proving that the motion of uniformly accelerated bodies has certain properties and then experimentally verifying that the motion of falling bodies has those properties. Thus Galileo says that his hypothesis is "confirmed mainly by the consideration that experimental results are seen to agree with and exactly correspond to those properties which have been, one after another, demonstrated by us" (1638, p. 160; cf., p. 167). Similarly, Huygens, in the preface to his *Treatise on Light* (1690), says of that book:

> There will be seen in it demonstrations of those kinds which do not produce as great a certitude as those of geometry, and which even differ much therefrom, since, whereas the geometers prove their propositions by fixed and incontestable principles, here the principles are verified by the conclusions to be drawn from them; the nature of these things not allowing of this being done otherwise. It is always possible to attain thereby to a degree of probability which very often is scarcely less than complete proof.

Let us now examine this assumption that hypotheses are confirmed by verifying their consequences. To do so, we will first express the assumption in terms of our explicata. Although the scientists' statements of the assumption do not mention background evidence, a careful analysis of their deductions would show that when they say evidence is logically implied by a hypothesis, often this is only true given some background evidence. This suggests that the scientists' assumption might be stated in terms of our explicata thus: if E is a logical consequence of $H.D$, then $C(H, E, D)$. The following theorem says that this assumption is true with a couple of provisos.[2]

THEOREM 1. *If E is a logical consequence of $H.D$, then $C(H, E, D)$, provided that $0 < p(H|D) < 1$ and $p(E|{\sim}H.D) < 1$.*

If the provisos do not hold then – as can be proved – $C(H, E, D)$ might be false even though E is a logical consequence of $H.D$. The provisos make sense intuitively; if H is already certainly true or false, or if the evidence is certain to obtain even if H is false, then we do not expect E to confirm H. Nevertheless, the need for these provisos is apt to be overlooked when there is no precise statement of the assumption or any attempt to prove it.

Another assumption that is made intuitively by scientists is that a hypothesis is confirmed more strongly the more consequences are verified. This can be expressed more formally by saying that if E_1 and E_2 are different logical consequences of $H.D$, then $E_1.E_2$ confirms H more than E_1 does, given D. The corresponding statement in terms of our explicata is that if E_1 and E_2 are different logical consequences of $H.D$, then $M(H, E_1.E_2, E_1, D)$. The following theorem shows that this is correct, with a few provisos like those of Theorem 1.

2 Proofs of all theorems except Theorem 4 are given in section 3.12.

THEOREM 2. *If E_2 is a logical consequence of H.D, then M(H, $E_1.E_2$, E_1, D), provided that $0 < p(H|E_1.D) < 1$ and $p(E_2|{\sim}H.E_1.D) < 1$.*

This theorem holds whether or not E_1 is a logical consequence of H.

Comparing E_1 and E_2 separately, it is also intuitive that E_1 confirms H better than E_2 does if E_1 is less probable given $\sim H$ than E_2 is. This thought may be expressed in terms of our explicata by saying that if E_1 and E_2 are both logical consequences of $H.D$ and $p(E_1|{\sim}H.D) < p(E_2|{\sim}H.D)$, then $M(H, E_1, E_2, D)$. The next theorem shows that this is also correct, with a natural proviso.

THEOREM 3. *If E_1 and E_2 are logical consequences of H.D and $p(E_1|{\sim}H.D) < p(E_2|{\sim}H.D)$, then M(H, E_1, E_2, D), provided that $0 < p(H|D) < 1$.*

3.5 Probabilities for Two Properties

Often, data about some sample is held to confirm predictions about other individuals. In order to deal with these kinds of cases, we need to specify the language L and the function p more completely than we have done so far. This section will give those further specifications and subsequent sections will apply them to questions about confirmation from samples.

Let us stipulate that our language L contains two primitive predicates "F" and "G" and infinitely many individual constants "a_1," "a_2,". . . .[3] A predicate followed by an individual constant means that the individual denoted by the individual constant has the property denoted by the predicate. For example, "Fa_3" means that a_3 has the property F.

Let us define four predicates "Q_1" to "Q_4" in terms of "F" and "G" by the following conditions. Here, and in the remainder of this chapter, "a" stands for any individual constant:

$$Q_1a = Fa.Ga; \qquad Q_2a = Fa.{\sim}Ga; \qquad Q_3a = {\sim}Fa.Ga; \qquad Q_4a = {\sim}Fa.{\sim}Ga.$$

A *sample* is a finite set of individuals. A *sample description* is a sentence that says, for each individual in some sample, which Q_i applies to that individual. For example Q_3a_1 is a sample description for the sample consisting solely of a_1, while $Q_1a_2.Q_4a_3$ is a sample description for the sample consisting of a_2 and a_3. We will count any logically true sentence as a sample description for the "sample" containing no individuals; this is artificial but convenient.

Let us also stipulate that L contains a sentence I which means that the properties F and G are statistically independent. Roughly speaking, this means that in a very large sample of individuals, the proportion of individuals that have both F and G is close to the proportion that have F multiplied by the proportion that have G. A more precise definition is given in Maher (2000, p. 65).

3   The restriction to two primitive predicates is partly for simplicity, but also because we do not currently have a satisfactory specification of p for the case in which there are more than two predicates (Maher, 2001).

So far, we have not specified the values of the function p beyond saying that they satisfy the axioms of probability. There are many questions about confirmation that cannot be answered unless we specify the values of p further. Maher (2000) states eight further axioms for a language such as L. The following theorem characterizes the probability functions that satisfy these axioms. In this theorem, and in the remainder of this chapter, "$\overline{F}$" is the predicate that applies to a iff "$\sim Fa$" is true, and similarly for "$\overline{G}$." (This theorem is a combination of Theorems 3, 5, and 6 of Maher, 2000.)

THEOREM 4. *There exist constants λ, γ_1, γ_2, γ_3, and γ_4, with $\lambda > 0$ and $0 < \gamma_i < 1$, such that if E is a sample description for a sample that does not include a, n is the sample size, and n_i is the number of individuals to which E ascribes Q_i, then*

$$p(Q_i a | E . \sim I) = \frac{n_i + \lambda \gamma_i}{n + \lambda}.$$

If $\gamma_F = \gamma_1 + \gamma_2$, $\gamma_{\overline{F}} = \gamma_3 + \gamma_4$, $\gamma_G = \gamma_1 + \gamma_3$, and $\gamma_{\overline{G}} = \gamma_2 + \gamma_4$, then $\gamma_1 = \gamma_F \gamma_G$, $\gamma_2 = \gamma_F \gamma_{\overline{G}}$, $\gamma_3 = \gamma_{\overline{F}} \gamma_G$, and $\gamma_4 = \gamma_{\overline{F}} \gamma_{\overline{G}}$. Also, if n_ϕ is the number of individuals to which E ascribes property ϕ, then

$$p(Q_1 a | E . I) = \frac{n_F + \lambda \gamma_F}{n + \lambda} \frac{n_G + \lambda \gamma_G}{n + \lambda},$$

$$p(Q_2 a | E . I) = \frac{n_F + \lambda \gamma_F}{n + \lambda} \frac{n_{\overline{G}} + \lambda \gamma_{\overline{G}}}{n + \lambda},$$

$$p(Q_3 a | E . I) = \frac{n_{\overline{F}} + \lambda \gamma_{\overline{F}}}{n + \lambda} \frac{n_G + \lambda \gamma_G}{n + \lambda},$$

$$p(Q_4 a | E . I) = \frac{n_{\overline{F}} + \lambda \gamma_{\overline{F}}}{n + \lambda} \frac{n_{\overline{G}} + \lambda \gamma_{\overline{G}}}{n + \lambda}.$$

I will now comment on the meaning of this theorem. In what follows, for any sentence H of L I use "$p(H)$" as an abbreviation for "$p(H|T)$," where T is any logically true sentence of L.

The γ_i represent the initial probability of an individual having Q_i, γ_F is the initial probability of an individual having F, and so on. This is stated formally by the following theorem.

THEOREM 5. *For $i = 1, \ldots, 4$, $\gamma_i = p(Q_i a)$. Also $\gamma_F = p(Fa)$, $\gamma_{\overline{F}} = p(\sim Fa)$, $\gamma_G = p(Ga)$, and $\gamma_{\overline{G}} = p(\sim Ga)$.*

As the sample size n gets larger and larger, the probability of an unobserved individual having a property moves from these initial values towards the relative frequency of the property in the observed sample (Maher, 2000, Theorem 10). The meaning of the factor λ is that it controls the rate at which these probabilities converge to the observed relative frequencies; the higher λ, the slower the convergence.

The formulas given $\sim I$ and I are similar; the difference is that when we are given $\sim I$ (F and G are dependent) we use the observed frequency of the relevant Q_i, and when we are given I (F and G are independent) we use the observed frequencies of F (or $\bar{F}$) and G (or $\bar{G}$) separately.

To obtain numerical values of p from Theorem 4, we need to assign values to γ_F, γ_G, λ, and $p(I)$. (The other values are fixed by these. For example, $\gamma_1 = \gamma_F\gamma_G$.) I will now comment on how these choices can be made.

The choice of γ_F and γ_G will depend on what the predicates "F" and "G" mean, and may require careful deliberation. For example, if "F" means "raven" then, since this is a very specific property and there are vast numbers of alternative properties that seem equally likely to be exemplified *a priori*, γ_F should be very small, surely less than 1/1,000. A reasoned choice of a precise value would require careful consideration of what exactly is meant by "raven" and what the alternatives are.

Turning now to the choice of λ, Carnap (1980, pp. 111–19) considered the rate of learning from experience that different values of λ induce and came to the conclusion that λ should be about 1 or 2. Another consideration is this: so far as we know *a priori*, the statistical probability (roughly, the long-run relative frequency) of an individual having F might have any value from 0 to 1. In the case in which $\gamma_F = 1/2$, it is natural to regard all values of this statistical probability as equally probable *a priori* (more precisely, to have a uniform probability distribution over the possible values), and it can be shown that this happens if and only if $\lambda = 2$. For this reason, I favor choosing $\lambda = 2$.

Finally, we need to choose a value of $p(I)$, the *a priori* probability that F and G are statistically independent. The alternatives I and $\sim I$ seem to me equally plausible *a priori* and for that reason I favor choosing $p(I) = 1/2$.

I have made these remarks about the choice of parameter values to indicate how it may be done but, except in examples, I will not assume any particular values of the parameters. What will be assumed is merely that $\lambda > 0$, $0 < \gamma_i < 1$, and $0 < p(I) < 1$.

3.6 Reasoning by Analogy

If individual b is known to have property F, then the evidence that another individual a has both F and G would normally be taken to confirm that b also has G. This is a simple example of reasoning by analogy. The following theorem shows that our explication of confirmation agrees with this. (From here on, "a" and "b" stand for any distinct individual constants.)

THEOREM 6. $C(Gb, Fa.Ga, Fb)$.

Let us now consider the case in which the evidence is that a has G but not F. It might at first seem that this evidence would be irrelevant to whether b has G, since a and b are not known to be alike in any way. However, it is possible for all we know that the property F is statistically irrelevant to whether something has G, in which case the fact that a has G should confirm that b has G, regardless of whether a has F. Thus I think educated intuition should agree that the evidence does confirm that b has G in this case too, although the confirmation will be weaker than in the preceding case. The following theorems show that our explication of confirmation agrees with both these judgments.

THEOREM 7. *C(Gb, ~Fa.Ga, Fb)*.

THEOREM 8. *M(Gb, Fa.Ga, ~Fa.Ga, Fb)*.

There are many other aspects of reasoning by analogy that could be investigated using our explicata, but we will move on.

3.7 Universal Generalizations

We are often interested in the confirmation of scientific laws that assert universal generalizations. In order to express such generalizations in *L*, we will stipulate that *L* contains an individual variable "*x*" and the universal quantifier "*(x)*," which means "for all *x*." Then, for any predicate "*φ*" in *L*, the generalization that all individuals have *φ* can be expressed in *L* by the sentence "*(x)φx*." Here, the variable "*x*" ranges over the infinite set of individuals $a_1, a_2, \ldots$. We now have the following:

THEOREM 9. *If φ is any of the predicates*

$$F, \overline{F}, G, \overline{G}, Q_1, Q_2, Q_3, Q_4$$

and if $p(E) > 0$, then $p[(x)φx|E] = 0$.

I do not know whether this result extends to other universal generalizations, such as $(x)(Fx \supset Gx)$ or $(x)\sim Q_i x$.

A corollary of Theorem 9 is the following:

THEOREM 10. *If φ is as in Theorem 9 and T is a logical truth then, for all positive integers m, $\sim C[(x)φx, φa_1 \ldots φa_m, T]$.*

On the other hand, we ordinarily suppose that if many individuals are found to have *φ*, with no exceptions, then that confirms that all individuals have *φ*. Thus our explicatum *C* appears to differ from its explicandum, the ordinary concept of confirmation, on this point. However, the discrepancy depends on the fact that the variables in *L* range over an infinite set of individuals. The following theorem shows that the discrepancy does not arise when we are concerned with generalizations about a finite set of individuals.

THEOREM 11. *If φ is as in Theorem 9 and T is a logical truth then, for all positive integers m and n, $C(φa_1 \ldots φa_n, φa_1 \ldots φa_m, T)$.*

We could modify our explicata to allow universal generalizations about infinitely many individuals to be confirmed by sample descriptions. However, that would add considerable complexity and the empirical generalizations that I will discuss in what follows can naturally be taken to be concerned with finite populations of individuals. Therefore, instead of modifying the explicata, I will in what follows restrict attention to universal generalizations about finite populations.

3.8 Nicod's Condition

Jean Nicod (1923) maintained that a law of the form "All F are G" is confirmed by the evidence that some individual is both F and G. Let us call this *Nicod's condition*.

For the reason just indicated, I will take the generalization "All F are G" to be about individuals $a_1, \ldots, a_n$, for some finite $n > 1$. I will denote this generalization by A, so we have

$$A = (Fa_1 \supset Ga_1) \ldots (Fa_n \supset Ga_n).$$

Nicod's condition, translated into our explicata, is then the claim that $C(A, Fa.Ga, D)$, where a is any one of $a_1, \ldots, a_n$ and D remains to be specified.

Nicod did not specify the background evidence D for which he thought that his condition held. We can easily see that Nicod's condition does not hold for any background evidence D. For example, we have:

THEOREM 12. $\sim C(A, Fa.Ga, Fa.Ga \supset \sim A)$.

On the other hand,

THEOREM 13. $C(A, Fa.Ga, Fa)$.

Thus Nicod's condition may or may not hold, depending on what D is. So let us now consider the case in which D is a logically true sentence, representing the situation in which there is no background evidence. Some authors, such as Hempel (1945) and Maher (1999), have maintained that Nicod's condition holds in this case. I will now examine whether this is correct according to our explicata.

In order to have a simple case to deal with, let $n = 2$, so

$$A = (Fa_1 \supset Ga_1).(Fa_2 \supset Ga_2).$$

In section 3.5, I said that I favored choosing $\lambda = 2$ and $p(I) = 1/2$, while the values of γ_F and γ_G should depend on what "F" and "G" mean and may be very small. Suppose, then, that we have

$$\gamma_F = 0.001; \qquad \gamma_G = 0.1; \qquad \lambda = 2; \qquad p(I) = 1/2.$$

In this case calculation shows that, with $a = a_1$ or a_2,

$$p(A|Fa.Ga) = 0.8995 < 0.9985 = p(A).$$

Table 3.1 Probabilities of $Q_i b$ in the counterexample to Nicod's condition

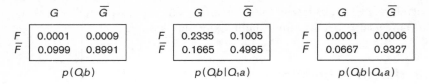

	G	$\overline{G}$		G	$\overline{G}$		G	$\overline{G}$
F	0.0001	0.0009	F	0.2335	0.1005	F	0.0001	0.0006
$\overline{F}$	0.0999	0.8991	$\overline{F}$	0.1665	0.4995	$\overline{F}$	0.0667	0.9327
	$p(Q_i b)$			$p(Q_i b \mid Q_1 a)$			$p(Q_i b \mid Q_4 a)$	

So in this case $\sim\!C(A, Fa.Ga, T)$, which shows that according to our explications Nicod's condition does not always hold, even when there is no background evidence.

Since Nicod's condition has seemed very intuitive to many, this result might seem to reflect badly on our explicata. However, the failure of Nicod's condition in this example is intuitively intelligible, as the following example shows.

According to standard logic, "All unicorns are white" is true if there are no unicorns. Given what we know, it is almost certain that there are no unicorns and hence "All unicorns are white" is almost certainly true. But now imagine that we discover a white unicorn; this astounding discovery would make it no longer so incredible that a nonwhite unicorn exists and hence would disconfirm "All unicorns are white."

The above numerical counterexample to Nicod's condition is similar to the unicorn example; initially, it is improbable that Fs exist, since $\gamma_F = 0.001$, and the discovery of an F that is G then raises the probability that there is also an F that is not G, thus disconfirming that all F are G. Table 3.1 shows $p(Q_i b)$ and $p(Q_i b \mid Q_1 a)$ for $i = 1, 2, 3,$ and 4. This shows how the evidence $Q_1 a$ raises the probability that b is Q_2 and hence a counterexample to the generalization that all F are G. Initially, that probability is 0.0009, but given $Q_1 a$ it becomes 0.1005.[4]

3.9 The Ravens Paradox

The following have all been regarded as plausible:

(a) Nicod's condition holds when there is no background evidence.
(b) Confirmation relations are unchanged by substitution of logically equivalent sentences.
(c) In the absence of background evidence, the evidence that some individual is a nonblack nonraven does not confirm that all ravens are black.

However, (a)–(c) are inconsistent. For (a) implies that a nonblack nonraven confirms "all nonblack things are nonravens" and the latter is logically equivalent to "all ravens are black." Thus there has seemed to be a paradox here (Hempel, 1945).

---

4 Good (1968) argued that Nicod's condition could fail when there is no background evidence for the sort of reason given here. Maher (1999, section 4.6) showed that Good's stated premises did not entail his conclusion. The present discussion shows that Good's reasoning goes through if we use a probability function that allows for analogy effects between individuals that are known to differ in some way. The probability functions used by Maher (1999) do not allow for such analogy effects.

We have already seen that (a) is false, which suffices to resolve the paradox. But let us now use our explicanda to assess (b) and (c).

In terms of our explicanda, what (b) says is that if $H = H'$, $E = E'$, and $D = D'$, then $C(H, E, D)$ iff $C(H', E', D')$. It follows from Axiom 5 and Definition 1 that this is true. So we accept (b).

Now let F mean "raven" and G mean "black." Then, in terms of our explicata, what (c) asserts is $\sim C(A, \sim Fa.\sim Ga, T)$. But, using the same parameter values as in the numerical example of the preceding section, we find that:

$$p(A|\sim Fa.\sim Ga) = 0.9994 > 0.9985 = p(A).$$

So in this case $C(A, \sim Fa.\sim Ga, T)$. This is contrary to the intuitions of many, but when we understand the situation better it ceases to be unintuitive, as I will now show.

The box on the right in table 3.1 shows $p(Q_ib|Q_4a)$, which on the current interpretation is the probability of Q_ib given that a is a nonblack nonraven. We see that

$$p(Q_2b|Q_4a) = 0.0006 < 0.0009 = p(Q_2b)$$

and so Q_4a reduces the probability that b is a counterexample to "All ravens are black." This should not be surprising. In addition, Q_4a tells us that a is not a counterexample to "All ravens are black," which *a priori* it might have been. So, for these two reasons together, it ought not to be surprising that a nonblack nonraven can confirm that all ravens are black.

Thus our response to the ravens paradox is to reject both (a) and (c). Neither proposition holds generally according to our explicata and the reasons why they do not hold make intuitively good sense.

3.10 Projectability

Goodman (1979, p. 74) defined the predicate "grue" by saying that "it applies to all things examined before t just in case they are green but to other things just in case they are blue." Goodman claimed, and it has generally been accepted, that "grue" is not "projectable" – although most discussions, including Goodman's own, do not say precisely what they mean by "projectable." Goodman's discussion is also entangled with his mistaken acceptance of Nicod's condition. Our explications allow us to clarify this confused situation.

One precise concept of projectability is the following:

DEFINITION 3. Predicate ϕ is absolutely projectable iff $C(\phi b, \phi a, T)$.

The basic predicates in L are projectable in this sense, that is:

THEOREM 14. *The predicates F, $\overline{F}$, G, and $\overline{G}$ are absolutely projectable.*

Now let us define a predicate G' as follows:

DEFINITION 4. $G'a = (Fa.Ga) \vee (\sim Fa.\sim Ga)$.

If "F" means "observed before t" and "G" means "green," then "G'" has a meaning similar to "grue."

THEOREM 15. *G' is absolutely projectable.*

This is contrary to what many philosophers seem to believe, but careful consideration will show that our explications here again correspond well to their explicanda. Theorem 15 corresponds to this statement of ordinary language: the justified degree of belief that an individual is grue, given no evidence except that some other individual is grue, is higher than if there was no evidence at all. If we keep in mind that we do not know of either individual whether it has been observed before t, then this statement should be intuitively acceptable.

Philosophers regularly claim that if "green" and "grue" were both projectable, then the same evidence would confirm both that an unobserved individual is green and that it is not green. It is a demonstrable fact about our explicata that the same evidence cannot confirm both a sentence and its negation, so this claim is definitely false when explicated as above. When philosophers say things like this, they are perhaps assuming that we know of each individual whether or not it has been observed before t; however, the concept of absolute projectability says nothing about what is true with this background evidence. So let us now consider a different concept of projectability.

DEFINITION 5. Predicate ϕ is projectable across predicate ψ iff $C(\phi b, \phi a.\psi a, \sim \psi b)$.

THEOREM 16. *G is, and G' is not, projectable across F.*

We saw that if "F" means "observed before t" and "G" means "green," then "G'" has a meaning similar to "grue." With those meanings, Theorem 16 fits the usual views of what is and is not projectable.

However, we could specify that "G" means "observed before t and green or not observed before t and not green." Then, with "F" still meaning "observed before t," "G'" would mean "green"; in that case, Theorem 16 would be just the opposite of the usual views of what is projectable. This shows that the acceptability of our explicata depends on the meanings assigned to the primitive predicates "F" and "G" in the language L. We obtain satisfactory results if the primitive predicates express ordinary concepts such as "green" and we may not obtain satisfactory results if some primitive predicates express gerrymandered concepts such as "grue."

3.11 Conclusion

The predicate "C" is a good explicatum for confirmation, because it is similar to its explicandum and theoretically fruitful. This predicate was defined in terms of probability. In that sense, probability captures the logic of scientific confirmation.

Patrick Maher

3.12 Proofs

3.12.1 Propositions

This section states and proves some propositions that will later be used in the proofs of the theorems.

PROPOSITION 1. *If H is a logical consequence of E, then* $p(H|E) = 1$.

Proof: Suppose that H is a logical consequence of E. Then

$$
\begin{aligned}
1 &= p(E|E), &\text{by Axiom 2}\\
&= p(H.E|E), &\text{by Axiom 5}\\
&= p(H|E)p(E|H.E), &\text{by Axiom 4}\\
&= p(H|E)p(E|E), &\text{by Axiom 5}\\
&= p(H|E), &\text{by Axiom 2.}
\end{aligned}
$$

PROPOSITION 2. *If E is inconsistent, then* $p(H|E) = 1$.

Proof: If E is inconsistent, then H is a logical consequence of E and so, by Proposition 1, $p(H|E) = 1$.

PROPOSITION 3. *If E is consistent and E.H is inconsistent, then* $p(H|E) = 0$.

Proof: Suppose that E is consistent and $E.H$ is inconsistent. Then $\sim H$ is a logical consequence of E and so

$$
\begin{aligned}
p(H|E) &= 1 - p(\sim H|E), &\text{by Axiom 3}\\
&= 1 - 1, &\text{by Proposition 1}\\
&= 0.
\end{aligned}
$$

PROPOSITION 4. $p(E|D) = p(E.H|D) + p(E.\sim H|D)$, *provided that D is consistent*.

Proof: Suppose that D is consistent. If $E.D$ is consistent, then

$$
\begin{aligned}
p(E|D) &= p(E|D)[p(H|E.D) + p(\sim H|E.D)], &\text{by Axiom 3}\\
&= p(E.H|D) + p(E.\sim H|D), &\text{by Axiom 4.}
\end{aligned}
$$

If $E.D$ is inconsistent, then Proposition 3 implies that $p(E|D)$, $p(E.H|D)$, and $p(E.\sim H|D)$ are all zero, and so again $p(E|D) = p(E.H|D) + p(E.\sim H|D)$.

PROPOSITION 5 (law of total probability). *If D is consistent, then*

$$
p(E|D) = p(E|H.D)p(H|D) + p(E|\sim H.D)p(\sim H|D).
$$

Proof: Suppose that D is consistent. Then:

$$p(E|D) = p(E.H|D) + p(E.\sim H|D), \qquad \text{by Proposition 4}$$
$$= p(E|H.D)p(H|D) + p(E|\sim H.D)p(\sim H|D), \qquad \text{by Axiom 4.}$$

PROPOSITION 6. *If D is consistent and $p(E|D) > 0$, then*

$$p(H|E.D) = \frac{p(E|H.D)p(H|D)}{p(E|D)}.$$

Proof: Suppose that D is consistent and $p(E|D) > 0$. Then

$$p(H|E.D) = \frac{p(E.H|D)}{p(E|D)}, \qquad \text{by Axiom 4}$$
$$= \frac{p(E|H.D)p(H|D)}{p(E|D)}, \qquad \text{by Axiom 4.}$$

PROPOSITION 7 (Bayes's theorem). *If D is consistent and $p(E|D) > 0$, then*

$$p(H|E.D) = \frac{p(E|H.D)p(H|D)}{p(E|H.D)p(H|D) + p(E|\sim H.D)p(\sim H|D)}.$$

Proof: Immediate from Propositions 5 and 6.

PROPOSITION 8. *If E is a logical consequence of $H.D$, then $p(E|D) \geq p(H|D)$.*

Proof: If D is consistent, then

$$p(E|D) = p(E|H.D)p(H|D) + p(E|\sim H.D)p(\sim H|D), \qquad \text{by Proposition 5}$$
$$\geq p(E|H.D)p(H|D), \qquad \text{by Axiom 1}$$
$$= p(H|D), \qquad \text{by Proposition 1.}$$

If D is inconsistent then, by Proposition 2, $p(E|D) = p(H|D) = 1$, so again $p(E|D) \geq p(H|D)$.

3.12.2 Proof of Theorem 1

Suppose that E is a logical consequence of $H.D$, $0 < p(H|D) < 1$, and $p(E|\sim H.D) < 1$. By Proposition 2, D is consistent. So, by Axiom 3, $p(\sim H|D) > 0$. Since $p(E|\sim H.D) < 1$, it then follows that

$$p(E|\sim H.D)p(\sim H|D) < p(\sim H|D). \qquad (1)$$

Since $p(H|D) > 0$, it follows from Proposition 8 that $p(E|D) > 0$. So

$$p(H|E.D) = \frac{p(E|H.D)p(H|D)}{p(E|H.D)p(H|D) + p(E|{\sim}H.D)p({\sim}H|D)}, \qquad \text{by Proposition 7}$$

$$= \frac{p(H|D)}{p(H|D) + p(E|{\sim}H.D)p({\sim}H|D)}, \qquad \text{by Proposition 1}$$

$$> \frac{p(H|D)}{p(H|D) + p({\sim}H|D)}, \qquad \text{by (1) and } p(H|D) > 0$$

$$= p(H|D), \qquad \text{by Axiom 3.}$$

So, by Definition 1, $C(H, E, D)$.

3.12.3 Proof of Theorem 2

Suppose that E_2 is a logical consequence of $H.D$. It follows that E_2 is a logical consequence of $H.E_1.D$. Suppose further that $0 < p(H|E_1.D) < 1$ and $p(E_2|{\sim}H.E_1.D) < 1$. It then follows from Theorem 1 that $C(H, E_2, E_1.D)$. By Definition 1, this means that $p(H|E_1.E_2.D) > p(H|E_1.D)$. By Definition 2, it follows that $M(H, E_1.E_2, E_1, D)$.

3.12.4 Proof of Theorem 3

Suppose that E_1 and E_2 are logical consequences of $H.D$, $p(E_1|{\sim}H.D) < p(E_2|{\sim}H.D)$, and $0 < p(H|D) < 1$. By Proposition 2, D is consistent. So, by Axiom 3, $p({\sim}H|D) > 0$. Since $p(E_1|{\sim}H.D) < p(E_2|{\sim}H.D)$, it then follows that

$$p(E_1|{\sim}H.D)p({\sim}H|D) < p(E_2|{\sim}H.D)p({\sim}H|D). \tag{2}$$

Since $p(H|D) > 0$, it follows from Proposition 8 that $p(E_1|D) > 0$ and $p(E_2|D) > 0$. So

$$p(H|E_1.D) = \frac{p(E_1|H.D)p(H|D)}{p(E_1|H.D)p(H|D) + p(E_1|{\sim}H.D)p({\sim}H|D)}, \qquad \text{by Proposition 7}$$

$$= \frac{p(H|D)}{p(H|D) + p(E_1|{\sim}H.D)p({\sim}H|D)}, \qquad \text{by Proposition 1}$$

$$> \frac{p(H|D)}{p(H|D) + p(E_2|{\sim}H.D)p({\sim}H|D)}, \qquad \text{by (2) and } p(H|D) > 0$$

$$= \frac{p(E_2|H.D)p(H|D)}{p(E_2|H.D)p(H|D) + p(E_2|{\sim}H.D)p({\sim}H|D)}, \qquad \text{by Proposition 1}$$

$$= p(H|E_2.D), \qquad \text{by Proposition 7.}$$

So, by Definition 2, $M(H, E_1, E_2, D)$.

3.12.5 Proof of Theorem 5

$$p(Q_1a) = p(Q_1a|I)p(I) + p(Q_1a|{\sim}I)p({\sim}I), \qquad \text{by Proposition 5}$$

$$= \gamma_1 p(I) + \gamma_1 p({\sim}I), \qquad \text{by Theorem 4}$$

$$= \gamma_1, \qquad \text{by Axiom 3.}$$

Similarly, $\gamma_2 = p(Q_2a)$, $\gamma_3 = p(Q_3a)$, and $\gamma_4 = p(Q_4a)$.

$$
\begin{aligned}
p(Fa) &= p(Q_1a) + p(Q_2a), \qquad \text{by Proposition 4} \\
&= \gamma_1 + \gamma_2, \qquad \text{as just shown} \\
&= \gamma_F(\gamma_G + \gamma_{\overline{G}}), \qquad \text{by Theorem 4} \\
&= \gamma_F(\gamma_1 + \gamma_2 + \gamma_3 + \gamma_4), \qquad \text{by definition of } \gamma_G \text{ and } \gamma_{\overline{G}} \\
&= \gamma_F[p(Q_1a) + p(Q_2a) + p(Q_3a) + p(Q_4a)], \qquad \text{as just shown} \\
&= \gamma_F[p(Fa) + p(\sim Fa)], \qquad \text{by Proposition 4} \\
&= \gamma_F, \qquad \text{by Axiom 3.}
\end{aligned}
$$

Similarly, $\gamma_{\overline{F}} = p(\sim Fa)$, $\gamma_G = p(Ga)$, and $\gamma_{\overline{G}} = p(\sim Ga)$.

3.12.6 Propositions used in the proof of Theorems 6–8

This section states and proves some propositions that will be used in proving the theorems in section 3.6.

PROPOSITION 9. $p(I|Q_ia) = p(I)$ and $p(\sim I|Q_ia) = p(\sim I)$ for $i = 1, \ldots, 4$.

Proof:

$$
\begin{aligned}
p(I) &= \frac{\gamma_1 p(I)}{\gamma_1}, \qquad \text{trivially} \\
&= \frac{p(Q_1a|I)p(I)}{\gamma_1}, \qquad \text{by Theorem 4} \\
&= \frac{p(Q_1a|I)p(I)}{p(Q_1a)}, \qquad \text{by Theorem 5} \\
&= p(I|Q_1a), \qquad \text{by Proposition 6.}
\end{aligned}
$$

Similarly, $p(I|Q_ia) = p(I)$ for $i = 2, 3$, and 4. It follows from Axiom 3 that $p(\sim I|Q_ia) = p(\sim I)$.

PROPOSITION 10. $\gamma_F + \gamma_{\overline{F}} = \gamma_G + \gamma_{\overline{G}} = 1$.

Proof: By Theorem 5, $\gamma_F = p(Fa)$ and $\gamma_{\overline{F}} = p(\sim Fa)$. By Axiom 3, the sum of these is 1. Similarly, $\gamma_G + \gamma_{\overline{G}} = 1$.

PROPOSITION 11. *If ϕ is F, $\overline{F}$, G, or $\overline{G}$, then $0 < \gamma_\phi < 1$.*

Proof: By Theorem 4, $\gamma_F = \gamma_1 + \gamma_2 > 0$. Also

$$
\begin{aligned}
\gamma_F &= 1 - \gamma_{\overline{F}}, \qquad \text{by Proposition 10} \\
&< 1, \qquad \text{since } \gamma_{\overline{F}} > 0.
\end{aligned}
$$

The argument for $\overline{F}$, G, and $\overline{G}$ is similar.

PROPOSITION 12. *If ϕ is F, $\bar{F}$, G, or $\bar{G}$, then $\gamma_\phi < (1 + \lambda\gamma_\phi)/(1 + \lambda)$.*

Proof: By Proposition 11, $\gamma_\phi < 1$. Adding $\lambda\gamma_\phi$ to both sides and then dividing both sides by $1 + \lambda$ gives the proposition.

PROPOSITION 13. $\gamma_G < (1 + \lambda\gamma_1)/(1 + \lambda\gamma_F)$.

Proof: Substitute G for ϕ and $\lambda\gamma_F$ for λ in Proposition 12.

PROPOSITION 14. $p(Ga|Fa) = \gamma_G$.

Proof:

$$
\begin{aligned}
\gamma_F\gamma_G &= \gamma_1, && \text{by Theorem 4}\\
&= p(Q_1 a), && \text{by Theorem 5}\\
&= p(Fa)p(Ga|Fa), && \text{by Axiom 4}\\
&= \gamma_F\, p(Ga|Fa), && \text{by Theorem 5.}
\end{aligned}
$$

Dividing both sides by γ_F gives the proposition.

PROPOSITION 15.

$$
p(Gb|Fa.Ga.Fb) = \frac{1+\lambda\gamma_G}{1+\lambda}\,p(I) + \frac{1+\lambda\gamma_1}{1+\lambda\gamma_F}\,p(\sim I).
$$

Proof:

$$
\begin{aligned}
p(Q_1 b|Q_1 a) &= p(Q_1 b|Q_1 a.I)p(I|Q_1 a) + p(Q_1 b|Q_1 a.\sim I)p(\sim I|Q_1 a), && \text{by Proposition 5}\\
&= p(Q_1 b|Q_1 a.I)p(I) + p(Q_1 b|Q_1 a.\sim I)p(\sim I), && \text{by Proposition 9}\\
&= \frac{1+\lambda\gamma_F}{1+\lambda}\frac{1+\lambda\gamma_G}{1+\lambda}\,p(I) + \frac{1+\lambda\gamma_1}{1+\lambda}\,p(\sim I), && \text{by Theorem 4.} \qquad (3)
\end{aligned}
$$

Similarly,

$$
p(Q_2 b|Q_1 a) = \frac{1+\lambda\gamma_F}{1+\lambda}\frac{\lambda\gamma_{\bar{G}}}{1+\lambda}\,p(I) + \frac{\lambda\gamma_2}{1+\lambda}\,p(\sim I). \qquad (4)
$$

$$
\begin{aligned}
p(Fb|Q_1 a) &= p(Q_1 b|Q_1 a) + p(Q_2 b|Q_1 a), && \text{by Proposition 4}\\
&= \frac{1+\lambda\gamma_F}{1+\lambda}\frac{1+\lambda\gamma_G + \lambda\gamma_{\bar{G}}}{1+\lambda}\,p(I) + \frac{1+\lambda\gamma_1 + \lambda\gamma_2}{1+\lambda}\,p(\sim I), && \text{by (3) and (4)}\\
&= \frac{1+\lambda\gamma_F}{1+\lambda}\,p(I) + \frac{1+\lambda\gamma_F}{1+\lambda}\,p(\sim I), && \text{by Proposition 10 and the}\\
&&& \text{definition of } \gamma_F\\
&= \frac{1+\lambda\gamma_F}{1+\lambda}, && \text{by Axiom 3.} \qquad (5)
\end{aligned}
$$

By Axioms 4 and 5,

$$p(Q_1b|Q_1a) = p(Fb|Q_1a)p(Gb|Fa.Ga.Fb).$$

Substituting (3) and (5) in this gives:

$$\frac{1+\lambda\gamma_F}{1+\lambda}\frac{1+\lambda\gamma_G}{1+\lambda}p(I) + \frac{1+\lambda\gamma_1}{1+\lambda}p(\sim I) = \frac{1+\lambda\gamma_F}{1+\lambda}p(Gb|Fa.Ga.Fb).$$

Dividing both sides by $(1 + \lambda\gamma_F)/(1 + \lambda)$ gives the proposition.

PROPOSITION 16.

$$p(Gb|\sim Fa.Ga.Fb) = \frac{1+\lambda\gamma_G}{1+\lambda}p(I) + \gamma_G p(\sim I).$$

Proof: Similar to Proposition 15.

3.12.7 Proof of Theorem 6

$$\begin{aligned}
p(Gb|Fa.Ga.Fb) &= \frac{1+\lambda\gamma_G}{1+\lambda}p(I) + \frac{1+\lambda\gamma_1}{1+\lambda\gamma_F}p(\sim I), &&\text{by Proposition 15} \\
&> \gamma_G p(I) + \gamma_G p(\sim I), &&\text{by Propositions 12 and 13} \\
&= \gamma_G, &&\text{by Axiom 3} \\
&= p(Gb|Fb), &&\text{by Proposition 14.}
\end{aligned}$$

So, by Definition 1, $C(Gb, Fa.Ga, Fb)$.

3.12.8 Proof of Theorem 7

$$\begin{aligned}
p(Gb|\sim Fa.Ga.Fb) &= \frac{1+\lambda\gamma_G}{1+\lambda}p(I) + \gamma_G p(\sim I), &&\text{by Proposition 16} \\
&> \gamma_G p(I) + \gamma_G p(\sim I), &&\text{by Proposition 12 and } p(I) > 0 \\
&= \gamma_G, &&\text{by Axiom 3} \\
&= p(Gb|Fb), &&\text{by Proposition 14.}
\end{aligned}$$

So, by Definition 1, $C(Gb, \sim Fa.Ga, Fb)$.

3.12.9 Proof of Theorem 8

$$\begin{aligned}
p(Gb|Fa.Ga.Fb) &= \frac{1+\lambda\gamma_G}{1+\lambda}p(I) + \frac{1+\lambda\gamma_1}{1+\lambda\gamma_F}p(\sim I), &&\text{by Proposition 15} \\
&> \frac{1+\lambda\gamma_G}{1+\lambda}p(I) + \gamma_G p(\sim I), &&\text{by Proposition 13 and } p(I) < 1 \\
&= p(Gb|\sim Fa.Ga.Fb), &&\text{by Proposition 16.}
\end{aligned}$$

So, by Definition 2, $M(Gb, Fa.Ga, \sim Fa.Ga, Fb)$.

3.12.10 Propositions used in the proof of Theorem 9

PROPOSITION 17. *If $E_1, \ldots, E_n$ are pairwise inconsistent and D is consistent, then*

$$p(E_1 \vee \cdots \vee E_n | D) = p(E_1 | D) + \ldots + p(E_n | D).$$

Proof: If $n = 1$, then the proposition is trivially true. Now suppose that the proposition holds for $n = k$ and let $E_1, \ldots, E_{k+1}$ be pairwise inconsistent propositions. Then

$$
\begin{aligned}
p(E_1 \vee \cdots \vee E_{k+1} | D) &= p((E_1 \vee \cdots \vee E_{k+1}).\sim E_{k+1} | D) + \\
&\quad p((E_1 \vee \cdots \vee E_{k+1}).E_{k+1} | D), \qquad \text{by Proposition 4} \\
&= p(E_1 \vee \cdots \vee E_k | D) + p(E_{k+1} | D), \qquad \text{by Axiom 5} \\
&= p(E_1 | D) + \ldots + p(E_{k+1} | D), \qquad \text{by assumption.}
\end{aligned}
$$

Thus the proposition holds for $n = k + 1$. So, by mathematical induction, the proposition holds for all positive integers n.

PROPOSITION 18. *If $\lambda > 0$ and $0 < \gamma < 1$, then*

$$\prod_{i=0}^{\infty} \frac{i + \lambda\gamma}{i + \lambda} = 0.$$

Proof: Let $\bar{\gamma} = 1 - \gamma$. Then, for all $i \geq 0$,

$$\frac{i + \lambda\gamma}{i + \lambda} = 1 - \frac{\lambda\bar{\gamma}}{i + \lambda}. \tag{6}$$

Also $0 < \lambda\bar{\gamma}/(i + \lambda) < 1$ for all $i \geq 0$. Now

$$\int_{x=0}^{\infty} \frac{\lambda\bar{\gamma}}{x + \lambda} dx = \lambda\bar{\gamma}[\ln(x + \lambda)]_{x=0}^{\infty} = \infty.$$

So, by the integral test for convergence of infinite series (Flatto, 1976, Theorem 5.10),

$$\sum_{i=0}^{\infty} \frac{\lambda\bar{\gamma}}{i + \lambda} = \infty.$$

Hence, by (6) and Theorem 5.32(2) of Flatto (1976),

$$\prod_{i=0}^{\infty} \frac{i + \lambda\gamma}{i + \lambda} = 0.$$

PROPOSITION 19. *If ϕ is any of the predicates F, $\overline{F}$, G, or $\overline{G}$, then, for all positive integers n,*

$$p(\phi a_1 \ldots \phi a_n) = \prod_{i=0}^{n-1} \frac{i + \lambda \gamma_\phi}{i + \lambda}.$$

Proof: Let E be any sample description for $a_1, \ldots, a_n$ that ascribes either Q_1 or Q_2 to each of $a_1, \ldots, a_n$. Let n_i be the number of individuals to which E ascribes Q_i. Then

$p(Fa_{n+1}|Fa_1 \cdots Fa_n.I)$

$\displaystyle = \sum_E p(Fa_{n+1}|E.I)p(E|Fa_1 \cdots Fa_n.I),$ by Proposition 5

$\displaystyle = \sum_E [p(Q_1a_{n+1}|E.I) + p(Q_2a_{n+1}|E.I)]p(E|Fa_1 \cdots Fa_n.I),$ by Proposition 4

$\displaystyle = \sum_E \frac{n + \lambda \gamma_F}{n + \lambda}\left[\frac{n_G + \lambda \gamma_G}{n + \lambda} + \frac{n_{\overline{G}} + \lambda \gamma_{\overline{G}}}{n + \lambda}\right]p(E|Fa_1 \cdots Fa_n.I),$ by Theorem 4

$\displaystyle = \sum_E \frac{n + \lambda \gamma_F}{n + \lambda}\, p(E|Fa_1 \cdots Fa_n.I),$ by Proposition 10 and $n_G + n_{\overline{G}} = n$

$\displaystyle = \frac{n + \lambda \gamma_F}{n + \lambda} \sum_E p(E|Fa_1 \cdots Fa_n.I),$ since n is the same for all E

$\displaystyle = \frac{n + \lambda \gamma_F}{n + \lambda}\, p\Big(\bigvee_E E|Fa_1 \cdots Fa_n.I\Big),$ by Proposition 17

$\displaystyle = \frac{n + \lambda \gamma_F}{n + \lambda},$ by Proposition 1.   (7)

$p(Fa_{n+1}|Fa_1 \cdots Fa_n.\sim I)$

$\displaystyle = \sum_E p(Fa_{n+1}|E.\sim I)p(E|Fa_1 \cdots Fa_n.\sim I),$ by Proposition 5

$\displaystyle = \sum_E [p(Q_1a_{n+1}|E.\sim I) + p(Q_2a_{n+1}|E.\sim I)]p(E|Fa_1 \cdots Fa_n.\sim I),$ by Proposition 4

$\displaystyle = \sum_E \left[\frac{n_1 + \lambda \gamma_1}{n + \lambda} + \frac{n_2 + \lambda \gamma_2}{n + \lambda}\right]p(E|Fa_1 \cdots Fa_n.\sim I),$ by Theorem 4

$\displaystyle = \sum_E \frac{n + \lambda \gamma_F}{n + \lambda}\, p(E|Fa_1 \cdots Fa_n.\sim I),$ since $n_1 + n_2 = n$

$\displaystyle = \frac{n + \lambda \gamma_F}{n + \lambda},$ using Proposition 17 as above.   (8)

$p(Fa_{n+1}|Fa_1 \cdots Fa_n)$

$\displaystyle = p(Fa_{n+1}|Fa_1 \cdots Fa_n.I)p(I|Fa_1 \cdots Fa_n) +$
$\displaystyle \quad p(Fa_{n+1}|Fa_1 \cdots Fa_n.\sim I)p(\sim I|Fa_1 \cdots Fa_n),$ by Proposition 5

$\displaystyle = \frac{n + \lambda \gamma_F}{n + \lambda}[p(I|Fa_1 \cdots Fa_n) + p(\sim I|Fa_1 \cdots Fa_n)],$ by (7) and (8)

$\displaystyle = \frac{n + \lambda \gamma_F}{n + \lambda},$ by Axiom 3.   (9)

$$p(Fa_1 \cdots Fa_n) = p(Fa_1) \prod_{i=1}^{n-1} p(Fa_{i+1}|Fa_1 \cdots Fa_i), \qquad \text{by Axiom 4}$$

$$= \prod_{i=0}^{n-1} \frac{i + \lambda\gamma_F}{i + \lambda}, \qquad \text{by Theorem 5 and (9).}$$

So, by mathematical induction, the proposition holds for $\phi = F$. Parallel reasoning shows that it also holds for $\phi = \overline{F}$, $\phi = G$, and $\phi = \overline{G}$.

3.12.11 Proof of Theorem 9

Case (i): ϕ is F, $\overline{F}$, G, or $\overline{G}$. Let $\varepsilon > 0$. By Theorem 4, $\lambda > 0$ and $0 < \gamma_\phi < 1$, so by Proposition 18 there exists an integer N such that

$$\prod_{i=0}^{N-1} \frac{i + \lambda\gamma_\phi}{i + \lambda} < \varepsilon.$$

Now

$$p[(x)\phi x] \le p(\phi a_1 \cdots \phi a_N), \qquad \text{by Proposition 8}$$

$$= \prod_{i=0}^{N-1} \frac{i + \lambda\gamma_\phi}{i + \lambda}, \qquad \text{by Proposition 19}$$

$$< \varepsilon, \qquad \text{by choice of } N.$$

Hence $p[(x)\phi x] = 0$. Also,

$$p[(x)\phi x|E] = \frac{p[(x)\phi x.E]}{p(E)}, \qquad \text{by Axiom 4 and } p(E) > 0$$

$$\le \frac{p[(x)\phi x]}{p(E)}, \qquad \text{by Proposition 8}$$

$$= 0, \qquad \text{since } p[(x)\phi x] = 0.$$

Case (ii): ϕ is Q_1, Q_2, Q_3, or Q_4.

$$p[(x)Q_1 x|E] \le p[(x)Fx|E], \qquad \text{by Proposition 8}$$

$$= 0, \qquad \text{from case (i).}$$

Similar reasoning shows that the result also holds if ϕ is Q_2, Q_3, or Q_4.

3.12.12 Proof of Theorem 10

Using Proposition 19 and Axiom 5, $p(\phi a_1 \ldots \phi a_m.T) > 0$. Also, $p(T) > 0$. So by Theorem 9,

$$p[(x)\phi x|\phi a_1 \cdots \phi a_m.T] = p[(x)\phi x|T] = 0.$$

So, by Definition 1, $\sim C[(x)\phi x,\ \phi a_1 \ldots \phi a_m,\ T]$.

3.12.13 Proof of Theorem 11

If $m \geq n$, then

$$p(\phi a_1 \cdots \phi a_n | \phi a_1 \cdots \phi a_m) = 1, \qquad \text{by Proposition 1}$$
$$> p(\phi a_1 \cdots \phi a_n), \qquad \text{by Proposition 19.}$$

By Proposition 19, $0 < p(\phi a_1 \ldots \phi a_m) < 1$. So if $m < n$, then

$$p(\phi a_1 \cdots \phi a_n | \phi a_1 \cdots \phi a_m) = \frac{p(\phi a_1 \cdots \phi a_n)}{p(\phi a_1 \cdots \phi a_m)}, \qquad \text{by Axiom 4}$$
$$> p(\phi a_1 \cdots \phi a_n).$$

Thus $p(\phi a_1 \ldots \phi a_n | \phi a_1 \ldots \phi a_m) > p(\phi a_1 \ldots \phi a_n)$ for all m and n. So, by Definition 1, $C(\phi a_1 \ldots \phi a_n,\ \phi a_1 \ldots \phi a_m,\ T)$.

3.12.14 Proof of Theorem 12

Let $D = Fa.Ga \supset \sim A$. Then $Fa.Ga.D$ is consistent and $Fa.Ga.D.A$ is inconsistent, so by Proposition 3, $p(A|Fa.Ga.D) = 0$. By Axiom 1, $p(A|D) \geq 0$ and so, by Definition 1, $\sim C(A|Fa.Ga,\ D)$.

3.12.15 Proof of Theorem 13

$$p(A|Fa.Ga.Fa) = p(A|Fa.Ga), \qquad \text{by Axiom 5}$$
$$> p(A|Fa.Ga)p(Ga|Fa),$$
$$\quad \text{since } p(Ga|Fa) < 1 \text{ by Propositions 11 and 14}$$
$$= p(A|Fa.Ga)p(Ga|Fa) + p(A|Fa.\sim Ga)p(\sim Ga),$$
$$\quad \text{since } p(A|Fa.\sim Ga) = 0 \text{ by Proposition 3}$$
$$= p(A|Fa), \qquad \text{by Proposition 5.}$$

So, by Definition 1, $C(A,\ Fa.Ga,\ Fa)$.

3.12.16 Proof of Theorem 14

Let ϕ be F, $\overline{F}$, G, or $\overline{G}$. Then

$$p(\phi b|\phi a) = \frac{p(\phi a.\phi b)}{p(\phi a)}, \qquad \text{by Axiom 4}$$
$$= \frac{1 + \lambda \gamma_\phi}{1 + \lambda}, \qquad \text{by Proposition 19}$$
$$> \gamma_\phi, \qquad \text{by Proposition 12}$$
$$= p(\phi b), \qquad \text{by Theorem 5.}$$

So, by Definition 1, $C(\phi b,\ \phi a,\ T)$. Hence, by Definition 3, ϕ is absolutely projectable.

3.12.17 Proof of Theorem 15

$$p(G'a.G'b|I) = p(Q_1a.Q_1b|I) + p(Q_1a.Q_4b|I) + p(Q_4a.Q_1b|I) +$$
$$p(Q_4a.Q_4b|I), \qquad \text{by Proposition 4}$$

$$= \gamma_1 \frac{1+\lambda\gamma_F}{1+\lambda} \frac{1+\lambda\gamma_G}{1+\lambda} + 2\gamma_1\gamma_4 \frac{\lambda^2}{(1+\lambda)^2} +$$
$$\gamma_4 \frac{1+\lambda\gamma_{\bar{F}}}{1+\lambda} \frac{1+\lambda\gamma_{\bar{G}}}{1+\lambda}, \qquad \text{by Axiom 4 and Theorem 4}$$

$$= \gamma_1\left(\gamma_F + \frac{\gamma_{\bar{F}}}{1+\lambda}\right)\left(\gamma_G + \frac{\gamma_{\bar{G}}}{1+\lambda}\right) + 2\gamma_1\gamma_4 \frac{\lambda^2}{(1+\lambda)^2} +$$
$$\gamma_4\left(\gamma_{\bar{F}} + \frac{\gamma_F}{1+\lambda}\right)\left(\gamma_{\bar{G}} + \frac{\gamma_G}{1+\lambda}\right)$$

$$= (\gamma_1 + \gamma_4)^2 + \frac{1}{1+\lambda}\left(\gamma_1\gamma_2 + \gamma_1\gamma_3 + \gamma_2\gamma_4 + \gamma_3\gamma_4 - 4\gamma_1\gamma_4\frac{\lambda}{1+\lambda}\right)$$

$$> (\gamma_1 + \gamma_4)^2 + \frac{1}{1+\lambda}(\gamma_1\gamma_2 + \gamma_1\gamma_3 + \gamma_2\gamma_4 + \gamma_3\gamma_4 - 4\gamma_1\gamma_4)$$

$$= (\gamma_1 + \gamma_4)^2 + \frac{1}{1+\lambda}[\gamma_G\gamma_{\bar{G}}(\gamma_F - \gamma_{\bar{F}})^2 + \gamma_F\gamma_{\bar{F}}(\gamma_G - \gamma_{\bar{G}})^2]$$

$$\geq (\gamma_1 + \gamma_4)^2. \tag{10}$$

$$p(G'a.G'b|{\sim}I) = p(Q_1a.Q_1b|{\sim}I) + p(Q_1a.Q_4b|{\sim}I) + p(Q_4a.Q_1b|{\sim}I) +$$
$$p(Q_4a.Q_4b|{\sim}I), \qquad \text{by Proposition 4}$$

$$= \gamma_1 \frac{1+\lambda\gamma_1}{1+\lambda} + 2\gamma_1\gamma_4 \frac{\lambda}{1+\lambda} + \gamma_4 \frac{1+\lambda\gamma_4}{1+\lambda},$$
$$\text{by Axiom 4 and Theorem 4}$$

$$= \gamma_1\left(\gamma_1 + \frac{1-\gamma_1}{1+\lambda}\right) + 2\gamma_1\gamma_4\left(1 - \frac{1}{1+\lambda}\right) + \gamma_4\left(\gamma_4 + \frac{1-\gamma_4}{1+\lambda}\right)$$

$$= (\gamma_1 + \gamma_4)^2 + \frac{(\gamma_1 + \gamma_4)(\gamma_2 + \gamma_3)}{1+\lambda}$$

$$> (\gamma_1 + \gamma_4)^2, \qquad \text{by Theorem 4.} \tag{11}$$

$$p(G'b|G'a) = \frac{p(G'a.G'b)}{p(G'a)}, \qquad \text{by Axiom 4}$$

$$= \frac{p(G'a.G'b|I)p(I) + p(G'a.G'b|{\sim}I)p({\sim}I)}{p(G'a)}, \qquad \text{by Proposition 5}$$

$$> \frac{(\gamma_1 + \gamma_4)^2 p(I) + (\gamma_1 + \gamma_4)^2 p({\sim}I)}{p(G'a)}, \qquad \text{by (10) and (11)}$$

$$= \frac{(\gamma_1 + \gamma_4)^2}{p(G'a)}, \qquad \text{by Axiom 3}$$

$$= p(G'b), \qquad \text{since } p(G'a) = p(G'b) = \gamma_1 + \gamma_4.$$

So, by Definition 1, $C(G'b, G'a, T)$. Hence, by Definition 3, G' is absolutely projectable.

3.12.18 Proof of Theorem 16

Interchanging F and $\sim F$ in Theorem 7 gives $C(Gb, Fa.Ga, \sim Fb)$. The proof of this is the same, *mutatis mutandis*, as the proof of Theorem 7. So, by Definition 5, G is projectable across F.

$$p(G'b|Fa.G'a.\sim Fb) = \frac{p(G'b.\sim Fb|Fa.G'a)}{p(\sim Fb|Fa.G'a)}, \qquad \text{by Axiom 4}$$

$$= \frac{p(\sim Gb.\sim Fb|Fa.Ga)}{p(\sim Fb|Fa.Ga)}, \qquad \text{by Definition 4 and Axiom 5}$$

$$= p(\sim Gb|Fa.Ga.\sim Fb), \qquad \text{by Axiom 4}$$

$$= 1 - p(Gb|Fa.Ga.\sim Fb), \qquad \text{by Axiom 3}$$

$$< 1 - p(Gb|\sim Fb), \qquad \text{since } G \text{ is projectable across } F$$

$$= p(\sim Gb|\sim Fb), \qquad \text{by Axiom 3}$$

$$= \frac{p(\sim Gb.\sim Fb)}{p(\sim Fb)}, \qquad \text{by Axiom 4}$$

$$= \frac{p(G'b.\sim Fb)}{p(\sim Fb)}, \qquad \text{by Definition 4 and Axiom 5}$$

$$= p(G'b|\sim Fb), \qquad \text{by Axiom 4.}$$

So, by Definition 1, $\sim C(G'b, Fa.G'a, \sim Fb)$. Hence, by Definition 5, G' is not projectable across F.

Bibliography

Carnap, R. 1950: *Logical Foundations of Probability*. Chicago: University of Chicago Press. Second edition 1962.

— 1980: A basic system of inductive logic, part II. In R. C. Jeffrey (ed.), *Studies in Inductive Logic and Probability*, vol. 2. Berkeley, CA: University of California Press, 7–155.

Flatto, L. 1976: *Advanced Calculus*. Baltimore, MD: Williams & Wilkins.

Galileo Galilei 1638: *Discorsi e Dimostrazioni Matematiche intorno à due nuove scienze*. Leiden: Elsevier. (References are to the English translation, *Dialogues Concerning Two New Sciences*, by H. Crew and A. de Salvio. New York: Dover, 1954.)

Good, I. J. 1968: The white shoe *qua* herring is pink. *British Journal for the Philosophy of Science*, 19, 156–7.

Goodman, N. 1979: *Fact, Fiction, and Forecast*, 3rd edn. Indianapolis, IN: Hackett.

Hempel, C. G. 1945: Studies in the logic of confirmation. *Mind*, 54. (Page references are to the reprint in Hempel, 1965.)

Hempel, C. G. 1965: *Aspects of Scientific Explanation*. New York: The Free Press.

Huygens, C. 1690: *Traite de la Lumiere*. Leiden: Pierre vander Aa. (English translation, *Treatise on Light*, by S. P. Thompson, New York: Dover, 1962.)

Maher, P. 1999: Inductive logic and the ravens paradox. *Philosophy of Science*, 66, 50–70.

— 2000: Probabilities for two properties. *Erkenntnis*, 52, 63–91.

— 2001: Probabilities for multiple properties: the models of Hesse and Carnap and Kemeny. *Erkenntnis*, 55, 183–216.

Nicod, J. 1923: *Le Problème Logique de l'Induction*. Paris: Alcan. (English translation in Nicod, 1970.)

— 1970: *Geometry and Induction*. Berkeley and Los Angeles: University of California Press. (English translation of works originally published in French in 1923 and 1924.)

von Wright, G. H. 1957: *The Logical Problem of Induction*, 2nd edn. Oxford: Blackwell.

Further reading

Corfield, D. and Williamson, J. (eds.), 2001: *Foundations of Bayesianism*. Dordrecht: Kluwer.

Earman, J. 1992: *Bayes or Bust?* Cambridge, MA: The MIT Press.

Horwich, P. 1982: *Probability and Evidence*. Cambridge, UK: Cambridge University Press.

Howson, C. and Urbach, P. 1993: *Scientific Reasoning: The Bayesian Approach*, 2nd edn. Chicago: Open Court.

Jaynes, E. T. 2003: *Probability Theory: The Logic of Science*. Cambridge, UK: Cambridge University Press.

Maher, P. 1990: How prediction enhances confirmation. In J. M. Dunn and A. Gupta (eds.), *Truth or Consequences: Essays in Honor of Nuel Belnap*. Dordrecht: Kluwer, 327–43.

— 1996: Subjective and objective confirmation. *Philosophy of Science*, 63, 149–73.

CHAPTER FOUR

Why Probability does not Capture the Logic of Scientific Justification

Kevin T. Kelly and Clark Glymour

4.1 Introduction

Here is the usual way philosophers think about science and induction. Scientists do many things – aspire, probe, theorize, conclude, retract, and refine – but successful research culminates in a published research report that presents an argument for some empirical conclusions. In mathematics and logic there are sound deductive arguments that fully justify their conclusions, but such proofs are unavailable in the empirical domain, because empirical hypotheses outrun the evidence adduced for them. Inductive skeptics insist that such conclusions cannot be justified. But "justification" is a vague term – if empirical conclusions cannot be established fully, as mathematical conclusions are, perhaps they are justified in the sense that they are partially supported or *confirmed* by the available evidence. To respond to the skeptic, one merely has to *explicate* the concept of confirmation or partial justification in a systematic manner that agrees, more or less, with common usage and to observe that our scientific conclusion are confirmed in the explicated sense. This process of explication is widely thought to culminate in some version of Bayesian confirmation theory.

Although there are nearly as many Bayesianisms as there are Bayesians, the basic idea behind Bayesian confirmation theory is simple enough. At any given moment, a rational agent is required to assign a unique degree of belief to each proposition in some collection of propositions closed under "and," "or," and "not." Furthermore, it is required that degrees of belief satisfy the axioms of probability. *Conditional probability* is defined as follows:

$$P(h|e) = \frac{P(h \text{ and } e)}{P(e)}$$

Confirmation can then be explicated like this:

evidence *e confirms* hypothesis *h* (for agent *P*) if and only if $P(h|e) > P(h)$.

In other words, confirmation is just positive statistical dependence with respect to one's degrees of belief prior to their modification in light of *e*. In a similar spirit, the *degree* of confirmation can be explicated as the difference $P(h|e) - P(h)$.

So defined, confirmation depends on the structure of the prior probability function *P* to the extent that, for some choice of *P*, the price of tea in China strongly confirms that the moon is green cheese. Personalists embrace this subjectivity, whereas objective Bayesians impose further restrictions on the form of *P* to combat it (cf., Patrick Maher's chapter 3 in this volume).

Confirmation theory's attractiveness to philosophers is obvious. First, it responds to the skeptic's challenge not with a proof, but with a conceptual analysis, conceptual analysis being the special skill claimed by analytic philosophers. Second, confirmation theorists seem to derive the pure "ought" of scientific conduct from the "is" of manifest practice and sentiment. No further argument is required that confirmation helps us *accomplish* anything, like finding the truth, for confirmation is (analytically) justification of belief and justification of belief that *h* is justification of belief that *h* is true. Third, in spite of the dependence on prior probability, Bayesian confirmation provides a simple unification of a variety of qualitative judgements of evidential relevance, some of which will be described below (cf., Patrick Maher's chapter 3). Fourth, explications are hard to refute. Celebrated divergences between human behavior and the Bayesian ideal can be chalked up as "fallacies" due to psychological foible or computational infeasibility. What matters is that the explication provides a unified, simple explanation of a wide range of practice and that when violations are called to attention, the violator (or at least you, as a third party), will agree that the violation should be corrected (Savage, 1951). Fifth, there are unexpected, *a priori*, arguments in favor of Bayesian principles called Dutch book arguments (DeFinetti, 1937; Teller, 1973). The basic idea is that you can't guard against possible disasters when you bet on the future, but at least you can guard against *necessary* disasters (i.e., combinations of bets in which one loses on matter what). It is then argued that Bayesian methodology is the unique way to avoid preferences for sure-loss bets. That isn't what anybody ever thought scientific method is for, but who ever said that philosophy can't make novel discoveries? Finally, confirmation theory restricts philosophical attention to a tractable subdomain of scientific practice. Confirmation stands to science as proof stands to mathematics. There are interesting psychological questions about how we find proofs, but regardless of their intrinsic psychological and sociological interest, such issues are irrelevant to the resulting proof's validity. Similarly, science is an ongoing social process that retracts, repairs, and revises earlier theories; but the philosophical relevance of these social, psychological, and historical details is "screened off" by confirmation (Hempel, 1965; Laudan, 1980).

So what's not to like? One might dwell upon the fact that scientists from Newton through Einstein produced, modified, and refined theories without attaching probabilities to them, even thought they were capable of doing so, had they so desired (Glymour, 1980); or on the fact that the sweeping consistency conditions implied by

Bayesian ideals are computationally and mathematically intractable even for simple logical and statistical examples (Kelly and Schulte, 1995); or on the fact that, in Bayesian statistical practice, the selection of prior probabilities often has more to do with mathematical and computational tractability than with anyone's genuine degree of belief (Lee, 1989); or on gaps in the Bayesians' pragmatic Dutch book arguments (Kyburg, 1978; Maher, 1997; Levi, 2002); or on the fact that Bayesian ideals are systematically rejected by human subjects, even when cognitive loading is not at issue (e.g., Allais, 1953; Ellsberg, 1961; Kahneman and Tversky 1972); or on the fact that a serious attempt to explicate real scientific practice today would reveal classical (non-Bayesian) statistics texts on the scientist's bookshelf and classical statistical packages running on her laboratory's desktop computer.

Our objection, however, is different from all of the above. It is that Bayesian confirmation is not even the right *sort* of thing to serve as an explication of scientific justification. Bayesian confirmation is just a change in the current output of a particular strategy or method for updating degrees of belief, whereas scientific justification depends on the truth-finding *performance* of the methods we use, whatever they might be. The argument goes like this:

1 Science has many aims, but its most characteristic aim is to find true answers to one's questions about nature.
2 So scientific justification should reflect how intrinsically difficult it is to find the truth and how efficient one's methods are at finding it. Difficulty and efficiency can be understood in terms of such cognitive costs as errors or retractions of earlier conclusions prior to convergence to the truth.
3 But Bayesian confirmation captures neither: conditional probabilities can fluctuate between high and low values any number of times as evidence accumulates, so an arbitrarily high degree of confirmation tells us nothing about how many fluctuations might be forthcoming in the future or about whether an alternative method might have required fewer.
4 Therefore, Bayesian confirmation cannot explicate the concept of scientific justification. It is better to say that Bayesian updating is just one method or strategy among many that may or may not be justified depending on how efficiently it answers the question at hand.

The reference to truth in the first premise is not to be taken too seriously. We are not concerned here with the metaphysics of truth, but with the problem of induction. Perhaps science aims only at theories consistent with all future experience or at theories that explain all future experience of a certain kind, or at webs of belief that don's continually have to be repaired as they encounter novel surface irritations in the future. Each of these aims outruns any finite amount of experience and, therefore, occasions skeptical concerns and a confirmation-theoretic response. To keep the following discussion idiomatic, let "truth" range over all such finite-evidence-transcending cognitive goals.

The second premise reflects a common attitude toward procedures in general: means are justified insofar as they efficiently achieve our ends. But immature ends are often infeasible – we want everything yesterday, with an ironclad warranty. Growing up is the painful process of learning to settle for what is possible and learning to achieve it efficiently. Truth is no exception to the rule: there is no foolproof, mechanical process that terminates in true theories. Faced with this dilemma, one must either give up on finding the truth or abandon the infeasible requirement that inductive procedures must *halt* or signal success when they succeed. Confirmation theorists adopt the first course, substituting confirmation, which can be obtained for sure, for truth, which cannot. We prefer the latter option, with retains a clear connection between method and truth-finding (Kelly, 2000). As William James (1948) wryly observed, no bell rings when science succeeds, but science may, nonetheless, slip across the finish line unnoticed.[1] Although one cannot demand that it do so smoothly and without a hitch, one may hope that due diligence will at least minimize the number of ugly surprises that we might encounter, as well as the elapsed time to their occurrence. Empirical justification is not a static Form in Plato's heaven, waiting to be recollected through philosophical analysis. If it is anything at all worth bothering about, it is grounded in the intrinsic difficulty of finding the truths that we seek and in the relative efficiency of our means for doing so.

The third premise is the crucial one. Confirmation theory encourages hope for more than efficient convergence to the truth, by promising some sort of "partial justification" in the short run. The terms "partial justification" and "partial support" hint at something permanent, albeit incomplete. But high degrees of belief are not permanent at all: arbitrarily high confirmation can evaporate in a heartbeat and can fluctuate between extremes repeatedly, never providing a hint about how many bumps might be encountered in the future or about whether some other method could guarantee fewer. Hence, Bayesian confirmation, or any other notion of confirmation that can be arbitrarily high irrespective of considerations of truth-finding efficacy, cannot explicate scientific justification.

Shifting the focus from confirmation relations to the feasibility and efficiency of truth-finding turns traditional, confirmation-based philosophy of science on its head. Process, generation, refinement, and retraction – the topics relegated to the ash-heap of history (or sociology or psychology) by confirmation theorists – are placed squarely in the limelight. Confirmation, on the other hand, is demoted to the status of a cog in the overall truth-finding process that must earn its keep like all the other parts. At best, it is a useful heuristic or defeasible pattern for designing efficient methods addressed to a wide range of scientific questions. At worst, rigid adherence to a preconceived standard of confirmation may prevent one from finding truths that might have been found efficiently by other means (Kelly and Schulte, 1995; Osherson and Weinstein, 1988).

4.2 Inductive Performance and Complexity

Our approach focuses on problems (ends) rather than on methods (means). An *empirical problem* is a pair (q, k) consisting of a *question q*, which specifies a unique, correct

---

1 The fallible convergence viewpoint on inquiry was urged in philosophy by James, Peirce, Popper, Von Mises, Reichenbach, and Putnam. Outside of philosophy, it shows up in classical estimation theory, Bayesian convergence theorems, and computational learning theory. Several of the points just made (e.g., noncumulativity) were argued on purely historical grounds by Kuhn and others.

answer for each possible world, and a *presupposition k* that restricts the range of possible worlds in which success is required. Let the presupposition be that you are watching a light that is either blue or green at each stage, and that is otherwise unconstrained. The question is what color the light will be at the very next stage. Then there is an easy method for *deciding* the problem with certainty: simply wait and report what you see. This procedure has the attractive property that it halts with the right answer, whatever the answer happens to be. But that is clearly because the next observation entails the right answer, so the problem is not really inductive when the right answer is obtained.

Next, consider the properly inductive question of whether the color will remain green forever. The obvious procedure guesses that the color won't change until it does, and then halts with the certain output that it does. This procedure is guaranteed to converge to the right answer whatever it is, but never yields certainty if the color never changes. It simply keeps waiting for a possible color change. We may say that such a method *refutes* the unchanging color hypothesis with certainty. This "one-sided" performance is reminiscent of Karl Popper's (1959) "anti-inductivist" philosophy of science – we arrive at the truth, but there is no such thing as accumulated "support" for our conviction, aside from the fact that we must leap, after a sufficiently long run of unchanging colors, to the conclusion that the color will never change *if* we are to converge to the right answer in the limit. In other words, the inductive leap is not pushed upward or supported by evidence; it is *pulled* upward by the aim of answering the question correctly.

One would prefer a "two-sided" decision procedure to the "one-sided" refutation procedure just described, but no such procedure exists for the problem at hand. For suppose it is claimed that a given method can decide the question under consideration with certainty. Nature can then feed the method constantly green experience until it halts with the answer that experience will always remain green (on pain of not halting with the true answer in that case). The method's decision to halt cannot be reversed, but Nature remains free to present a color change thereafter. In the stream of experience so presented, the method converges to the wrong answer "forever green," which contradicts the *reductio* hypothesis that the method converges to the truth. So, by *reductio and absurdum*, no possible method decides the question with certainty. This is essentially the classical argument for inductive skepticism, the view that induction is unjustified. We recommend the opposite conclusion: since no decision procedure is feasible in this case, a refutation procedure yields the best feasible sort of performance and, hence, is justified in light of the intrinsic difficulty of the problem addressed. Of course, this "best-we-can-do" justification is not as satisfying as a "two-sided" decision procedure would be, but *that* kind of performance is impossible and the grown-up attitude is to obtain the best possible performance as efficiently as possible, rather than to opine the impossible.

Next, suppose that the question is whether the color will never change, changes exactly once, or changes at least twice. The obvious method here is to say "never" until the color changes, "once" after it changes the first time, and "at least twice" thereafter. The first problem (about the color tomorrow) requires no retractions of one's initial answer. The second question (about unchanging color) requires one, and this question requires two. It is easy to extend the idea to questions requiring three, four, . . . retractions. Since retractions, or noncumulative breaks in the scientific tradition, are the observable signs of the problem of induction in scientific inquiry, one can measure the *intrinsic difficulty* or *complexity* of an empirical question by the least number of retraction that Nature could exact from an arbitrary method that converges to the right answer.

Some problems are not solvable under any fixed, finite bound on retractions. For example, suppose that it is *a priori* that the color will change only finitely often and the questions is how many times it will changes. Then Nature can lead us to change our minds any number of times by adding another color change just after the point at which we are sure we will never see another one. But the question is still *decidable in the limit* in the sense that it is possible to converge to the right answer, whatever it might be (e.g., by concluding at each stage that the color will never change again).

There are also problems for which no possible method can even converge to the truth in the limit of inquiry. One of them is a Kantian antinomy of pure reason: the question of whether matter is infinitely divisible. Let the experiment of attempting to cut matter be successively performed (failures to achieve a cut are met with particle accelerators of ever higher energy). Nature can withhold successful cuts until the method guesses that matter is finitely divisible. Then she can reveal cuts until the method guesses that matter is infinitely divisible. In the limit, matter is infinitely divisible (new cuts are revealed in each "fooling cycle"), but the method does not converge to "infinitely divisible." Nonetheless, it is still possible to converge to "finitely divisible" if and only if matter is only finitely divisible: just answer "finitely divisible" while no new cuts are performed and "infinitely divisible" each time a new cut is performed. Say that this method *verifies* finite divisibility in the limit. The same method may be said to *refute* infinite divisibility in the limit, in the sense that it converges to the alternative hypothesis just in case infinite divisibility is false. These "one-sided" concepts stand to decision in the limit as verificiation and refutation with certainty stand to decision with certainty. Of course, we would prefer a two-sided, convergent, solution to this problem, but none exists, so one-sided procedures are justified insofar as they are the best possible.

There are even questions that are neither refutable in the limit nor verifiable in the limit, such as whether the limiting relative frequency of green observations exists. This problem has the property that a method could output "degree of belief" or "confirmation values" that converge to unity if and only if the limiting relative frequency exists, but no possible method of this kind converges to unity if and only if the limiting relative frequency does not exist. And then there are problems that have neither of these properties. At this point, it starts to sound artificial to speak of convergent success in any sense.

Changes in background information can affect solvability. For example, let the question be whether a sequence of observed colors will converge to blue. Absent further background knowledge, the best one can do is to verify convergence to blue in the limit (the analysis is parallel to that of the finite divisibility example). But if we know *a priori* that the sequence of colors will eventually stabilize (e.g., the current color corresponds to which magnet a damped pendulum is nearest to), then the question is decidable in the limit: just respond "yes" while the color is blue and respond "no" otherwise. This shows that extra assumption may make a problem intrinsically easier to solve without giving the game away altogether.

In mathematical logic and computability theory, it is a commonplace that formal problems have *intrinsic complexities* and that a problem's intrinsic complexity determines the best possible sense in which a procedure can solve it. For example, the unavailability of a decision procedure for the first-order predicate calculus justifies the use of a one-sided verification procedure for inconsistency and of a one-sided refutation procedure for consistency. The notion that background presuppositions can make a problem easier is also familiar, for the predicate calculus is decidable if it is known in advance that all encountered instances will involve only monadic (one-place) predicates. All we have done so far is to apply this now-familiar computational perspective, which has proven so salutary in the philosophy of deductive reasoning, to empirical problems.

Philosophers of science are accustomed to think in terms of confirmation and underdetermination rather than in terms of methods and complexity, but the ideas are related. Bayesian updating, on our view, is just one method among the infinitely many possible methods for attaching numbers to possible answers. Underdetermination is a vague idea about the difficulty of discerning the truth of the matter from data. We propose that a problem's intrinsic complexity is a good explication for this vague notion, since it determines the best possible sense in which the problem is solvable.

Using retractions to measure inductive complexity is a more natural idea than it might first appear. First, the concepts of refutability and verifiability have a long standing in philosophy, and these concepts constititute just the first step in a retraction hierarchy (verifiability is success with one retraction starting with initial guess $\neg h$ and refutability is success with one retraction starting with h). Second, Thomas Kuhn (1970) emphasized that science is not cumulative, because in episodes of major scientific change some content of rejected theories is lost. These are retractions. Kuhn also emphasized the tremendous cost of cognitive retooling that these retractions occasion. Unfortunately, he did not take the next logical step of viewing the minimization of retractions as a natural aim that might provide alternative explanations of features of scientific practice routinely explained along confirmation-theoretic lines. Third, by generalizing resource bounds in a fairly natural way (Freivalds and Smith, 1903; Kelly, 2002), one can obtain the equation that each retraction is worth infinitely many errors (ω many, to be precise), so that the aim of minimizing the number of errors committed prior to convergence generates exactly the same complexity classes as minimizing retractions. Fourth, the concept of minimizing retractions is already familiar in logic and computability. In analysis, retraction complexity is called *difference* complexity (Kuratowski, 1966), and in computability it is known as "n-trial" complexity, a notion invented by Hilary Putnam (1965). Finally, the idea has been extensively studied in empirical applications by computational learning theorists (for an extended summary and bibliograhy, see Jain et al., 1999).

4.3 Explanations of Practice

Bayesian confirmation theorists have some foundational arguments for their methods (e.g., derivation from axioms of "rational" preference, Dutch book theorems) but one gets the impression that these are not taken too seriously, even by the faithful. What really impresses confirmation theorists is that Bayesian updating provides a unified, if highly idealized, explanation of a wide range of short-run judgments of evidential relevance. Such explanations are facilitated by *Bayes's theorem*, a trivial logical consequence of the definition of conditional probability:

$$P(h|e) = \frac{P(e|h)P(h)}{P(e)}.$$

It follows immediately, for example, that initial plausibility of h is good ($P(h)$ is upstairs), that prediction of e is good and refutation by e is bad ($P(e|h)$ is upstairs), and that surprising predictions are good ($P(e)$ is downstairs). Successive confirmation by instances has diminishing returns simply because the sum of the increases is bounded by unity. These explanations are *robust*: they work for any prior probability assignment such that $P(e) > 0$. Other explanations depend on prior probability. For example, it seems that black ravens confirm "all ravens are black" better than white shoes, and under some plausible assignments of prior probability, this judgment is accommodated. Under others, it isn't (cf., Patrick Maher's chapter 3).

The trouble with this naively hypothetico-deductive case for Bayesianism (even by Bayesian standards) is that it ignores competing explanations. In particular, it ignores the possibility that some of the same intuitions might follow from truth-finding efficiency itself, rather than from the details of a particular method. For an easy example, consider the maxim that scientific hypotheses should be consistent with the available evidence. If we assume that the date are true (as the Bayesian usually does), then any method that produces a refuted answer obviously hasn't converged to the truth *yet* and it is possible to do better (Schulte, 1999a,b). To see how, suppose that a method produces an answer h that is inconsistent with current evidence e but the method eventually converges to the truth. Since the answer is inconsistent with true data e and the method converges to the truth, the method eventually converges to an answer other than h in each world compatible with e. Let n be the least stage by which the method converges to the true answer h' (distinct from h) in some world w compatible with e. Now construct a new method that returns h' in w from the end of e onward. This method converges to the truth immediately in w, but converges no more slowly in any other world, so in decision-theoretic jargon one says that it *weakly dominates* the inconsistent method in convergence time, or that the inconsistent method is *inadmissable* with respect to convergence time. Since science is concerned primarily with finding the truth, avoiding needless delays is a natural and direct motive for consistency.

The preceding argument does not take computability into account. In some problems, a computable method can maintain consistency at each stage only by timidly failing to venture substantive answers, so it fails to converge to the truth in some worlds (Kelly and Schulte, 1995; Kelly, 1996). In other words, computable methods may have to produce refuted theories if they are to converge to the truth. In that case, a committed truth-seeker could rationally side with convergence over consistency, so the Bayesian's blanket insistence on idealized consistency as a necessary condition for "rationality" is too strong.

Consider next the maxim that it is better to predict the data than to merely accommodate them. Recall the example of whether the color will change no times, exactly once, or at least twice, and suppose that we have seen ten green observations. The only answer that predicts the next datum in light of past data is "the color never changes." Neither of the other answers is refuted, however, so why not choose one of them instead? Here is a reason based on efficiency: doing so could result in a needless retraction. For Nature could continue to present green inputs until, on pain of converging to the wrong answer, we cave in and conclude that "the color never changes." Thereafter, Nature could exhibit one color change followed by constant experience until we revise to "the color changes exactly once" and could then present another color change to make us revise again to "the color changes at least twice," for a total of three retractions. Had we favored "the color never changes" on constant experience, we could have succeeded with just two retractions in the worst case. Furthermore, after seeing a color change, we should prefer the answer "one color change," which is the only answer compatible with experience that entails the data until another color change occurs. To do otherwise would result in the possibility of two retractions from that point onward, when one retraction should have sufficed in the worst case.

Consider the question of whether "all ravens are black," and suppose that Nature is obligated to show us a black raven, eventually, if one exists. Then the most efficient possible method (in terms of retractions and convergence time) is to assume that the hypothesis is true until a counterexample is encountered and to conclude the contrary thereafter, since this method uses at most one retraction and is not weakly dominated in convergence time by any other method. Now suppose that one were to filter shoes out of the data stream and to reject "all ravens are black" as soon as a nonblack raven is encountered. We would succeed just as soon and with no more retractions than if we were to look at the unfiltered data. If one were to filter out ravens, however, no possible method could converge to the truth, even in the limit. More generally, a kind of datum is *irrelevant* (for the purposes of efficient inquiry) if systematically filtering data of that kind does not adversely affect efficiency. Suppose that we know in advance that all observed ravens are within one meter of a white paint can. Then, by the time the sphere with that radius is filled with positive instances, "all ravens are black" is conclusively refuted and an efficient method must reject it immediately. No mystery there: different problems call for different solutions. Indeed, the performance viewpoint explains what background information "is for": extra background constraints tend to make an empirical problem easier to solve.

4.4 Ockham's Razor and Efficiency

In this section, we show how efficiency explains one of the great mysteries of scientific method better than Bayesian confirmation can. A quick survey of the major scientific revolutions (e.g., the Copernican, the Newtonian, the Lavoisierian, the Darwinian, and so on) reveals an unmistakable pattern, described already by William Whewell (1840). The received theory of some domain achieves broad, shallow coverage over a range of phenomena by positing a large number of free parameters and then tweaking them until the various phenomena are accounted for. Then another, narrower, but more *unified* explanation is proposed that involves fewer parameters. The new theory appears implausible to those trained in the older tradition, but its ability to unify previously unrelated facts makes it ultimately irresistable.

Twenty years ago, one of us (Glymour, 1980) proposed that the unified theory is better *confirmed* because it is cross-tested in more different ways than the disunified theory by the same data. This has a tough, Popperian ring: the simpler or more unified theory survives a more rigorous, self-inflicted, cross-testing ordeal. But a theory is not a long-distance runner who needs training and character development to win – it just has to be *true*. Since reality might be disunified and complex (indeed, it *is* more complex than we used to suspect), how is the quest for *truth* furthered by presuming the true theory to be simple and severely cross-testable? If there is no clear answer to this question, then science starts to look like an extended exercise in sour grapes (if the world isn't they way I what it to be, I don't *care* what it is like) or in wishful thinking (I like simplicity, so the world must be simple).

Of course, Bayesians have no trouble *accommodating* simplicity biases: just assign greater prior probability to simpler hypotheses (e.g, Jeffreys, 1985). But that approach evidently presupposes the very bias whose special status is to be explained. One doesn't have to favor the simple theory outright, however; one need only "leave the door open" to it (by assigning it nonzero prior probability) and it *still* wins against a strong *a priori* bias toward its complex competitor (cf., Rosencrantz, 1983). For suppose that one merely assigns nonzero prior probability to the simple, unified theory s, which entails evidence e without extra assumptions, so that $P(e|s) = 1$. The complex, logically incompatible, competitor $c = \exists\theta.q(\theta)$ has a free continuous parameter θ to wiggle in order to account for future data. In the strongest possible version of the argument, there is a unique value θ_0 of the parameter such that $P(e|q(\theta_0)) = 1$ and at every other value of θ, $P(e|q(\theta)) = 0$. Finally, a "free continuous parameter" isn't really free if we have sharp *a priori* ideas about the best way to set it, so suppose that $P(q(\theta)|c)$ is zero for each value of θ including θ_0. Given these assumptions, $P(e|c) = \int P(e|q(\theta))P(q(\theta)|c)d\theta = 0$. Hence,

$$\frac{P(c|e)}{P(s|e)} = \frac{P(c)}{P(s)}\frac{P(e|c)}{P(e|s)} = \frac{P(c)}{P(s)}\frac{0}{1} = 0.$$

So the simple theory trounces its complex competitor, as long as $P(s) > 0$. This accounts for the temptation to say that it would be *a miracle* if the parameters of the complex theory were carefully adjusted by nature to reproduce the effects of s. If the hard-edged assumptions of the preceding argument are softened a bit, then the complex theory may end up victorious, but only if it is assigned a much greater prior probability than the simple theory.

This improved argument merely postpones the objectionable circularity of the first version, however. For focus not on the contest between c and s, but on the contest between $q(\theta_0)$ and s. Since both of these theories account for e equally well, there is no external or objective reason to prefer one to the other. In fact, the only reason s wins is because $P(s) > 0$, whereas $P(q(\theta_0)) = P(q(\theta_0)|c))P(c) + P(q(\theta_0)|s))P(s) = 0 \cdot p(c) + 0 \cdot P(s) = 0$. In other words, probabilistic "fairness" to s in the contest against c necessarily induces an infinite bias for s in the contest against $q(\theta_0)$. But one could just as well insist upon "fairness" in the contest between s and $q(\theta_0)$. Since we don't think

$q(\theta)$ is more plausible *a priori* than $q(\theta')$, for any other value θ', it follows that $P(q(\theta)) = P(s) = 0$. Then $P(c) = 1 - P(s) = 1$, so the complex theory c wins *a priori* (because it covers so many more possibilities). The moral is that *neither* Bayesian prior is really "open minded" – we are being offered a fool's choice between two extreme biases. What open-mindedness really dictates in this case is to reject the Bayesian's forced choice between prior probabilities altogether; but then the Bayesian explanation of Ockham's razor evaporates, since it is grounded entirely in prior probability.

Here is an alternative, efficiency-based explanation that presupposes no prior bias for or against simplicity and that doesn't even mention prior probabilities. Recall the question of whether the observed color will change zero times, once, or at least twice. Three different intuitions about simplicity lead to the same simplicity ranking over these answers. First, the hypothesis that the color never changes is more *uniform* than the hypotheses that allow for color changes. Secondly, the hypothesis that there are no color changes is the most *testable*, since it is refutable in isolation, whereas the other answers are refutable only given extra auxiliary hypotheses (e.g., the hypothesis that the color changes exactly once is refutable only under the extra assumption that the color will change at least once). Thirdly, the theory that there are no color changes has *no free parameters*. The theory that there is one color changes has one free parameter (the time of the change). The theory that there are at least two has at least two parameters (one for each change). So the constant-color hypothesis seems to carry many of the intuitive marks of simplicity. Now suppose that we prefer the needlessly complex theory that the color will change prior to seeing it do so. Nature can withhold all color changes until, on pain of converging to the wrong answer, our method outputs "no changes." Then Nature can exact two more retractions, for a total of three, when two would have sufficed had we always sided with the simplest hypothesis compatible with experience (once for the first color change and another for the second).

Easy as it is, this argument suggests a general, performance-based understanding of the role of simplicity in science that sheds new light on the philosophical stalemate over scientific realism. The anti-realist is right that Ockham's razor doesn't point at or *indicate* the truth, since the truth might be simple or complex, whereas Ockham's razor points at simplicity no matter what. But the realist is also right that simplicity is more than an arbitrary, subjective bias that is washed out, eventually, by future experience. It is something *in between*: a necessary condition for minimizing the number of surprises prior to convergence. So choosing the simplest answer compatible with experience is better justified (in terms of truth-finding efficacy) than choosing competing answers, but such justification provides *no security whatever* against multiple, horrible surprises in the future. The two theses are consistent, so the realism debate isn't a real debate. It is a situation. Our situation.

Suppose that one were to ask whether the "grolor" changes no times, once, or at least twice, where the "grolor" of an observation is either "grue" or "bleen," where "grue" means "green" prior to n_0 and blue thereafter" and "bleen" means "blue prior to n_0 and green thereafter" (Goodman, 1983). The preceding argument now requires that one guess "no grolor change" until a grolor change occurs. But there is no grolor change only if there is a color change, so it seems that the whole approach is inconsistent. The right moral, however, is that simplicity is relative to the problem addressed.

That is as it must be, for if simplicity is to facilitate inquiry over a wide range of problems, simplicity must somehow adapt itself to the contours of the particular problem addressed. There must be some general, structural concept of simplicity that yields distinct simplicity rankings in different problems.[2]

Here it is. Recall the problem in which it is known in advance that the observed color will change at most three times and the question is how many times it will change. Intuitively, the simplicity of a world in this problem is lower insofar as it presents more color changes: the zero color change worlds are most uniform and the three color change worlds are least uniform (with respect to the question at hand, which concerns color rather than grolor). Extrapolating from this example, let us suppose that worlds have discrete degrees of simplicity starting with zero and that worlds with lower simplicity degrees are less simple than those with higher degrees. So we would like worlds presenting three color changes to have simplicity degree zero in this problem. A characteristic property of such worlds is that they eventually present inputs verifying the answer "three color changes" for sure relative to the problem's presupposition that it changes at most three times. Next, we would like worlds that present two color changes to have simplicity degree 1, since they are apparently the next-to-least-simple worlds in the problem. Such worlds verify no answer to the original problem, but if we *add* the assumption that the color won't change three times (i.e., that the world's simplicity degree exceeds 0), then such worlds verify the answer "color changes twice." Worlds that present one color change should have the next simplicity degree, 2, and they verify the answer "color changes once" given the strengthened assumption that the color changes at most once (i.e., that the world's simplicity degree exceeds 1). Finally, worlds that present no color changes (the simplest, or uniform worlds) should have the maximum simplicity degree 3 in this problem and they verify the answer "no color changes" given that there are fewer than one color changes (i.e., given the assumption that the world has simplicity greater than 2).

Generalizing from the example, we define simplicity degrees inductively as follows. A world has simplicity degree 0 in a problem just in case it eventually presents inputs that verify some answer to the problem relative to the problem's original presupposition. A world has simplicity degree no greater than $n + 1$ just in case it eventually verifies some answer to the problem *given* the assumption that the world has simplicity strictly greater than n.[3] A world has simplicity degree *exactly* n just in case the world has simplicity degree no greater than n but greater than $n - 1$. Inductions are hard to stop once they get started. A world has simplicity degree no greater than (possibly infinite) ordinal number β if it verifies an answer to the problem given the strengthened presupposition that the world's simplicity exceeds each lower ordinal degree.

2 Goodman's own response to this issue was that there is a special of family of *projectible* predicates out of which confirmable gneralizations may be formulated. That approach grounds simplicity and justification in personal sentiment, which strikes us as wrong-headed. For us, sentiment is relevant only to the selection of problems. Justification then supervenes on objective efficiency with respect to the problems that sentiment selects. Hence, our approach involves a middle term (problems) that "screens off" sentiment from justification, allowing us to give an objective proof of the truth-finding efficacy of Ockham's razor over a broad range of problems.

3 This really is an induction, because if you already know what "no greater than n" means, then you know what it means to say that this is not the case: namely, that the world has simplicity degree strictly greater than n.

Sometimes, there is not even a transfinite ordinal bound on the simplicity degree of a world. Then we say that the world has indefinite simplicity degree. For example, if we know only that the color will change at most finitely often, then every world has indefinite simplicity degree, since no answer "n changes" is verified in any world (more color changes are always possible). Indeed, this will be the case in any problem in which no answer is eventually verified in any world, because then the backward induction can't get started. So the question arises whether there is a natural, epistemically motivated, sufficient condition for nontrivial application of the theory of simplicity degrees just proposed. And there is: every world has a definite simplicity degree if it is merely possible to decide the truth of the presupposition of the problem in the limit relative to the trivial presupposition that the data may arrive however they please (Kelly, 2002). In other words, if you could have converged to the truth about your presuppositions to begin with (else where did you get them, anyway?), then you do in fact assign a well-defined, ordinal simplicity degree to each world satisfying the presupposition of your problem. This isn't obvious,[4] but it is true (the proofs of the claims in this section are presented in detail in Kelly, 2002).

In this chapter, we will stick with problems in which worlds have finitely bounded simplicity degrees.[5] The simplicity degree of an answer is then definable as the maximum simplicity degree over worlds satisfying the answer (simple answers are true of simple worlds) and the *structural complexity* of a problem is definable as the maximum simplicity degree taken over worlds satisfying the problem's presupposition (complex problems have worlds of high simplicity degree because simplicity is a matter of Nature foregoing "dirty tricks" she is entitled to in the problem).

Ockham's razor can now be stated in the obvious way: never output an answer unless it is the uniquely simplest answer compatible with current experience. The important point is this: it is a mathematical theorem that any violation of Ockham's razor implies either that one fails to decide the question at hand in the limit or that one uses more retractions than necessary in the subproblem entered when Ockham's razor is violated. Hence, *efficiency in each subproblem requires that one follow Ockham's razor at each stage of inquiry.* Furthermore, one can show that following Ockham's razor is *equivalent* to minimizing errors prior to convergence, assuming that the method converges to the truth in the limit and that success under a (transfinite) error bound is possible at all.

To see how the idea applies in a different context, consider an idealized version of the problem of inferring conservation laws in particle physics (Schulte, 2000). The standard practice in this domain has been to infer the most restrictive conservation laws compatible with the reactions observed so far (Ford, 1963). Retraction efficiency demands this very practice, for suppose that one were to propose looser conservation laws than necessary. Then Nature could withhold the unobserved reactions incompatible with the most restrictive laws until we give in (on pain of converging to the wrong laws) and propose the most restrictive laws. Thereafter, Nature can exhibit reactions excluded by these laws, forcing a retraction, and so forth for the remaining

4 The proof is akin to the proof of the Cantor–Bendixson theorem in set theory.
5 The general theory of infinite simplicities is also worked out in Kelly (2002). The definition of simplicity degree is a bit more complicated than the one presented here.

degrees of restrictiveness. Notice that tighter conservation laws are "simpler" in our general sense than are looser laws, for in worlds in which the tighter laws hold, Nature forever reserves her right to exhibit reactions violating these laws, but in worlds in which looser laws are true, eventually Nature has to reveal reactions refuting simpler laws, assuming that all the reactions are observable.

4.5 Statistical Retractions

There is something admittedly artificial about examples involving ravens and discrete color changes. Both the world and the measurements we perform on it are widely thought to involve chance, and where chance is involved, nothing is strictly verified or refuted: it is always *possible* for a fair coin to come up heads every time or for a measurement of weight to be far from the true value due to a chance conspiracy of disturbances. In this section, we extend the preceding efficiency concepts, for the first time, to properly statistical problems. Doing so illustrates clearly how the problem of induction arises in statistical problems and allows one to derive Ockham's razor from efficiency in statistical settings, with applications to curve fitting and causal inference. Readers who are willing to take our word for it are invited to skip to the next section, in which the applications are sketched.

In a statistical problem concerning just one continuous, stochastic measurement X, each possible statistical world w determines a *probability density* function p_w over possible values of X. If we repeatedly sample values of X for n trials, we arrive at a sample sequence $(X_1 = x_1, \ldots, X_n = x_n)$ in which $X_i = x_i$ is the outcome of the ith trial. If the sampling process is independent and identically distributed, then samples are distributed according to the product density:

$$p_w^n(X_1 = x_1, \ldots, X_n = x_n) = p_w(X_1 = x_1) \cdots p_w(X_n = x_n).$$

Increasing sample size will serve as the statistical analog of the notion of accumulating experience through time.

A *statistical question* partitions the possible statistical worlds into mutually incompatible potential answers, a *statistical presupposition* delimits the set of worlds under consideration, and a *statistical problem* consists of a question paired with a presupposition. A *statistical method* is a rule that responds to an arbitrary sample of arbitrary size with some guess at the correct answer to the question or with "?," which indicates a refusal to commit at the current time. In a familiar, textbook example, the background presupposition is that the observed value of X is normally distributed with known variance σ^2 and unknown mean μ. Then each possible value of the mean μ determines the normal sampling density with mean μ and variance σ^2. The question might be whether $\mu = 0$. A method for this problem returns $\mu = 0$, $\mu \neq 0$, or "?" for an arbitrary sample of arbitrary size.

There is always some small probability that the sample will be highly unrepresentative, in which case the most sensible of statistical methods will produce spurious results. Hence, there is no way to guarantee that one's method actually converges to the truth. It is better to focus on how the probability of producing the right answer evolves as the sample size increases. Accordingly, say that *M solves* a problem *in the limit* (in probability) just in case in each world satisfying the problem's presupposition, the probability of producing the right answer for that world approaches unity as the sample size increases.

Statistical retractions occur when a method's chance of producing some answer drops from a high to a low value. Let $1 > \gamma > 0.5$. Method *M* γ-*retracts h* between stages *n* and *n'* in *w* just in case $P_w^n(M = h) > \gamma$ and $P_w^{n'}(M = h) < 1 - \gamma$. Then *M* γ-*retracts at least k times* in *w* iff there exist $n_0 < n_1 < \ldots < n_k$ such that *M* γ-retracts some answer to the question between n_0 and n_1, between n_1 and n_2, and so on. Also, *M* γ-retracts exactly *k* times iff *k* is the greatest *k'* such that *M* γ-retracts at least *k'* times in *w*. Moreover, *M solves* a given statistical problem *with at most k* γ-*retractions* just in case *M* solves the problem in the limit in probability and γ-retracts at most *k* times in each world. Finally, *M solves* a given statistical problem *with at most k* γ-*retractions starting with h* just in case *M* solves the problem with at most *k* retractions and in each world in which which *M* uses all *k* γ-retractions, *h* is the first answer *M* produces with probability $> \gamma$.

The γ-*retraction complexity* of a statistical problem *starting with h* is the least γ-retraction bound under which some method can solve the problem starting with *h*. The γ-*retraction complexity* of a statistical problem is the least γ-retraction bound under which some method can solve it. As before, γ-*verifiability* is solvability with one γ-retraction starting with $\neg h$, γ-*refutability* is solvability with one γ-retraction starting with *h*, and γ-*decidability* is solvability with zero γ-retractions.

To see how it all works, recall the textbook problem described earlier, in which observed variable *X* is known to be normally distributed with variance σ^2 and the question is whether or not *h* is true, where *h* says that the mean of *X* is zero. Let M_α^n be the standard statistical test of the point null hypothesis *h* at sample size *n* and significance level α (where nonrejection is understood as acceptance). In this test, one reject *h* if the average of the sampled values of *X* deviates sufficiently from zero. The significance level of the test is just the probability of mistakenly rejecting *h* when *h* is true (i.e., when the true sampling distribution is p_μ). It won't do to hold the significance level fixed over increasing samples, for then the probability of producing *h* when *h* is true will not go to unity as *n* increases. That is readily corrected, however, by "tuning down" α according to a monotone schedule $\alpha(n)$ that decreases so slowly that the successive tests $M_{\alpha(n)}^n$ have ever-narrower acceptance zones.

It is a familiar fact that $M_{\alpha(n)}^n$ solves the preceding problem in the limit (in probability), but the current idea is to attend to γ-retractions as well, where $1 > \gamma > 0.5$. Suppose that the initial significance level is low – less than $1 - \gamma$. Then $M_{\alpha(n)}^n$ starts out producing *h* with high probability in *w*, where $\mu_x = 0$. Also, since the significance level drops monotonically to zero as the sample size increases, the probability of producing *h* rises monotonically to unity in *w*, so there are no γ-retractions of *h*. If *w'* satisfies $\neg h$, then since the sample mean's density peaks monotonically around the true mean in *w'* and the acceptance zone shrinks monotonically around *w*, the probability that $M_{\alpha(n)}^n$ produces $\neg h$ approaches unity monotonically. If *w'* is very close to *w*, then the method may start out producing *h* with high probability in *w'*, because p_w will be very similar to $p_{w'}$. But since the probability of producing $\neg h$ rises monotonically in *w'*, *h* is γ-retracted just once. Far from *w* there are no γ-retractions at all, because h is never produced with high probability. So $M_{\alpha(n)}^n$ succeeds with one γ-retraction starting with h and, hence, *h is γ-refutable, for arbitrary γ such that $1 > \gamma > 0.5$.*

That doesn't suffice to justify the proposed method. We still have to argue that *no possible method* can decide the problem in the more desirable, two-sided sense. For this, it suffices to show that no possible method γ-verifies h in probability when $1 > \gamma > 0.5$. Suppose, for *reductio*, that M solves the problem with one retraction starting with $\neg h$. Suppose that w satisfies h. Since M succeeds in the limit and $\gamma < 1$, there exists an n_0 in w such that $P_w^{n_0}(M = h) > \gamma$. There exists a small, open interval I around w such that, for each world w' in I, $P_{w'}^{n_0}(M = h) > \gamma$.[6] Choose $w' \neq w$ in I. Then since M succeeds in the limit and w' does not satisfy h, there exists $n_1 > n_0$ such that $P_{w'}^{n_1}(M = \neg h) > \gamma$. Since w' is in I, we also have that $P_{w'}^{n_0}(M = h) > \gamma$, so at least one γ-retraction occurs in w'. By the *reductio* hypothesis, the first answer output with probability $> \gamma$ in w' is $\neg h$. So another retraction occurs by stage n_0, for a total of two retractions. This contradicts the *reductio* hypothesis and closes the proof. A corollary is that *no possible method solves the problem with zero γ-retractions if $1 > \gamma > 0.5$,* for any such method would count as a γ-verifier of h. Hence, h is γ-refutable but is not γ-verifiable or γ-decidable.

This asymmetry is not so surprising in light of the familiar, statistical admonition that rejections of tests are to be taken seriously, whereas acceptances are not. Less familiar is the question of whether statistical problems can require more than one retraction, so that *neither* side of the question is refutable (in probability). In fact, such problems are easy to construct. Suppose that we have two independent, normally distributed variables X and Y, and we want to know which of the variables has zero mean (both, one or the other or neither). *This problem is solvable with two γ-retractions starting with "both zero," but is not solvable with two γ-retractions starting with any other answer, as long as $1 > \gamma > 0.5$.* The negative claim can be shown by the following extension of the preceding argument. Let h_S be the answer that exactly the variables in S have zero means, so we have possible answers h_ϕ, $h_{\{X\}}$, $h_{\{Y\}}$, and $h_{\{X,Y\}}$. Suppose, for *reductio*, that M solves the problem with two γ-retractions starting with some answer other than $h_{\{X,Y\}}$. Each possible world corresponds to a possible value of the joint mean (x, y). Since M succeeds in the limit and $\gamma < 1$, there exists an n_0 such that $P_{(0,0)}^{n_0}(M = h_{\{X,Y\}}) > \gamma$. There exists a small open disk B_0 around world $(0, 0)$ such that, for each world w' in B_0, $P_{w'}^{n_0}(M = h_{\{X,Y\}}) > \gamma$.[7] Choose $(0, r)$ in B_0 so that $(0, r)$ satisfies $h_{\{X\}}$. Since M succeeds in the limit, there exists an $n_1 > n_0$ such that $P_{(0,r)}^{n_1}(M = h_{\{X\}}) > \gamma$. Again, there exists a small open disk B_1 around $(0, r)$ such that, for each world w' in B_1, $P_{w'}^{n_1}(M = h_{\{X\}}) > \gamma$. Choose (r', r) in B_1 (and hence in B_0) so that (r', r) satisfies h_ϕ. Since M succeeds in the limit, there exists an $n_2 > n_1$ such that $P_{(r',r)}^{n_2}(M = h_\phi) > \gamma$. Since world (r', r) is in both B_0 and B_1, we also have that $P_{(r',r)}^{n_0}(M = h_{\{X,Y\}}) > \gamma$ and that $P_{(r',r)}^{n_1}(M = h_{\{X\}}) > \gamma$, for a total of at least two retractions in (r', r). By the *reductio* hypothesis, the first answer output with probability $> \gamma$ in w' is not $h_{\{X,Y\}}$. But then another retraction occurs by n_0, for a total of three. This contradicts the *reductio* hypothesis and closes the proof. It follows as an immediate corollary that *no possible method solves the problem with one γ-retraction if $1 > \gamma >*

6   This follows from the Lebesgue convergence theorem (cf., Royden, 1988, p. 267).
7   Again by the Lebesgue convergence theorem.

0.5, for any such method would count as succeeding with two retractions starting with an arbitrary answer different from $h_{\{X,Y\}}$.

So the best one can hope for is two retractions starting with the hypothesis that all the means are zero. The following, natural strategy does as well as possible. Choose the usual statistical tests for $\mu_X = 0$ and for $\mu_Y = 0$. Tune down the significance levels to make both tests γ-refute their respective hypotheses, as in the preceding example. Let M produce $h_\emptyset$ if both tests reject, $h_{\{Y\}}$ if only the X test rejects, $h_{\{X\}}$ if the Y test rejects, and $h_{\{X,Y\}}$ if neither test rejects. The probability of producing the right answer rises monotonically toward unity in each test, as was described above. The probability that M produces the right answer is the product of the marginal probabilities that the component tests are right. In the worst case, the right answer is $h_\emptyset$ and the actual world (r, r') has the property that r is quite small and r' is even smaller. In such a world, the probability that the X test rejects will rise late and the probability that the Y test rejects will rise later. Then, at worst, there is a time at which both tests probably accept, followed by a time at which the X test probably rejects and the Y test probably accepts, followed by a time after which both tests probably reject. Since the joint probability is the product of the marginal probabilities, M γ-retracts at most twice. Furthermore, each worst-case world in which two retractions occur has $h_{\{X,Y\}}$ as the first output produced with probability $> \gamma$. Hence, this problem's complexity is exactly "two retractions starting with $h_{\{X,Y\}}$." It is clear that each new variable added to the problem would result in an extra retraction, so there is no limit to the number of retractions that a statistical problem can require.

The multiple mean problem has suggestive features. In order to minimize retractions, one must start out with the hypothesis that both means are zero. This is the most uniform hypothesis (if a mean distinct from zero is close to zero, that fact will become apparent only at large sample sizes, resulting in a "break" in the signal from the environment as sample size increases). It is also the most testable hypothesis (the reader may verify that this answer is γ-refutable but none of the alternative answers is). Finally, $h_{\{X,Y\}}$ has no free parameters (both μ_X and μ_Y are fixed at zero), whereas $h_{\{X\}}$ allows adjustment of μ_Y and $h_\emptyset$ allows for adjustment of both μ_X and μ_Y. These features suggest that retraction efficiency should explain intuitive simplicity preferences in more interesting statistical problems, such as curve fitting, model selection, and causal inference.

4.6 Curves and Causes

Suppose that we know that the true law is of the form

$$y = \alpha x^3 + \beta x^2 + \gamma x + \varepsilon,$$

where ε is normally distributed measurement error and the question is whether the law is linear, quadratic, or cubic. Simplicity intuitions speak clearly in favor of linearity, but why should we agree? Minor variants of the preceding arguments show that the problem requires at least two retractions, and requires more if the method starts with a nonlinear answer. Moreover, any method that probably outputs a law of higher order than necessary in a world that is simplest in some subproblem uses more retractions than necessary in the subproblem. We conjecture that the usual, nested sequence of tests succeeds with two retractions starting with linearity.

Another sort of simplicity is minimal causal entanglement. The key idea behind the contemporary theory of causal inference is to axiomatize the appropriate connection between the true causal network and probability rather than to attempt to reduce the former to the latter (Pearl, 2000; Spirtes et al., 2000). The principal axiom is the *causal Markov condition*, which states that each variable is probabilistically independent of its noneffects given its immediate causes. A more controversial assumption is *faithfulness*, which states that every conditional probabilistic independence follows from causal structure and the causal Markov condition alone (i.e., is not due to causal pathways that cancel one another out exactly). If all variables are observable and no common causes have been left out of consideration, it follows from the two axioms that there is a direct causal connection between two variables (one way or the other) just in case the two variables are statistically dependent conditional on each subset of the remaining variables.

The preceding principles relate the (unknown) causal truth to the (unknown) probabilistic truth. The methodological question is what to infer now, from a sample of the current size. Spirtes et al. have proposed the following method (which we now oversimplify – the actual method is much more efficient in terms of the number of tests performed). For each pair of variables X, Y, and for each subset of the remaining variables, perform a statistical test of independence of X and Y conditional on the subset. If every such test results in rejection of the null hypothesis of independence, add a direct causal link between X and Y (without specifying the direction). Otherwise, conclude provisionally that there is no direct causal connection. In other words, presume against a direct causal connection until rejections by tests verify that it should be added.

The proposed method is, again, a Boolean combination of standard statistical tests, because the edges in the output graph result from rejections by individual tests and missing edges correspond to acceptances. The procedure can be implemented on a laptop computer and it has been used with success in real problems. However, it is neither Bayesian nor Neyman–Pearsonian: the significance levels and powers of the individual tests do not really pertain to the overall inference problem. For some years, the principal theoretical claim for the method has been that it solves the causal inference problem in the limit (in probability); a rather weak property. But now one can argue, as we have done above several times, that (a) the problem of inferring immediate causal connections requires as many probable retractions as there are possible edges in the graph, and that (b) an extra retraction is required in the current subproblem if the method ever probably outputs a complex graph in one of the simplest worlds in the subproblem. Furthermore, we conjecture that the proposed method (or some near variant thereof) succeeds under the optimal retraction bound.

4.7 Confirmation Revisited

Bayesian methods assign numbers to answers instead of producing answers outright. This hedging is thought to be an especially appropriate attitude in the face of possible, nasty surprises in the future. It is, rather, a red herring, for there is a natural sense in which hedgers retract just as much and as painfully as methods that leap straight for the answers themselves. Say that Bayesian $P(.|.)$ *solves* a statistical problem just in case for each $\varepsilon < 1$ and that, for each statistical world w satisfying the problem's presupposition, there exists a stage n such that for each stage $m \geq n$, $P_w^m(P(h_w|.) > \varepsilon) > \varepsilon$, where h_w is the correct answer in w to the statistical question posed by the problem. Let γ be strictly between 0.5 and unity and say that confirmation method $P(.|.)$ *γ-retracts* answer h between n and n' in w just in case $P_w^n(P(h|.) > \gamma) > \gamma$ and $P_w^{n'}(P(h|.) < 1 - \gamma) < 1 - \gamma$. Also, $P(.|.)$ *starts with h* (relative to k, γ) iff in each world in which k is realized, the first answer assigned more than γ probability by $P(.|.)$ with chance $>\gamma$ is h.

Given these concepts, one can re-run all the preceding arguments for exacting retractions from a statistical method, only now one forces the Bayesian's credence to drop by a large amount (from γ to $1 - \gamma$) with high chance (from γ to $1 - \gamma$). Indeed, you are invited to run your favorite choices of Bayesian prior probabilities through the negative arguments of the preceding section. Be as tricky as you like; assign point mass to simple answers or assign continuous priors. Either your agent fails to converge to the truth in probability in some world or it realizes the worst-case retraction bound in some world.

That doesn't mean Bayesian methods are *bad*, for the same arguments apply to *any strategy* for attaching partial credences to answers in light of samples or for inferring answers from samples. We are not like the classical statisticians who reject Bayesian methodology unless the prior corresponds to a known chance distribution. Nor are we like the idealistic extremists in the Bayesian camp, who call their arbitrary prior distributions "knowledge," even when nothing is known. We advocate the middle path of letting problems speak for themselves and of solving them as efficiently as possible by whatever means. Bayesian means may be as good as any others, but they are not and cannot be better than the best.

The moral for confirmation theory is that good Bayesian methods are *only* good methods. The probabilities that they assign to hypotheses are just the current outputs of good methods which, at best, converge to the truth with the minimum of surprises. The same efficiency could be had by attaching the numbers in other ways or by dispensing with the numbers and producing theories outright, as scientists have always done until fairly recently. There is no special aptness about softening one's views in the face of uncertainty: Nature can wreak as much havoc on a high confirmation value as on outright acceptance of an answer. High confirmation provides no guarantee or partial guarantee of a smooth inductive future. There are just smooth problems, bumpy problems, methods that add extra bumps and methods that avoid all the avoidable ones.

Acknowledgments

We are indebted to Oliver Schulte for comments on a draft of this chapter, and to Joseph Ramsey, Richard Scheines, and Peter Spirtes for helpful discussions. We also thank Peter Tino for some very helpful corrections.

Bibliography

Allais, M. 1953: Le comportement de l'homme rationel devant le risque: critiques des postulats et axiomes de l'école americaine. *Econometrika*, 21, 503–46.

DeFinetti, B. 1937: Foresight: its logical laws, its subjective sources. Reprinted in H. Kyburg and H. Smokler (eds.), *Studies in Subjective Probability*, 2nd edn. New York: John Wiley, 1980, 53–118.

Ellsburg, D. 1961: Risk, ambiguity, and the savage axioms. *Quarterly Journal of Economics*, 75, 643–69.

Freivalds, R. and Smith, C. 1993: On the role of procrastination in machine learning. *Information and Computation*, 107, 237–71.

Ford, K. 1963: *The World of Elementary Particles*. New York: Blaisdell.

Glymour, C. 1980: *Theory and Evidence*. Princeton, NJ: Princeton University Press.

Goodman, N. 1983: *Fact, Fiction, and Forecast*, 4th edn. Cambridge, MA: Harvard University Press.

Hempel, C. G. 1965: Studies in the logic of confirmation. In *Aspects of Scientific Explanation*. New York: The Free Press, 3–51.

Jain, S., Osherson, D., Royer, J., and Sharma, A. 1999: *Systems that Learn*, 2nd edn. Cambridge, MA: The MIT Press.

James, W. 1948: The will to believe. In A. Castell (ed.), *Essays in Pragmatism*. New York: Collier.

Jeffreys, H. 1985: *Theory of Probability*, 3rd edn. Oxford: The Clarendon Press.

Kahneman, D. and Tversky, A. 1972: Subjective probability: a judgment of representativeness. *Cognitive Psychology*, 2, 430–54.

Kelly, K. 1996: *The Logic of Reliable Inquiry*. New York: Oxford University Press.

—— 2000: The logic of success. *British Journal for the Philosophy of Science*, 51, 639–66.

—— 2002: A close shave with realism: how Ockham's razor helps us find the truth. CMU Philosophy Technical Report 137.

Kelly, K. and Schulte, O. 1995: The computable testability of theories with uncomputable predictions. *Erkenntnis*, 42, 29–66.

Kuhn, T. 1970: *The Structure of Scientific Revolutions*. Chicago: The University of Chicago Press.

Kuratowski, K. 1966: *Topology*, vol. 1. New York: Academic Press.

Kyburg, H. 1978: Subjective probability: criticisms, reflections and problems. *Journal of Philosophical Logic*, 7, 157–80.

Laudan, L. 1980: Why abandon the logic of discovery? In T. Nickles (ed.), *Scientific Discovery, Logic, and Rationality*. Boston: D. Reidel, 173–83.

Lee, P. 1989: *Bayesian Statistics: An Introduction*. London: Edward Arnold.

Levi, I. 1993: Money pumps and diachronic books. *Philosophy of Science*. Supplement to vol. 69, s236–7.

Maher, P. 1997: Depragmatized Dutch book arguments. *Philosophy of Science*, 64, 291–305.

Osherson, D. and Weinstein, S. 1988: Mechanical learners pay a price for Bayesianism. *Journal of Symbolic Logic*, 53, 1245–52.

Pearl, J. 2000: *Causation*. New York: Oxford University Press.

Popper, K. 1959: *The Logic of Scientific Discovery*. New York: Harper.

Putnam, H. 1965: Trial and error predicates and a solution to a problem of Mostowski. *Journal of Symbolic Logic*, 30, 49–57.

Rosencrantz, R. 1983: Why Glymour is a Bayesian. In J. Earman (ed.), *Testing Scientific Theories*. Minneapolis: University of Minnesota Press, 69–97.

Royden, H. 1988: *Real Analysis*, 3rd edn. New York: Macmillan.

Savage, L. J. 1951: *The Foundations of Statistics*. New York: John Wiley.

Schulte, O. 1999a: The logic of reliable and efficient inquiry. *The Journal of Philosophical Logic*, 28, 399–438.

—— 1999b: Means–ends epistemology. *The British Journal for the Philosophy of Science*, 51, 151–3.

—— 2000: Inferring conservation laws in particle physics: a case study in the problem of induction. *The British Journal for the Philosophy of Science*, 51, 771–806.

Spirtes, P., Glymour, C., and Scheines, R. 2000: *Causation, Prediction, and Search*. Cambridge, MA: The MIT Press.

Teller, P. 1973: Conditionalization and observation. *Synthese*, 26, 218–58.

Whewell, W. 1840: *Philosophy of the Inductive Sciences*. London: Parker.

Further reading

Earman, J. 1992: *Bayes or Bust? A Critical Examination of Bayesian Confirmation Theory*. Cambridge, MA: The MIT Press.

Glymour, C. 1980: *Theory and Evidence*, Princeton, NJ: Princeton University Press.

Hempel, C. G. 1965: Studies in the logic of confirmation. In *Aspects of Scientific Explanation*. New York: The Free Press, 3–51.

Howson, C. and Urbach, P. 1989: *Scientific Reasoning: The Bayesian Approach*. New York: Open Court.

Jain, S., Oshersion, D., Royer, J., and Sharma, A. 1999: *Systems that Learn*. Cambridge, MA: The MIT Press.

Kelly, K. 1996: *The Logic of Reliable Inquiry*. New York: Oxford University Press.

Lee, P. 1989: *Bayesian Statistics: An Introduction*. London: Edward Arnold.

Martin, E. and Osherson, D. 1998: *Elements of Scientific Inquiry*. Cambridge, MA: The MIT Press.

Popper, K. 1959: *The Logic of Scientific Discovery*. New York: Harper.

Spirtes, P., Glymour, C., and Scheines, R. 2000: *Causation, Prediction, and Search*. Cambridge, MA: The MIT Press.

CAN A THEORY'S PREDICTIVE SUCCESS WARRANT BELIEF IN THE UNOBSERVABLE ENTITIES IT POSTULATES?

Many theories posit entities that cannot be directly observed: atoms, quarks, magnetic fields, genes, mental representations, and so on. By their very nature, it is never possible to confirm the existence of such entities by direct observation. (There is, however, a question about what counts as a "direct" observation – is looking through a microscope allowed?) Nonetheless, theories that posit such entities often do make predictions that can be tested by observation, and some theories are highly successful in this enterprise. Does this sort of predictive success give us reason to believe that the posited entities really exist? One argument, often dubbed the "miracle argument," claims that the empirical success of a theory would be a *miracle*, or at least a coincidence of cosmic proportions, if the theory were wrong about the basic entities underlying the phenomena in question. Jarrett Leplin and also André Kukla and Joel Walmsley, reject the miracle argument in its broadest form. Nonetheless, Leplin argues for the realist position that we are sometimes warranted in believing in the existence of the unobservable entities postulated by science. First, there is a certain presumption in favor of belief in these entities, and none of the standard anti-realist arguments are successful in dislodging this belief, once admitted. Secondly, one specific kind of predictive success, the prediction of *novel* phenomena, *does* provide particularly strong warrant for the belief in unobservables. Kukla and Walmsley criticize this argument, and other refinements of the miracle argument, on the grounds that they presuppose a certain conception of what it would be to *explain* the predictive success of a scientific theory. They arrive at the strong conclusion that no version of the miracle argument could possibly give us grounds for believing in unobservable entities.

CHAPTER FIVE

A Theory's Predictive Success can Warrant Belief in the Unobservable Entities it Postulates

Jarrett Leplin

5.1 The Burden of Argument

Theoretical entities are the unobservable entities that scientific theories posit to explain or predict empirical results. Inaccessible to experience, their claim to conviction derives from the acceptability of the theories in which they figure. Electrons, fields, and genes are examples normally thought to be real entities. Phlogiston and the electromagnetic ether, though once confidently embraced, have turned out not to be real. The fate of the strings and gravitons advanced by theories at the frontiers of physics is unresolved. The ontological status of theoretical entities is frequently uncertain, disputed, revised. The existence of such an entity may be denied because, with changes in theory, something different, with a related explanatory role, is thought to exist instead. Thus oxygen replaces phlogiston in the chemical theory of combustion. Or an entity may be rejected because a new theory denies it any continuing explanatory utility. The crystalline spheres of ancient and early modern astronomy are simply obviated by Newtonian gravity. In general, the question within science is not *whether* theoretical entities exist but *which* theoretical entities exist. Answers to this question change with the fortunes of theory.

Scientific realism is the position that this question can be answered on evidentially probative grounds. Many posited entities turn out not to exist and the status of many others remains unsettled. But in some cases science develops, through the testing and application of its theories, adequate reason to believe that certain theoretical entities are real. Further, according to realism, the success of theories warrants some beliefs about the nature – the properties and behavior – of these entities. For, as they are unobservable, the mere assertion of their existence, without an account of their nature, is insufficient to serve their explanatory and predictive purposes. All such realist beliefs are defeasible; new evidence could undermine them, as new evidence frequently forces changes in theory. Nevertheless, according to scientific realism, such beliefs are epistemically justifiable.

Anti-realism claims, as a matter of philosophical principle, that there can never be adequate reason to invest credence beyond the range of the observable. Anti-realism regards all theoretical entities, regardless of the ontological status assigned to them within science – regardless of the available evidence – as conceptual tools whose roles are pragmatic, not epistemic. Anti-realism need not deny the reality of unobservable entities; it denies that their reality is ever warrantedly assertable by us. Anti-realism proclaims a sweeping agnosticism with respect to theory, independently of the evidence used to evaluate theories. Scientific evidence attests not to the truth of theories, nor to the existence of the unobservable entities they posit, but only to their explanatory and predictive utility.

If the dispute between realism and anti-realism is a dispute between science and philosophy, anti-realism loses. There is no *a priori* stance from which philosophy can presume to dictate the standards and methods for acquiring knowledge. How knowledge is best acquired depends on the nature of the objects of knowledge, on what the world is like, and is therefore itself knowledge to be acquired, in the way that any empirical knowledge is acquired, through scientific investigation of the world. Abstract reasoning, conceptual or linguistic analysis, appeal to common sense or intuition – or any distinctively philosophical mode of inquiry – are notoriously unreliable as a determinant of the nature or scope of scientific knowledge. Nor does autonomous philosophy deliver a consistent verdict to compare with the conclusions that science reaches. No settled, dependable method of appraisal, such as operates within the natural sciences, is available to adjudicate among the indefinite number of competing positions that philosophers fashion.

Some realists have contended that their dispute with anti-realism could be won on just this basis. Realism sides with science; comparing the progress of science with the state of philosophical inquiry, the attraction of siding with science is an advantage that no philosophically generated anti-realism can overcome. This is essentially what realists are saying when they claim that but for their position science is mysterious, its successfulness an unprojectable accident. To reject realism, they suggest, is to reject science itself. Science's detractors gleefully agree. They think they can infer from their rejection of realism that the success and progressiveness of science are illusory, that its epistemic status is no better than that of any other social institution or practice.

Unfortunately, the resolution of the dispute over realism does not reduce to one's attitude toward science. Essentially, this is because science and philosophy are not autonomous. Philosophical assumptions are ineliminable from the reasoning by which science fixes its ontological commitments. And substantive scientific results often support philosophical limitations on science. It is open to the anti-realist to contend that realist beliefs cannot be read off the record of what science achieves, but must be read in via a certain, optional, philosophical interpretation. Science can, or even does, operate just as well without such beliefs.

For example, science operates just as well with a system of fundamental equations that admit of no coherent physical interpretation as it does with conceptually tractable laws, for no physical commitments at the theoretical level follow in any case. From a (consistent) anti-realist perspective, the quest for an understandable interpretation of quantum mechanics is misdirected. The anti-realist therefore disagrees with theoretical physicists as to the importance of interpreting quantum mechanics, even as he insists that his philosophy is consonant with scientific practice.

But this rejection of realism's scientific pedigree concedes that the priority of science would favor a philosophical position that science really did require. And this concession places the argumentative burden on anti-realism. Science claims to discover and to learn the nature of certain theoretical entities. "Electron," for example, is a purportedly referential term, and properties of electrons – mass, charge, and spin – are held to be well established. According to anti-realism, all claims to discover or learn such things are mistaken. Given the priority to which science is entitled, the opening move in the debate should be to argue that its characteristic theoretical hypotheses cannot be warranted.

Accordingly, I elect to begin with anti-realist arguments. In subsequent sections, I shall consider two significant challenges to realism. Although unsuccessful, they may be judged sufficiently compelling to shift the burden of argument back upon the realist. Therefore, I shall follow with development of an independent argument for realism. In summary, I contend that realism is the default position, and that the case to be made for switching to anti-realism is at best indecisive. At the same time, a compelling defense of realism is available.

5.2 Underdetermination

A major source of anti-realist argumentation is the simple, incontestable fact that observational evidence is logically inconclusive with respect to the truth of theory. That is, the falsity of any theory T is logically consistent with the truth of all observation statements O_i used in assessing T. T is, in this respect, *underdetermined* by the evidence; indeed, by all possible evidence. It follows that in affirming T on the basis of evidence, one must reason ampliatively; one must use forms of reasoning that are not truth-preserving. Eliminative induction, a kind of process of elimination in which the range of potential contending hypotheses is unrestricted, is such a form, as is inductive generalization. Another is abduction, in which the explanatory power of a hypothesis counts as evidence for its truth. In practice, scientific argumentation exhibits all the rationally cogent ampliative forms that philosophers have identified.

But this reliance on ampliative reasoning is not immediately an objection to affirming theories. For such reasoning is endemic to, and ineliminable from, ordinary inference that grounds common-sense beliefs about the observable world. Without it one could not ground even the belief that a plainly observable object continues to exist when unobserved. This belief is certainly abductive. Without ampliation, one gets not anti-realism about science but a sweeping skepticism that no party to the dispute over realism accepts.

The question for realism, therefore, is whether, *allowing* the rational cogency of standard forms of ampliative inference, theories are *still* underdetermined. It is usual to formulate this thesis of ampliative underdetermination, the only kind worth considering, by asserting the existence of rival theories to T that are equally well supported by the evidence. It is then said that theory-*choice* is underdetermined: a preference for T over its rivals must have some nonevidential, pragmatic basis.

Is ampliative underdetermination credible? Notice that the usual formulation of this thesis is not its minimal formulation. Minimally, the claim is that no body of observational evidence $\{O_i\}$ warrants any theory T; neither truth-preserving inference nor rational modes of ampliative inference can get you from $\{O_i\}$ to T. The existence of evidentially equivalent rival theories is a *further* assertion. It is easy to see why this further assertion is made. Without it, there is no particular reason to suppose, quite generally and abstractly, that T is not rationally reachable. Embedded into the very formulation of underdetermination, then, is an argument for it: there will always be alternative theoretical options that fare as well on the evidence as T does.

But now we discern a serious equivocation. Once the legitimacy of ampliation is conceded, the existence of the evidentially equivalent rivals can have no logical or otherwise *a priori* guarantee. For the sort of rivals to T whose existence the logical gap between $\{O_i\}$ and T guarantees are surely not defensible by ampliation. For example, the construction $\sim T \& \Pi\{O_i\}$, while both inconsistent with T (and to that extent a "rival") and consistent with the evidence for T, is certainly not supported by this evidence. ($\Pi\{O_i\}$ is the conjunction of all the propositions in the set $\{O_i\}$.) While the O_i might fall short in warranting T – this is the possibility that underdetermination declares realized invariably – the O_i can hardly be supposed to warrant $\sim T$. The situation at issue is not one in which T faces counter-evidence or disconfirmation, but one in which the evidence bearing on T is allowed to be as supportive as one likes; the claim is that *even then* T is underdetermined because some rival is supported equally well. $\sim T \& \Pi\{O_i\}$ is merely logically consistent with the evidence; it is not *supported* at all. What would support $\sim T$ is the *negation* of some O_i. Thus an immediate reason to declare $\sim T \& \Pi O_i$ unconfirmable is that confirmation of its first component requires refutation of the second.

This argument assumes that if the O_i support $\sim T \& \Pi\{O_i\}$ then they support $\sim T$. What makes this assumption reasonable is not some holistic conception of confirmation according to which any evidence for a theory supports equally, or even to any extent, all distinguishable components of the theory or all consequences of the theory. Such holism is implausible on its face and proves detrimental to realism. Rather, in the particular situation depicted, $\sim T$ is the only bit of theory around to be the subject of support. The O_i are suppliers of support, not objects of it. It makes little sense to speak of a relation of support between $\{O_i\}$ and itself. It could only be in virtue of supporting $\sim T$ that the O_i support $\sim T \& \Pi\{O_i\}$, as they have been hypothesized to do.

Of course $\sim T \& \Pi\{O_i\}$ makes empirical commitments and is truth-valuable. Why isn't this enough to make it confirmable? Although the O_i are not self-supporting, are they not confirmed by the facts? That is, do not the observations, as opposed to the observation sentences that report them, confirm $\sim T \& \Pi\{O_i\}$?

The answer is negative. We are not concerned with choices among rival theories that the evidence refutes. The underdetermination thesis applies to theories that the evidence supports; it is these that are supposed to have equally supported rivals. Therefore, the O_i must either be supposed true or supposed to instantiate generalizations of supposed truths if the supposed rival committed to them is even to be formulated. And a theory is not confirmed by an observation presupposed in its very formulation. Observational results obtainable from a theory only by defining the theory to include them do not support the theory. For it is not even logically possible for the theory to get them wrong, whereas a result can support a theory only if it is unlikely to be obtained if the theory is false.

Accordingly, $\sim T \& \Pi\{O_i\}$ is not confirmed by $\{O_i\}$. The only possible epistemic route to $\sim T \& \Pi\{O_i\}$ is indirect, via some further theory inconsistent with T that delivers all of T's observational consequences. But this further theory cannot be generated algorithmically, as $\sim T \& \Pi\{O_i\}$ was generated, by operating on T. Not even the existence of such a theory, let alone its confirmability, has any *a priori* guarantee.

Strictly speaking, of course, the anti-realist's claim is not that the O_i confirm $\sim T \& \Pi\{O_i\}$, but only that they support $\sim T \& \Pi\{O_i\}$ as well as they do T, that T and its rival are *equally* confirmed by the evidence. This common amount of confirmation cannot, however, be zero without repudiating ampliation. More generally, it is a reasonable constraint on theories that the thesis of underdetermination invokes as rivals that these be at least *amenable* to evidential support. A propositional structure that could not in principle be confirmed violates this constraint. Such a structure, crafted solely for logical consistency with the observational consequences of an existing theory, will not be entertained as an alternative to the existing theory because there is nothing to be done with it; it is rightly dismissed as dead in the water because its only possible support is derivative from some further, independent theory that would itself be the proper object of confirmation. Once theories are required to be defensible by ampliation, conditions such as confirmability in principle, explanatory power, and generality – conditions that standard forms of ampliative inference select for – become reasonable constraints on theoretical status.

Similar problems befall other candidates for T's rivals. Let T' affirm that T holds whenever observations are made but not otherwise, the universe instantaneously reverting, when observations recommence, to the conditions that, according to T, would have then prevailed had observation been continuous. Certainly T' is unconfirmable. It has a confirmable component, but not only do confirmations of this component not confirm the other component; *nothing* could confirm the other component. So again, it does not take holism to rule T' unconfirmable. But if T' is unconfirmable, it cannot be equally well supported as T, which *is* confirmable. Of course, T, as much as T', carries a commitment to conditions that prevail in the absence of observation. But T's commitment, unlike that of T', is ampliatively defensible; stability is a paradigm of ordinary, unproblematic ampliation. Moreover, the component T^*, common to T and T', that T holds during observation, is confirmable and exempt from the underdetermination that T' is supposed to create. T' offers no rival to T^*.

The equivocation of the underdetermination thesis, then, is this: while only the ampliative form of underdetermination challenges realism, it is the merely logical form of underdetermination that supplies the rival theories invoked to establish the underdetermination thesis. For the challenge to get off the ground, there will have to be some substantive, independent basis for supposing the rivals to exist at all.

What could this be? There are examples of theoretical rivalries that resist adjudication, but they do so for reasons that presuppose a substantial body of theory that further evidence could undermine. Thus, the possibility of an epistemically principled choice cannot be precluded. Within Newtonian theory, rival attributions of motion to the center of mass of the universe are unadjudicable, but Newtonian theory could prove to be wrong about the detectability of absolute motion. Any theory that both fixes one of its parameters and declares it unmeasurable, or defines a parameter but leaves it unspecified, generates rivals. But these rivals share the theory's fallible onto- logical and nomological commitments.

Even if such examples were compelling, argument by example is unlikely to moti- vate a perfectly general thesis about all theories and all possible bodies of evidence. Moreover, there are plenty of examples that point the other way, cases in which sci- entists are unable to produce even a *single* theory that makes sense of the mystify- ing empirical regularities their experiments have revealed. Astronomy has many such anomalies, such as the motion of stars at the periphery of the Milky Way, which is greater than known gravitational forces allow, and the "great chain" of galaxies, which violates the large-scale uniformity required by big bang cosmology. Far from unavoid- able, a multiplicity of theoretical options may not even be the norm.

I submit that any challenge to realism from underdetermination is vastly underdetermined.

5.3 Superseded Science

Lots of successful theories turn out to be wrong; the entities that they posit nonex- istent. Why is currently successful science any more entitled to credence than the once successful science we now reject? Does the history of theorizing not provide ample reason to distrust theories, regardless of the evidence that supports them?

An obvious answer is that past theories ultimately proved unsuccessful whereas current theories have not. This is why their temporal status differs. But with further testing and theoretical developments, might current theories not prove unacceptable, just as their temporarily accepted predecessors did? Indeed they might; realism admits the defeasibility of all theory. The question is whether there is reason to forecast this development. The major fundamental theories of current physics – general relativity, the basic laws of quantum mechanics, the standard model of elementary particles – are certainly the most severely tested theories ever, and they have proved flawless to a precision uncontemplated in the assessment of their predecessors. These theories are not thought to be the final word; their very multiplicity reflects limitations that a more fundamental, unifying theory will overcome. But there is no reason to expect a unifying theory to require their rejection. Why should the fact that earlier theories failed count against the different, better-tested theories that we have now?

The issue here is the status of second-order evidence. The challenge to realism is that our methods of developing and evaluating theories, our standards for investing credence, are demonstrably unreliable. They have led us to judge Newtonian gravity, phlogistic chemistry, and the electromagnetic ether to be firmly established, as well confirmed as a theoretical commitment could be expected to be. In the nineteenth century, Maxwell famously considered the ether the best-confirmed theoretical entity in natural philosophy. Lavoisier declared the material theory of heat to be no longer a hypothesis, but a truth (*Mémoires de Chimie*, vol. 1, section 2). The phenomena of heat, wrote Lavoisier, are inconceivable without "admitting that they are the result of a real, material substance, of a very subtle fluid, that insinuates itself throughout the molecules of all bodies and pushes them apart" (*Traité de Chimie*, vol. 1, sections

1–3). Chastened by such misjudgments, we distrust the methods that licensed them. Distrusting our methods, we distrust the theories that they now recommend, however much these theories excel under them. First-order evidence supports current theory, but there is second-order evidence against reliance on first-order evidence. The priority of second-order evidence challenges realism.

But why is second-order evidence privileged? As there could be no second-order evidence without first-order evidence to learn from, the relation would appear symmetric. We cannot very well infer from the conclusions to which first-order evidence leads that the conclusions to which it leads are untrustworthy, for if they are untrustworthy then they are no basis for inference. Yet the trustworthiness of conclusions we draw from first-order evidence depends on the evidential warrant of the standards of evidence and modes of inference used in drawing them. If the trustworthiness of each level of evidence is presupposed in assessing that of the other, neither is privileged.

Giving priority to second-order evidence raises quite general problems in epistemology. Not only do theories prove wrong; ordinary, paradigmatically justified beliefs that ampliative reasoning must be allowed to license prove wrong. That systems of ordinary beliefs have proven to contain errors is second-order evidence for the erroneousness of current belief systems, none of whose component beliefs is currently individually impeachable. Am I to induce, from my record of fallibility, that some of my present beliefs are false, although the evidence favors each of them and I have no grounds to doubt any? If so, I am lodged in paradox. For in addition to believing that some of my beliefs are false, I am entitled to believe that all of them are true by the principle that epistemic justification is closed under conjunction. If each of these propositions is justified, so, by further application of this closure principle, is their self-contradictory conjunction, which is absurd.

Partly for this reason, the closure principle for justification under conjunction is disputed within epistemology. But an anti-realism that purports to rationalize scientific practice cannot afford to dispute it. Without this principle, rational inference does not in general transmit epistemic warrant. For in general it is only in conjunction, not individually, that premises provide a basis for inference. And science grows as much by forging new inferential connections – by relating new ideas to what is already known – as by introducing new theories, hypotheses, empirical laws, and experimental results. Inference, often without prospect of independent empirical confirmation, is a frequent basis for extensions of science.

The presumption that second-order evidence trumps first-order evidence gives too simple a picture. Methods of evaluation depend on substantive developments in theory. What we expect of theories responds to what our best theories achieve. As current theories are more severely tested, current methods may reasonably be supposed more reliable. As our knowledge of the world improves, so does our knowledge of how such knowledge is obtained. It might be difficult to prove that methods improve, without assuming, impermissibly, the superiority of current theories. But the burden of argument here is squarely on the anti-realist. A challenge mounted from history is not entitled to presuppose that methods are stable. Unless the anti-realist can show that rejected theories were once warranted by the highest standards that the best current theories meet, the challenge fails.

It is also unclear that the failures of past theories are epistemic failures. Not only do we learn *from* our mistakes; it is an epistemic advance to learn *that we have been mistaken*. That a posited theoretical entity does not, after all, exist, or that a posited theoretical mechanism is not responsible for a certain effect, is important theoretical information. It is not clear that a consistent anti-realism can allow for this information. What makes it any more trustworthy than the information, equally a conclusion from first-order evidence, that a theoretical entity *does* exist? The indispensability of auxiliary hypotheses in generating observable predictions from theories belies the apparent logical asymmetry between verification and falsification. If empirical evidence cannot establish theories, neither can it refute them. For the refutation of a theory requires that theoretical auxiliaries assumed in testing it be independently established.

As much as we see once successful theories rejected, we see once unsuccessful theories resurrected. Why is the anti-realist more impressed by the fall of an admired theory than by the rise of one scorned? Heliocentrism, the vision of ancient Greek astronomers eclipsed by Aristotle, and the checkered history of Prout's hypothesis are as compelling examples as phlogiston and nested spheres. Yet the correctness of the information that a posited theoretical entity does not exist after all is presupposed in pronouncing past theories wrong. There is a certain commitment to the correctness of current theory in impugning past theoretical commitments, for it is current theory that corrects them. Yet the anti-realist impugns past theory *so as* to induce that current theory is unfounded.

Anti-realists have several tactics for finessing this problem. Their essential theme is that inconsistencies among theories – either among historical theories or between a historical theory and current theory – guarantee that, one way or another, there will be examples of successful but failed theoretical entities. Thus we need not assume that current theory is right and past theories wrong; it suffices to assume that not all can be right.

The obvious rejoinder is that appeal to inconsistency is unpromising as a source of data for induction. Inductive strength depends on the preponderance of evidence, whereas inconsistencies necessarily generate evidence in different directions. Less obviously, the anti-realist is to be pushed as to the status of the inconsistencies he discerns. It is easy to take inconsistency for a purely formal relation identifiable independently of substantive empirical judgments. But inconsistencies that relate rival or successive scientific theories are rarely so straightforward. Theories may be thought inconsistent, only to be reconciled by a radical new idea. The anti-realist may be forced to endorse a particular theoretical perspective even to diagnose the inconsistency that he wants to supply his inductive evidence.

More reasonable than to read scientific successes as license to diagnose failures is to regard the record of theory change as constructive. In being rejected, theories are improved upon. Because science has been able to identify its mistakes and rectify them, its current commitments are all the more trustworthy.

5.4 Selective Confirmation

What are these commitments? What, for that matter, are the commitments of any successful theory? I have argued that because the ontological status of theoretical posits is inconstant and disputatious within science, a philosophy that imposes uniformity assumes the burden of dissenting from science. But if the status of theoretical entities is disputatious, what is one to be realist about? In particular, are the once successful posits, from whose eventual rejection the anti-realist induces the epistemic unreliability of current science, ones to which a realist should have been committed in the first place?

Success has many forms. Not only is scientific appraisal unstable; it is multidimensional. Theories and the entities they posit are assessed for their utility, heuristic power, explanatory value, mathematical tractability, experimental manipulability, and cohesion with background knowledge. Epistemic justification is neither the only, nor necessarily the most pressing, concern. Because of this complexity, the epistemic commitments of science cannot simply be read off of scientific practice. Whether or not the failures of past science are failures of a realist view of that science depends upon realism's criterion for deciding what theoretical posits to treat realistically. Deference to scientific practice is not a definitive criterion.

This problem is intractable if one takes a holistic view of confirmation. All manner of theoretical posits whose roles in a successful theory are not such as to accrue epistemic warrant then go along for the ride. Is the spacetime interval in Minkowski's interpretation of special relativity supposed to be a real entity, to whose existence the empirical evidence for relativity attests, or is it but a mathematical invariance of dispensable convenience? Are quarks a purely formal method of classification for hadrons, or are they their physical constituents? Is the mechanical ether an intuitive aid to picturing what happens in the space between the locations of charged bodies, or is it an entity that Maxwell's equations require to exist?

To treat the empirical success of a theory as confirmation of the theory as a whole is to obviate such questions; it is then the theory as such that evidence confirms, not this hypothesis or this theoretical posit over others. The loss in discrimination may appear innocuous from an overview of scientific practice, because the pragmatic goals of prediction and control are as well advanced by an entity's conceptual utility as by its existence. But the questions are pressing for the realist, who must discriminate entities whose existence is established by the evidence from those that can come or go with impunity.

To do this, the realist must adopt some criterion beyond mere participation in an empirically successful theory to identify theoretical posits to be treated realistically. There are both positive and negative criteria, and some of them are intuitively obvious in the abstract, if problematic in application. For example, a theoretical posit gets no epistemic support from the successful prediction of a result that it was artificially contrived to yield. Nor is it supported by a result that the theory predicts independently, a result the theory does not need it to obtain. The positive criterion would seem to be that the posit be used *essentially* in achieving the theory's empirical success, and that this success be unexplainable without it.

It may have been inconceivable to Maxwell that electrical phenomena could proceed without a mechanical medium to propagate electromagnetic waves. But his equations alone, without the ether hypothesis, generate the predictive success of his theory. Newton believed that the apparent motions, which his laws of mechanics governed, presupposed the existence of absolute frames of space and time. But these posits have no role in the use of Newton's laws to generate observable predictions. The igneous fluid that Lavoisier thought was necessary to push the molecules of a heated substance apart was not required to account for the phenomena of heat; molecular motion itself was the operative mechanism.

A pattern is evident in such examples. Like the hidden variables of intuitively picturable interpretations of quantum mechanics, an entity or structure is introduced to make physical sense of the laws used to predict empirical phenomena. But the particular properties attributed to this entity do not matter to the use of these laws in successful prediction. The center of mass of the universe is supposed, by Newton, to be at rest in absolute space, but it makes no difference to the use of Newton's laws to give it a positive constant absolute velocity. Maxwell thought there had to be some sort of ether to propagate waves, but he was free to give it all sorts of mechanical properties without affecting his laws. Lavoisier was vague by default as to the physical process by which his igneous fluid flowed. Entities such as phlogiston or the nested spheres of a geocentric universe are rejected because they give the wrong theoretical mechanism. They have identifiable successors in later theories – oxygen and gravity. In contrast, presuppositional posits in the conceptual background of successful laws may be rejected simply because their existence proves inconsistent with subsequent theory. They have no successors because there is no predictive role to continue to fill. We teach ourselves to regard them as metaphysically superfluous.

Thus, conceptual involvement, however fundamental, in a successful theory is not sufficient for the reality of a theoretical entity. What scientists believe the world must be like and how they make sense of the empirical phenomena owe too much to heuristic concepts that underlie their theory and to the entanglement of pragmatic among epistemic ends. Rather, the criterion must be that the theory owes its success to this entity. With this criterion, the realist can reject historical counterexamples of successful but nonexistent theoretical entities, and argue that entities meeting his criterion survive in current science.

5.5 An Argument from Novelty

Despite their inconclusiveness, the challenges to realism have pressured realists into independent lines of argumentation. For anti-realism has its own pedigree. It is embedded within a long-ascendant tradition of empiricist epistemology, and from the perspective of this tradition realism's epistemic commitments are excessive. Empiricism is certainly the philosophical root of anti-realism's indiscriminate suspicion of theoretical entities. If experience is the only possible source of knowledge of the world, beliefs as to the existence and nature of unobservable entities are inherently suspect and require a special defense. The predominant line of defense is explanationist: If theories owe their empirical success to unobservable entities, then we need realism to explain why theories are empirically successful. If there is no truth to theory, if theoretical entities are not real, then the predictive accuracy of theory is a coincidence too cosmic to accept.

The argument I shall construct is a descendant of this line. Lots of empirical success needs no explanation, and lots of it has nonrealist explanation. Lots of theory is unsuccessful. The problem is to identify a specific form of success that realism alone explains, and then to show that this virtue of realism is epistemically justificatory.

The form of success I propose is *novel* success; the successful prediction by a theory of an empirical result that is novel for it.

Intuitively, "novel" means "new," not just temporally, but also in the sense of different or unusual, and "unexpected." These attributes pose an explanatory challenge that it will take realism to meet. How does a theory manage to predict unusual and unexpected results correctly? That its predictions are correct is a matter of experience. The answer to the question *why* they are correct is that this is the way the world is found to be. That they are *its* predictions is a matter of deduction. The theory yields its predictions because of the inferential resources of its semantic content. Realism is not involved in answering these questions, but the question I have posed does not reduce to these. It is a second-order question about the success of the theory. That it be just the correct results that the theory predicts is not, if these results are novel, explained by the fact that the theory sustains certain logical relations to certain observation sentences which simply happen to be correct. The explanation must appeal to some property of the theory, something distinctive in its content that enables it reliably to forecast the unfamiliar and unexpected. Unless this content is interpreted realistically, the theory's novel success appears purely accidental.

To make clear why realism is required to explain the novel success of theories, let me be more precise about novelty. Classic examples of novelty are the prediction from general relativity of the gravitational deflection of starlight, and the prediction from Fresnel's transverse wave theory of light of the bright spot in the center of the shadow cast by a circular disk in spherical diffraction. These results were new, unknown, surprising, unanticipated independently of the theory predicting them, uninvolved in constructing this theory, and unlike results supporting rival theories. They present all of the features intuitively associated with novelty. But some of these features do not require realism and some need not be present for realism to be required. A result could be new and unknown, and yet instantiate a general law presupposed in constructing the theory that predicts it. Then the success of the prediction gives no epistemic support to the theory. A result could be well known yet unexplained, even contrary to the predictions of extant theories. Then its prediction by a new theory carries probative weight. The use of a result in constructing a theory might have been inessential, such that the theory would have predicted the result without its use. Even a theory expressly motivated by the need to explain a result can receive epistemic credit for doing so, if the result is not involved in its construction. The core conditions for an analysis of novelty are those under which novel predictive success depends on the existence of the mechanisms theorized to produce it and the accuracy of their description.

Two such conditions will be jointly sufficient for the prediction of a result R by a theory T to be novel for T. The more basic intuitively is *independence*: R must not instantiate any general law used essentially in constructing T. This captures the idea that R was not built-in, whether expressly or inadvertently, such that T would automatically have predicted R whether T is true or not. Secondly, the prediction must be *unique*: no viable rival of T provides an alternative basis for predicting R. For otherwise, R's evidential status is ambiguous. The uniqueness condition speaks to the intuition that a novel consequence of T must differ from the empirical consequences of other theories.

Classic examples of novelty exhibit these conditions. Young's law of interference could be used to obtain the positions and intensities of diffraction bands, but not Fresnel's predictions for spherical diffraction. And certainly no corpuscular theory of light could be made to yield the unexpected bright spot. A gravitational influence on light could be based on Newtonian theory, but only by suspending major theoretical developments since Newton: the incorporation of the wave theory of light into electromagnetic field theory, and the role of light in relativistic mechanics. No viable rival to Einstein's analysis of deflection was available.

My explanationist argument for realism is now straightforward. A theory's sustained record of novel predictive success is only explainable by supposing that the theory has correctly identified and described the entities or processes responsible for the observations predicted. The realist explanation of the theory's success is then epistemically unrivaled. If this success is uncompromised by failure, if the theory is free of disconfirming results and conceptual problems, then the realist explanation of its success is also epistemically undefeated. But an explanation that is neither rivaled nor defeated is justified, on pain of skepticism with respect to ordinary beliefs whose provenance is unavoidably abductive.

5.6 Consequences and Clarifications

Independence is a historical attribute; whether R satisfies the independence condition depends on how T was developed. If we imagine T having alternative provenances without overlapping reliance on common results, or generalizations of them, then all of T's predictions satisfy independence, for the use of none is essential to T's construction. This is appropriate, for alternative provenances constitute a form of *over*determination, which should be as much of an epistemic advantage as underdetermination is an epistemic liability. A theory's provenance provides reasons for thinking it plausible and taking it seriously as a potentially acceptable account of some domain of nature, pending empirical testing. Additional provenances represent additional grounding.

But, of course, predictions novel for T do not automatically support T. They must be established by observation with quantitative accuracy. A theory rich in novel consequences is rich in opportunities for epistemic support, none of which might materialize. Independent provenances are rare historically, and even if present are no guarantee that T merits realist interpretation.

Although novel for T, R might come to be explained or predicted by a rival T' of T that is developed after the fact. Rather than preempt R's novelty, the effect of the advent of T' is to challenge R's epistemic weight. The uniqueness condition is to be read as temporally indexed; it is not novel status that varies with historical developments but its epistemic significance. This interpretation accords with philosophical as well as scientific practice. The anti-realist wishes to credit superseded theories with empirical success. The most important form of empirical success is novel success, and it must be possible to diagnose such success however further developments affect T.

However, T' does not necessarily undermine R's support of T. If R would not have been novel for T' even if T were unavailable – if, for example, T' is expressly designed to yield R and provides no independent theoretical basis for explaining R – then T is still favored. For the correctness of T's explanation of R is still undefeated as an explanation of T's ability to predict R successfully.

What if, but for T, R would have been novel for T'? What if T' only *happens* to come later, its provenance owes nothing to T or R, and its prediction of R is epistemically as impressive as T's? Then, of course, R's support of T is undermined; R is no longer a reason to interpret T realistically. And if such sequences are historically common, the anti-realist has a new basis for skeptical induction.

Accordingly, realism predicts that such scenarios will *not* occur. It predicts that viable rivals to epistemically warranted theories will not arise. It predicts that theories that record novel success will continue to be successful, that the existence and properties of the theoretical entities they invoke to explain novel results will be upheld through further developments in science. As novel success is realism's standard of epistemic warrant, realism, if it is to be warranted, must itself meet this standard. The importance of the predictions in question is to deliver on this requirement.

No philosophical rival to realism provides a basis for these predictions, nor are they involved in the argument for or content of the realist thesis. Accordingly, they satisfy the independence and uniqueness conditions for novelty. Realism is self-referentially consistent, as any naturalistically defensible theory must be.

Scenarios that defeat realism leave us with no explanation of T's ability to predict R successfully. We must allow that some novel success may simply be chance. And of some novel success, the correct explanation may not be epistemically warrantable by us. But these possibilities do not prevent realism from being warranted where the explanation it offers is undefeated. Rather, they register the unavoidable defeasibility of realism and suggest that realism be embraced only where a substantial record of sustained novel success has been achieved. As a philosophical interpretation of the epistemic status of T, realism requires retrospective evaluation of the evidential situation; we should not expect to read realism off of scientific practice in real time.

5.7 The Importance of Novelty

I have rebuked anti-realism for insensitivity to differences in the status of theoretical entities. The positive argument that I have constructed for realism reveals a related failing. Anti-realism lacks the resources to distinguish novel predictive and explanatory success from a theory's routine empirical applications. There is no question that novel applications are especially compelling epistemically. From the use of Newtonian theory to discover the outer planets of the Solar System, to Mendel's backcross test of the genetic hypothesis, to the application of atomic theory to Brownian motion, to the bright spot discovered in spherical diffraction, to the conversion of matter and energy, to the gravitational deflection of starlight, novel results have provided a warrant for theories that mere conformity to observation cannot signify.

To account for this difference requires treating not just the observable results themselves, but a theory's success in predicting them, as a proper object of explanation. Where, but only where, the results are novel, the explanation of their successful prediction must be that the theory has correctly identified and described the theoretical mechanisms of their production. To treat all theoretical entities with agnostic indifference preempts this explanation. Anti-realism may acknowledge the possibility that a theory has posited the right entities. But in insisting that this possibility be epistemically inaccessible, anti-realism in effect reduces the theory to a mere predictive instrument: a theory's only permissible endorsement is that it is, unaccountably, a good predictor. And with respect to this attribute, all observable phenomena, from novel discoveries to programmed outcomes, are epistemically on a par. All of a theory's observed consequences are equal confirmation of its predictive accuracy.

The concept of novelty gives realism a criterion for epistemic commitment; it identifies the conditions under which theoretical belief is warranted. This criterion undercuts the holistic conception of confirmation, which is as unacceptably indiscriminate as anti-realism's dismissal of all theoretical entities. The portions of a theory that are to be interpreted realistically and expected to survive future theory-change are those responsible for the theory's novel success. The realist does not presuppose the correctness of current theory in allocating epistemic commitments to past theory. He identifies what was novel in the successes of past theories and determines how that novel success was achieved. His criterion of epistemic commitment is the same for past science as for present.

Nor does the realist presuppose the legitimacy of whatever inference is needed to close the logical gap between theory and evidence. He endorses a specific ampliative move to support a theory with novel success over its algorithmically generated rivals. I argued that T's rivals are not confirmable, while, by hypothesis, T is. The basis for the hypothesis was simply that ampliative inference as such may not be disallowed on pain of skepticism. Allowing ampliation as such is enough to make T a proper object of confirmation, but not T's rivals.

The positive argument that I have since constructed for realism affords greater specificity. The ampliative principle that warrants interpreting T realistically is an abductive inference from T's novel success. No algorithmically generated rival to T can possibly claim novel success. The rivals cannot make novel predictions at all, because their observational consequences violate both requirements for novelty. All their consequences are already consequences of T, which violates uniqueness. And all their consequences are used essentially in their construction; their semantic content is determined by specifying what their consequences are to be. This violates independence. It follows from my positive argument for realism that the rivals are ineligible for epistemic support. Their availability cannot, therefore, establish the underdetermination of T. This reasoning presupposes nothing more than the rejection of a sweeping skepticism that would deprive ordinary, paradigmatically unproblematic beliefs of their necessary grounding in explanatory inference.

5.8 Observation and Theory

To this point, I have assumed a clear and absolute distinction between observational and theoretical propositions. Observational propositions formulate the evidence by which theories are judged, and the question has been whether this evidence is ever, in principle, sufficient for epistemic commitment to theories. The distinction is necessary to engage the anti-realist, for without it anti-realism collapses into skepticism. Anti-realism requires an observational level exempt from problems of underdetermination and historical inconstancy to circumscribe its range of incredulity.

Realism, by contrast, makes no such requirement. Although traditionally formulated within the problematic of vouchsafing the inference from observation to theory, realism's essential message is that epistemic justification is not confined to the observable. Realism does not require that there even be an essential division of observation from theory; it requires only that if there is, it does not divide the justifiable from the unjustifiable. In particular, my own positive argument for realism does not depend upon the observational status of novel results. Results novel for a theory are consequences of it that satisfy independence and uniqueness. Any result, whether classifiable as observational or not, could be novel and, if epistemic justification is not restricted to an observational classification, could be independently established and evidentially probative.

As anti-realism requires an epistemically privileged category of observational propositions distinguished from theory and realism does not, it is appropriate to ask whether the requirement can be met. I am inclined to concede to the anti-realist at least a rough-and-ready distinction between observation and theory, although I think it is more contextual and variable than he can tolerate. The issue I wish to press is that of epistemic privilege: Could there be justified observational beliefs if no theoretical beliefs are justifiable?

Observation reports are certainly fallible, and often they are corrected or reinterpreted in ways that affect their evidential bearing on theory. In order to trust them as evidence, there must be reason to believe that conditions are not such as to undermine their reliability. But beliefs as to the conditions under which observation is reliable are not themselves classifiable as observational. We observe objects, not the accuracy of our observation of them. Should it be imagined that the accuracy of observation is itself somehow an object of observation, then it would become necessary to ask after the observability of the accuracy of observations of accuracy. That won't do. Unobservable entities and processes can interfere with our ability to represent objects accurately on the basis of observing them. We use theories about the nature of observation and the impediments to its veridicality in assessing the correctness and the evidential weight of observation reports. Only if these theories are justified are we entitled to our discrimination of trustworthy from untrustworthy reports.

Anti-realism assumes that epistemic justification has a rigidly foundationalist structure. There is a privileged class of judgments sanctioned by observation whose justification is unproblematic and automatic. This class must be identifiable independently of any theorizing, for it is assumed that its members are the common explananda of rival theories. Propositions outside this class are justifiable, if at all, only indirectly by inference from those inside. The more remote a proposition's content from paradigmatically observable situations, the more dubious is its epistemic status.

Given this foundationalist picture, doubts as to the existence of such a privileged class of propositions with the required properties are immediately skeptical doubts. If observation is fallible, if observational judgments may themselves be objects of justification, if their evaluation invokes judgments outside their class, then the entire structure of justification collapses. The anti-realist wants to be a skeptic only at the level of theory. But his epistemology makes this restriction untenable. For his arguments against theory can be repeated at the level of evidential reports used to judge theory. Such reports are revised, reinterpreted, and underdetermined by experience. In view of these liabilities, the consistent foundationalist can only retreat to a yet more basic epistemic level.

If there is such a level, then with respect to it the anti-realist's arguments fail. There will be a foundation for belief not itself in need of epistemic support but capable of providing it. The evidential judgments of science will be justified in terms of it. But if this works for the evidential judgments of science, why not for the theoretical judgments inferred in turn? There will be no reason to arrest epistemic assent short of theory. If there is no epistemically basic level, then anti-realist arguments succeed everywhere. There is no place to arrest epistemic descent into a sweeping skepticism that anti-realism eschews.

We are left with two nonskeptical options: repudiate foundationalism or allow epistemic support to extend throughout the structure of our belief system. The first option eliminates the anti-realist's grounds for denying the possibility of theoretical knowledge. On this option, whether or not a belief is justifiable has no automatic connection to a classification of beliefs into ontological kinds. The second option provides for the justification of theoretical beliefs. Their epistemic status may be less secure than that of beliefs closer to the foundations, but the difference is at most one of degree. Either way, scientific realism wins.

Bibliography

Boyd, R. 1984: The current status of scientific realism. In Leplin (1984a), op. cit., pp. 41–83.

Churchland, P. M. and Hooker, C. A. (eds.) 1985: *Images of Science*. Chicago: The University of Chicago Press.

Fine A. 1984: The natural ontological attitude. In Leplin (1984a), op. cit., pp. 83–108.

Kitcher, P. 1993: *The Advancement of Science*. New York: Oxford University Press.

Laudan, L. 1981: A confutation of convergent realism. *Philosophy of Science*, 48, 19–49.

Laudan, L. and Leplin, J. 1991: Underdetermination and the empirical equivalence of theories. *Journal of Philosophy*, 88, 449–72.

Leplin, J. (ed.) 1984a: *Scientific Realism*. Berkeley, CA: University of California Press.

— 1984b: Truth and scientific progress. In Leplin (1984a), op. cit., pp. 193–218.

— 1997: *A Novel Defense of Scientific Realism*. New York: Oxford University Press.

—Psillos, S. 1999: *Scientific Realism: How Science Tracks Truth*. New York: Routledge.

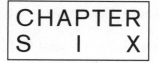

A Theory's Predictive Success does not Warrant Belief in the Unobservable Entities it Postulates

André Kukla and Joel Walmsley

6.1 Introduction

One problem facing the epistemology of science concerns the fact that many scientific theories consist of claims about unobservable entities such as viruses, radio waves, and quarks. If these entities are unobservable, how are we to know that the theories that make reference to them are true? The question we ask is "Does the success of a scientific theory ever give us reason to believe in the unobservable entities it postulates?" The existence of the unobservable entities postulated by a theory is, of course, entailed by the truth of that theory. If a theory is true, then the entities it postulates exist (if they didn't, the theory wouldn't be true). Accordingly, where we talk of the truth of a theory, we mean that to include the existence of the unobservable entities it postulates, as will be noted.

The fact is that scientists *do* hold theories about unobservable entities, and for the most part, they hold them to be *true*. At the risk of overgeneralization, then, we might say that most practicing scientists are realists, where realism is the thesis that scientific theories are true (with all that that entails). It must be said that this view is a particularly strong (and therefore fragile) form of realism. This chapter aims more ambitiously to unseat the argument from scientific success for the much weaker realist thesis that there are nomologically possible circumstances (circumstances which are possible given the laws of physics in this world) wherein we would be justified in believing in the unobservable entities postulated by a theory. Henceforth, we use the term "realism" to refer to this view – which Leplin has called "minimal epistemic realism" – that it is logically and nomologically possible to attain a state that warrants belief in a theory. It is worthy of note that there may be *other* good reasons for being a (minimal epistemic) realist – the aim of the present chapter is to show that the success of science isn't one of them: the *success of science* does not warrant even the weakest realist thesis.

We should start by saying what we mean by the "success of science." Let us say that by this phrase, we mean that our scientific theories allow us to make significantly more correct predictions than we could without them. A less circumscribed, but nonetheless consistent, version is Laudan's broadly pragmatic definition:

> A theory is successful provided it makes substantially more correct predictions, that it leads to efficacious interventions in the natural order, or that it passes a suitable battery of tests. (1984, p. 109)

Nothing substantial in our argument will hinge on the difference between these two formulations.

The argument for realism from the success of science is often referred to as the "no miracles" argument, following Putnam's (1979) contentious formulation. In his view, "[t]he positive argument for realism is that it is the only philosophy that doesn't make the success of science a miracle" (p. 73). In order to block Putnam's conclusion, then, we need to show that there is a coherent explanation for science's success, which neither requires nor entails realism. This is to say that even if we grant that realism does explain the success of science, Putnam's claim can be undermined by showing that there are comparably good anti-realist explanations.

6.2 Anti-Realist Explanations for the Success of Science

Leplin (1987, 1997) has conducted a thorough investigation of this topic and concludes that there are no viable anti-realist arguments for the success of science. He considers two categories of putative anti-realist explanations for the success of science: those that allude to evolutionary-type mechanisms of theory selection, and those that regard a theory's empirical adequacy (i.e., the extent to which the theory correctly describes observations and experimental results) to be an explanation of its success. According to Leplin, both strategies fail, so the only remaining explanation of scientific success is the truth of theories. His critique of the evolutionary alternatives is persuasive – more so than his critique of the empirical adequacy alternative – so let us consider them in that order.

The archetypal evolutionary explanation for the success of science can be traced to van Fraassen (1980), according to whom the success of our theories is due to their having been chosen by a mechanism of Darwinian selection:

> Species which did not cope with their enemies no longer exist. That is why there are only ones who do. In just the same way, I claim that the success of current scientific theories is no miracle. It is not even surprising to the scientific (Darwinist) mind. For any scientific theory is born into a life of fierce competition, a jungle red in tooth and claw. Only the successful theories survive – the ones which *in fact* latched on to actual regularities in nature. (p. 39)

This thesis often goes by the name of "evolutionary epistemology," and amounts to the view that theories are selected by the scientific community for predictive success and that this practice suffices to account for the success. It has to be defended against the charge that the competition among theories fails to satisfy some of the conditions necessary for a process of Darwinian selection to take place. Leplin's (1997) analysis, which we now summarize, shows that van Fraassen's account fails to provide an explanatory alternative to theoretical truth.

Leplin draws an analogy with the tennis-playing abilities of Wimbledon finalists: How do we explain the fact that Wimbledon finalists are so good at tennis? The question admits of two different interpretations. If we want to know why it is that Wimbledon finalists are in general good tennis players, it is appropriate to cite the stringency of the selection procedures for entry into the tournament. This answer, however, does not tell us why particular individuals – Sampras and Agassi, for example – are good at tennis. To answer *this* question, we would have to cite the relevant properties of the individuals, such as their training and genetic endowment. To make the analogy explicit, the problem with evolutionary epistemology is that the question it answers is not the same as the question answered by theoretical truth. The Darwinian account explains how we come to be in *possession* of successful theories, but it doesn't explain what it is about *these* theories, which we happen to possess, that makes them successful. Theoretical truth does not explain how we come to possess these particular theories. But it does answer the second type of question with regard to predictive success: it is an attribute of *these particular theories* that accounts for that success. We must conclude that evolutionary epistemology and theoretical truth are not explanatory rivals – van Fraassen's account does not give us an anti-realist explanation for the success of science and the "no miracles" argument remains unscathed.

Leplin notes that the same can be said for Laudan's (1984) elaboration of van Fraassen's Darwinian account, which specifies the mechanism by which theories are selected. According to Laudan, success is to be explained by the fact that scientists use appropriate experimental controls and methodologies in testing their theories. Indeed, the fact that our theories have been tested by such-and-such a methodology may explain why we possess successful theories, but it does not explain why quantum mechanics and not some other theory has survived all those tests. The hypothesis that quantum mechanics is *true*, by contrast, provides at least *an* explanation for its success.

Before we turn to his critique of the empirical adequacy alternative (i.e., the part with which we disagree), let us briefly set a few ground rules for the discussion. We will suppose, for now, that no other potential problems with the "no miracles" argument need to be worried about. For the sake of argument, let us grant the realist an awful lot, and assume that science *is* successful, that the realist argument from the success of science is not question-begging or invalid, that the notion of (mere) approximate truth is both meaningful and measurable, and that approximate truth *does* explain the success of science. Given these assumptions, it will make no difference whether we talk of a theory's truth or its approximate truth. We can, therefore, play fast and loose with the distinction as well as with the distinction between empirical adequacy and approximate empirical adequacy. For the sake of expository convenience, we speak of the truth of theories where we mean their truth or approximate truth, and of empirical adequacy where we mean empirical adequacy or approximate empirical adequacy. Provisionally, let us also define "T is (approximately) empirically adequate" as "T is (approximately) empirically equivalent to a true theory." (Note that this definition of "empirical adequacy" differs from the conventional one whereby a theory is empirically adequate if it gives a correct description of observations and experimental results.)

Leplin (1987) defines *surrealism* as the thesis that the explanation for the success of our theories is that they are empirically adequate in the sense just defined – that is, our theories are empirically equivalent to true theories. Surrealism holds that the world behaves *as if* our theories of the world were true, but it "makes no commitment as to the actual deep structure of the world." Surrealism "allows that the world has a deep structure," which is to say that it allows that there *is* a true theory of the world whose theoretical terms denote real entities, but it "declines to represent" these structures (Leplin, 1987, p. 520). Leplin attributes the development of this view to Fine (1986) but, as Leplin rightly notes, Fine does not actually endorse surrealism, he simply cites its coherence to discredit Putnam's "no miracles" argument.

Leplin presents two arguments designed to show that surrealism collapses into realism. The first is that the surrealist explanation for the success of science *presupposes* the realist explanation:

> To suppose that Surrealism explains success is to suppose that for the world to behave as if the theory is true *is* for the theory's predictions to be borne out. But this is to suppose that theoretical truth is normally manifested at the observational level, and thus to *pre*suppose the original, realist explanation of success rather than to provide a weaker alternative. (1987, p. 523)

If this is right, then surrealism does not qualify as an *alternative* to the realist explanation. But this claim merely equivocates between two senses of "explanation." Let us explain.

We need to disambiguate the phrase "A explains B," and do so without becoming entangled in the various philosophical questions concerning explanation that are peripheral to this chapter. To say that A is *an* explanation for B is to make a claim which is consistent with A being false and the possibility of there being some C, incompatible with A, which is also an explanation for B. To say that A is *the* explanation for B is clearly to imply that A is true, and that the truth of A is the reason why B is true. "A explains B" is ambiguous between these two options.

Returning to the issue at hand, to suppose that the empirical adequacy of a theory T explains its success (in either sense of "explains") is indeed to presuppose that T's truth must also qualify as *an* explanation of its success. But it is not to presuppose that T's truth is *the* (i.e., the correct) explanation of its success, or even that T is true at all. Both realists and anti-realists acknowledge that scientists frequently make use of theories that are known to be false in predicting observations. Scientists sometimes represent gas molecules as minute billiard balls. The explanation for why these patently false representations succeed is presumably surrealistic: they are approximately empirically equivalent to other theories about gases that are true. If we accepted Leplin's claim about surrealism's presupposition of realism, then we would have to conclude that gases *are* made up of tiny billiard balls.

One might argue that, whilst empirical adequacy does not presuppose truth *simpliciter*, it does presuppose approximate truth: the success of the billiard ball theory of gases does not entail that gas molecules are billiard balls, but it does entail that gas molecules are relevantly *like* billiard balls. Most realists would be content with the verdict that our best theories are approximately true, so the surrealist explanation of the success of science may presuppose the realist explanation after all. The temptation to think that approximate empirical adequacy entails approximate truth stems from the following considerations. For a theory to be approximately true means, let's say, that it gets most or many of its claims right, even though it may also get some things wrong. Similarly, for a theory to be approximately empirically adequate, it must get a lot of its *observational* claims right, even though it may also get some of them wrong. But to get a lot of observational claims right is to get a lot of claims right, *tout court*. Therefore, approximate empirical adequacy entails approximate truth.

The problem for realists, though, is that this notion of approximate truth has no connection with scientific realism. If getting *any* sizeable part of the story right is enough to count as approximate truth, then it is possible for an approximately true theory to get the *theoretical* part of the story *entirely* wrong despite getting the observational part right. Getting a lot right is certainly necessary for approximate truth, but if approximate truth is to have any bearing on the issue of realism, it should also be stipulated that approximately true theories get at least some things about *theoretical* entities right, not just observational statements. There is no need here to specify just what part or how much of the theoretical story an "approximately true" theory has to get right. But however slight that theoretical requirement may be, the result is a notion of approximate truth that is no longer presupposed by approximate empirical adequacy. The surrealist explanation does not presuppose the realist explanation because getting an arbitrarily large part of the observational story correct is logically compatible with getting the theoretical story totally wrong.

Let's call the hypothesis that every theory has empirically equivalent rivals EE. Leplin's second argument is that surrealism threatens to collapse into realism unless EE is true. For suppose that we have a successful theory T which has no empirically equivalent rival. Then the surrealist explanation for T's success – namely that T is empirically equivalent to a true theory – entails that T itself is true. This is to say that the surrealist explanation entails the realist explanation. To maintain its distinctiveness, surrealism must be *committed* to the truth of EE. Thus Leplin regards it as unproved that surrealism provides a genuine explanatory alternative, since it is an open question whether EE is true. The first response to make is that it is *not* an open question as to the universal availability of empirically equivalent rivals. Kukla (1998) argues for several algorithms for transforming any theory into an empirically equivalent rival theory (see chapter 5 in particular). We shan't insist on that point here – rather, let us concede, for the sake of argument, that the status of EE is uncertain, and hence that the distinctiveness of the surrealist explanation is also uncertain. We can nonetheless weaken surrealism in such a manner as to block the derivation of realism from the negation of EE whilst still leaving an available explanation for the success of science.

Let us here revert to the more traditional definition of empirical adequacy: T is empirically adequate if all of its empirical consequences are true. As before, the empirical adequacy of T entails that the observable world behaves as if the "deep structures" postulated by T do exist, but it is no longer required that the observable phenomena be consequences of any *real* "deep structures." This is to say that, in virtue of the fact that it drops the requirement that T's empirical consequences also be the consequences of a true theory, this notion is logically weaker than the previous one. We might give the name *weak surrealism* to the thesis that our best theories are empirically adequate in the new diminished sense, and *strong* surrealism to the thesis that our best theories are empirically equivalent to true theories.

Now consider the hypothesis that weak surrealism explains the success of science. In fact, such a hypothesis is immune to the second difficulty that Leplin makes for the strong surrealist explanation. Leplin claims, however, that such an escape from the collapse of strong surrealism merely lands us in another dilemma:

> The statement that things happen as they would happen were our theories true invites two directions of analysis. On one analysis, Surrealism simply tells us what goes on at the observational level. Then it is not simply metaphysically weaker than realism; it is not metaphysical at all. But neither is it explanatory. Rather, it restates the explanandum. (1987, p. 522)

The "direction of analysis" alluded to in this passage is the direction of weak surrealism. The other direction is strong surrealism. According to Leplin, in their choice between strong and weak surrealism, anti-realists are caught between the Scylla of realism (if they choose strong surrealism, their explanation threatens to be indistinguishable from that of the realists) and the Charybdis of vacuity (if they opt for weak surrealism, then their explanatory hypothesis amounts to no more than a recapitulation of the explanandum).

But there is a position between strong surrealism and the explanation that Leplin regards as vacuous, and so it seems his argument conflates three putative explanations of the success if science into two. Let's call T* the set of all empirical consequences of T. Weak surrealism is the view that T* is true. But one can hold that T* is true in two different ways. On the one hand, one may believe that the statement "The observable world behaves as if T is true" is *itself* a nomological truth about the world – a law of nature and part of T just like any other law-like statement about the world contained in T. On the other hand, one may believe in the truth of a set, X, of law-like empirical generalizations, along with the (nonnomological) hypothesis that, *as it happens*, X is coextensive with T*, perhaps by accident. In the latter case, one believes two things: (1) for each statement in X, that statement is a true law-like generalization; and (2) these generalizations exhaust T*. The conjunction of (1) and (2) amounts to saying that one does *not* believe in a further *law* that X is coextensive with T*; rather, one accepts this coextensiveness as a contingent accident.

The difference between the two can be brought out by considering what happens if some previously unconsidered empirical hypothesis E is found to be a consequence of T. If you believe that it is a law of nature that the world behaves as if T is true, then you will add E to your stock of beliefs. But if you think that the coextensiveness of T* and your empirical beliefs is merely accidental, then the same discovery will leave the status of E open; its derivation from T will not be a rationally compelling reason for its adoption into T*. Indeed, the discovery that E is a consequence of T might be a good reason for giving up the belief that the true empirical generalizations are coextensive with T*. We can, therefore, reserve the name "weak surrealism" for the first view, and call the second view *fragmentalism*. Fragmentalism is logically weaker than weak surrealism.

We thus have a series of progressively weakened views about theories: (1) realism (our theories are true); (2) strong surrealism (our theories are empirically equivalent to true theories); (3) weak surrealism (the observable world behaves as if our theories were true); and (4) fragmentalism (the true empirical generalizations happen to be coextensive with the empirical consequences of our theories). The further down the list we go, the less danger there is of a collapse into realism, but also the fewer resources we have for explaining the success of science. Leplin allows that strong surrealism explains the success of science but notes the danger of its collapse into realism. On the other hand, a fragmentalist account of success falls prey to the charge of vacuity, since according to that view, the ultimate truth about the world – or as much of it as we can ascertain – is given by a collection of empirical generalizations, and the truth of our empirical generalizations is exactly the explanandum in the success of science argument. But Leplin doesn't distinguish vacuous (4) from (3) – and (3) seems to avoid both Scylla and Charybdis. According to weak surrealism, the truth of our empirical generalizations is explained by the fact that (i) these generalizations are consequences of T and (ii) the observable world behaves as if T were true. Since this claim goes beyond what is contained in the explanandum, weak surrealism avoids the charge of vacuity. But there is no danger of it collapsing into realism, even if it turns out that some theories have no empirically equivalent rivals. Thus, weak surrealism is an anti-realist rival to the realist explanation for the success of science.

Of course, the existence of coherent and nonvacuous rivals to realism does not show that these explanations are *as good* as theoretical truth. The realist may abandon Putnam's own formulation of the "no miracles" argument and weaken her thesis to the claim that the realist explanation is *better* than the anti-realist explanation. The realist intuition is that the weak surrealist claim that the empirical adequacy of T is a fundamental fact about the world is unintelligible. We have no idea what could keep all the disparate empirical consequences of T in line, if not the causal mechanisms postulated by T itself, or at least, as strong surrealism asserts, the mechanisms posited by some other theory that is empirically equivalent to T: simply saying that they *are* kept in line doesn't explain anything. But what is the epistemic import of intelligibility?

It used to be thought that teleological explanations gave us understanding, but now we want to know the mechanism behind an appearance of teleology. Explanations that posit action at a distance have been considered intelligible and unintelligible by successive generations. The point is that the notion of "intelligibility" is at least relative if not vague. It would seem that the realists have in mind an archetypal explanation as one that satisfies our explanatory hunger, likely mechanistic in flavor. But it doesn't seem that the "intelligibility" of this kind of explanation signifies anything more than a psychological willingness to stop inquiry at that point. It would be, of course, premature to claim that there *cannot* be an epistemically relevant notion of intelligibility that brands weak surrealism as unintelligible, but at present there does not seem to be such a notion formulated. In fact, the only relevant discussion, to our knowledge, concludes that the intelligibility of scientific theories is a matter of "psychological acclimation" (Cushing, 1991, p. 346).

Does this critique break one of the ground rules? Recall that we granted the validity of the "no miracles" argument, and then went on to argue that weak surrealism is a coherent, nonvacuous, and distinct explanation of the success of science. If we are right, then realists need to show that theirs is a *better* explanation than that of the weak surrealists. They do not have to show that explanatory goodness has epistemic import, since that much was assumed when the validity of the "no miracles" argument was granted – if explanatory goodness were *not* epistemically relevant, then the conclusion of the "no miracles" argument wouldn't follow from its premises. It is precisely this principle that we call into question, but to grant that explanatory goodness is epistemically relevant is not to give the realists license to use any features of explanations they like in making a comparative evaluation. It would be cheating, for example, for the realist to claim that their theory is superior on the grounds that the positing of theoretical entities is an indicator of explanatory goodness.

The problem is that the assumption of the epistemic relevance of explanatory goodness doesn't tell us *which* theoretical properties are indicative of this epistemically relevant explanatory virtue. We can no longer deal with Leplin's problem in isolation from other problems faced by the "no miracles" argument. The realists' task is to find a property, p, of explanations which simultaneously satisfies this pair of criteria: (1) theoretical truth has more p than empirical adequacy; and (2) the possession of p is relevant to beliefworthiness. The realist must at the very least show that there exists some p such that theoretical truth has more of it than empirical adequacy, and that p is not demonstrably *ir*relevant to beliefworthiness. It is easy to satisfy either requirement by itself. On the one hand, theoretical truth is undoubtedly a more intelligible explanation of success than empirical adequacy. But if, indeed, intelligibility is relative, then it cannot play the role that realists need it to. On the other hand, an example of a property of explanations whose epistemic credentials cannot be denied is the *probability that it is true*. But since theoretical truth is logically stronger than empirical adequacy, it cannot be maintained that the realist explanation has more of this property than the surrealist explanation. The trick is to satisfy both requirements at once, and realists have not yet managed to perform this feat.

6.3 Two Realist Arguments from the Success of Science

Having shown that realism is not the *only* explanation for the success of science, it seems appropriate to return to the realist argument from the success of science and examine it in more detail. We might schematize Putnam's (1979) argument, which we hereafter refer to as SS, as the nonparenthetical reading in what follows:

(SS₁) The enterprise of science is (enormously) more successful than could be accounted for by chance.

(SS₂) The only (or best) explanation for this success is the truth (or approximate truth) of scientific theories (and therefore the existence of the unobservable entities postulated by those theories).

(SS3) Therefore, we should be scientific realists.

Notice that, as Putnam originally phrased it, the truth of theories is the *only* available explanation for science's success. This makes his view a particularly fragile version of the argument; our formulation of a single coherent alternative explanation, that of weak surrealism, is enough to block his conclusion. A stronger version of the argument would be to adopt the parenthetical reading of SS₂ and say that we should be scientific realists because the truth (or approximate truth) of theories is the *best* explanation of science's success, thus circumventing the issue of whether or not there is an anti-realist alternative.

The canonical anti-realist objections fall into three groups. First, one can follow the line suggested in section 6.2 and contend that even if truth or approximate truth does explain the success of science, there are equally good (or perhaps even better) anti-realist explanations. Secondly, one might claim that neither truth nor approximate truth is good enough to count as a passable explanation for the success of science. Thirdly, one can allege that even if truth or approximate truth were the only or the best explanation for science's success, the realist conclusion does not follow from the premises without additional (possibly question-begging) assumptions. We will examine the second and third of these claims in turn. But first, we should examine a fourth type of objection, conspicuous in its absence from the anti-realist literature.

It would appear that anti-realists have never tried to deny the truth of SS₁. But it is certainly not unthinkable to deny that science is successful. Indeed, the denial of SS₁ is central to many social constructionist and other epistemically relativist analyses of science. For example, Latour and Woolgar (1979) contend that the apparent success of science is explicable by the fact that scientists *construct* the data that confirm their own theories. Indeed, such a process of self-validation could be said to result in some kind of success – maybe even on Laudan's notion – for a self-validated theory *does* correctly predict the data that will be constructed to confirm it. But self-validating theories do not depend for their success on any of the *intrinsic* properties of the theory; our predictions would be just as successful with any other theory that was equivalently situated within our social milieu. So it is clear that neither realists nor anti-realists have this kind of performance in mind when they think of success for the purposes of SS. In fact, the characterization of success that we proposed in section 6.1 above gives the correct judgment in this case: self-validated theories are not successful because they don't enable us to make more correct predictions than it would be possible to make with some *other* theories which were also appropriately embedded in our social life.[1]

---

1 Note that this does not commit us to saying that a theory that has "empirically equivalent" rivals (i.e., rival theories that predict all and only the same observations) cannot be successful. It simply means that empirically equivalent theories must be regarded as equally successful. Since self-validation depends on social processes that are that are independent of the *intrinsic* properties of theories, the social constructionist theory must be that *all* theories are equally successful – which amounts to the assertion that the concept of success doesn't apply to theories.

Given the propensity of philosophers for occupying every nook and cranny of logical space, it is an interesting question as to why, despite the conceivability of doing so, anti-realists have never tried to undermine SS with the denial of SS_1. If the thesis of anti-realism *presupposed* the success of science, the absence of this strategy would be understandable. But there is no logical impropriety in endorsing either realism or its negation whilst simultaneously contending that all scientific theories suggested to date – or that will ever be suggested – are empirical failures. SS_1 is not entailed by anti-realism or by realism, or by a disjunction of the two. The supposition that both realists and anti-realists must make is that it is nomologically *possible* for us to have successful theories, for the issue at hand would not even make sense if empirical success were not possible – there is no point arguing about whether empirical success warrants belief in a theory, if there can be no such thing as empirical success. Such success does not have to be *actual* – in principle, the debate about scientific realism could have got started by cave-dwellers, long before the first remotely successful scientific theory was even conceived.

So, since SS_1 is independent of anti-realism, we might ask: Why have anti-realists not attempted to refute the realist argument from the success of science by asserting that science is not successful? Anti-realists may not have to concede that, in fact, there have ever been, or ever will be, successful theories. But they do have to concede that it would be at least *possible* for us to possess a successful theory. So one reason why anti-realists have not tried to deny SS_1 could be that although SS is about actual theories, it can be reformulated as an argument about *possible* theories, whilst maintaining all of its plausibility. And then the realist could claim that it is only the truth of such a theory that could account for its success. This new argument for realism, an argument from the possible success of science, which we call SS′, runs as follows:

(SS$_1$′) Some nomologically accessible theories would be successful if we used them to make predictions.

(SS$_2$′) The only (or best) explanation for this success would be the truth (or approximate truth) of the theories in question (and therefore the existence of the unobservable entities postulated by those theories).

(SS$_3$′) Therefore, there are nomologically possible circumstances – that is, the circumstance of being in possession of a successful theory – wherein we would be justified in believing a theory true.

Note that SS′ is an existential generalization of SS. SS′ is the stronger argument, and the negation of SS_1 would show SS to be unsound, but would not, by itself, adversely affect the status of SS′. This is to say that if SS is sound, then so is SS′, but that to refute SS is not to refute SS′.

It seems likely that relativists such as Latour and Woolgar would want to deny SS_1′ as well as SS_1 – and such a strategy would undermine SS′ as well as SS. The truth of SS′, however, is not up for grabs in the realist/anti-realist debate. But the fact that SS_1 is independent of both realism and anti-realism means that both sides are free to avail themselves of any argumentative strategies that involve its negation. One such strategy is deployed in a *realist* defense of SS′ and will be discussed in the next section.

6.4 Truth and Truthlikeness as Explanations for the Success of Science

Recall that we earlier argued that "A explains B" is ambiguous between two options – A's being *an* explanation of B, and A's being *the* explanation of B, where only the latter entails the truth of A. In what follows we will always use "A explains B" to mean (only) that A is *an* explanation for B.

The first anti-realist blow is that neither truth nor truthlikeness can be the explanation for the success of science (we use "truthlikeness" as a synonym for "approximate truth"). Such a view is developed at length by Laudan (1981). According to Laudan, a quick glance at the history of science shows that truth fails because the paradigmatically successful scientific theories of the past are now known not to be true. Truthlikeness fails for two similar reasons. First, approximate truth does not explain the success of science anyway – it is not even *an* explanation of success, let alone *the* explanation. Secondly, many of the successful theories of the past have not only failed to be true, but have also failed even to be approximately true.

As we shall see, the argument against theoretical truth as the explanation for the success of science preempts most of the issues relating to truthlikeness, so very little additional work will be needed to deal with the latter. If successful theories of the past were false, then it follows that either there must be some acceptable explanation for their success other than their truth, or that success sometimes has *no* acceptable explanation. Either way, SS is refuted by the repudiation of SS_2 – success is not a compelling ground for ascribing truth. Following such considerations, realists have mostly abandoned truth as an explanation for success and have taken refuge in approximate truth. But there is a rejoinder available for realists, and such a defense is similar to that used by McAllister (1993).

In fact, McAllister was not addressing the status of truth as an explanans for success. He was trying to defuse Laudan's parallel argument against the claim that *approximate* truth can explain success. Many of the successful theories of the past failed even to be approximately true. Theories of ether and phlogiston, for example, posited ontologies that are now completely rejected – there is nothing remotely like ether or phlogiston in our current world view. So Laudan's conclusion, as in the case of truth *simpliciter*, is that predictive success does not compel us to take our theories to be even approximately true.

There are two realist responses to this argument. First, the realist can insist that, in fact the successful theories of the past *were* in fact approximately true (see, e.g., Hardin and Rosenberg, 1982). This strategy fails because the more liberally we construe approximate truth, the more likely the strategy is to succumb to the charge that the approximate truth of our theories does not warrant a realist attitude towards them. Here, the truth of SS_2 is protected by weakening it to the point at which it ceases to entail SS_3. If phlogistic theories are to be counted as approximately true, then a theory may be approximately true despite the nonexistence of anything remotely like its theoretical entities. So, the strategy of maintaining that the theories of the past were in fact approximately true will not work. Secondly, the realist can admit that the theories cited by Laudan fail to be approximately true, but also deny that they are successful – this is the line taken by McAllister. According to McAllister, such judgments were made according to criteria for observational success that are now considered inadequate, even though, of course, these theories were *thought* to be successful at the time they were posited. Therefore, the examples given by Laudan do not actually show that theories far from the truth may nevertheless be successful and so do not warrant the anti-realist charge.

Interestingly, the same argument can be used to give another lease on life to truth itself (as opposed to mere truthlikeness) as the explanation for success. The fact that past theories were all false has no bearing on the truth-based version of SS if we deny that they were successful. Turning to our current theories, we might suppose that even the best of those is not successful. The assumption that no successful theory has ever been formulated contradicts premise SS_1. Hence it would undo SS. But, as noted above, it would not affect the status of the more general argument SS′, according to which the very possibility of having a successful theory is to be explained by the possibility of having a true theory. Here, the assumption that science is not successful can be used to reinforce the case for realism: it unseats Laudan's argument against the thesis that truth is the explanation of success, without harming SS′. Laudan's argument about truth doesn't make trouble for a minimal epistemic realist who claims only that it is possible to have a rationally warranted belief in theoretical entities. Laudan's argument does, however, create problems for stronger forms of realism; for example, the view that we are *already* entitled to believe in some theories. This may have been all that Laudan wished to claim; our point is that this is all he *can* claim.

Yet realists may be unwilling to pay what it costs for this argument to protect SS′ from Laudan's confutation. They have to assume that quantum mechanics (amongst other things) is empirically unsuccessful, and whilst this may be a possibility for the general skeptic about the possibility of there ever being a successful theory, as we've noted, anti-realists and realists alike must concede that empirical success is at least a possibility. Why doesn't quantum mechanics fit the bill for success? What more stringent standards for success could one possibly adopt than those which quantum mechanics has already passed? So let us assume that at least some our current theories are successful. If those successful theories are also *false*, then it follows either that there are other explanations of success besides truth, or that the success of science is inexplicable. In either case, the versions of SS and SS′ that refer to the truth of theories must fail.

What kind of a case can be made for the contention that our best current theories are false? The so-called "pessimistic meta-induction" bears on this question: all the theories of the past are now known to be false, so it is likely, by induction, that our present theories will turn out to be false as well. Indeed, the premise of the pessimistic meta-induction does provide us with *some* evidence for the view that our current theories are false but, as is the case with all inductions based on bare enumeration, the force of its conclusion is easy to resist. The fact that our past theories were false, *all by itself*, is a weak basis for projecting that falsehood onto our current theories. Of course, if we are progressing toward an attainable goal of absolute theoretical truth, our interim theories might nevertheless be false until we came very close to the end of this process of progress, but maybe we are close to that end *now*.

In fact, there are background-theoretical considerations that count against the pessimistic meta-induction. Recall that we are considering the realist contention that theoretical truth is the explanation for theoretical success. Laudan's objection is that this argument fails on the ground that the successful theories of the past were *not* true. Against Laudan, realists can claim that these false theories were not really successful either. But it is much more difficult to suppose that our current theories are also unsuccessful if the Laudan argument is run on our current theories as well as our past theories. On the other hand, it is not as obvious that our most successful current theories must be considered false – the only basis for that view is the pessimistic meta-induction. Finally, it has been conceded that reasonable realists have to admit that the best current theories are successful, but they still don't have to admit that the best theories of the past were successful. Moreover, realists are of the opinion that the inference to truth is warranted by success, so the data cited in support of the pessimistic meta-induction are consistent with the realist view. If it is accepted that the best theories of the past were unsuccessful whereas the best theories of the present are successful, then realists will rightly conclude that it is inappropriate to project that falsehood, which was possessed by theories of the past, onto theories of the present. The pessimistic meta-induction carries weight only if it is assumed that success is irrelevant to truth. So this anti-realist argument against the contention that truth is the explanation of success is a failure because the assumption begs the question against realists.

We now turn to the pair of counter-arguments relating to approximate truth. The realist argument here is that the approximate truth of our theories accounts for their success, and that approximate truth entails realism. The first of the two counter-arguments from Laudan is that there are successful theories that fail to be even approximately true. As noted above, anti-realists cannot secure the claim that there are successful theories that fail to be *true*, so the present, stronger claim can only be more problematic.

The second Laudanian counter-argument is that there is no reason to believe that approximate truth entails success. Laudan writes:

> [V]irtually all proponents of epistemic realism take it as unproblematic that if a theory is approximately true, it deductively follows that the theory is a relatively successful predictor and explainer of observable phenomena. (1984, p. 118)

But, of course, such an assumption is problematic, since it is conceivable that an arbitrarily minute error in our characterization of theoretical entities might result in a drastically incorrect account of observable phenomena. The root problem, as Laudan notes, is that realists have given us no clear account of how to assess claims of approximate truth. Until they do, the claim that approximate truth is an explanation of success will remain unwarranted.

Whilst this criticism of SS must be accepted, we should also note, however, that it does not constitute a *refutation* of SS. It does not follow from the vagueness of "approximate truth" that approximate truth does not *explain* success; what follows is that it is *hard to tell* whether approximate truth explains success. The moral that ought to be drawn is not that SS is unsound but, rather, that this particular issue – whether approximate truth explains success – is not the proper arena for a decisive confrontation over the status of SS.

To recap: the Laudanian counter-arguments are (1) that success does not warrant an inference to truth or truthlikeness because there are successful theories which are neither true nor approximately true, and (2) there is no reason to believe that truthlikeness is even a formally adequate explanation of success. Neither counter-argument is decisive – the first can be defused by insisting on a standard for success so high that all the theories whose truthlikeness is in doubt fail the test. Indeed, if necessary, the standard for success can be raised so high that all extant theories fail the test, rendering Laudan's conjunctive existential claim (that there are theories that are both successful and far from the truth) false. The second counter-argument, whilst sound, leaves the status of SS unsettled. So the fate of SS thus hinges on the results of the other pair of argumentative strategies deployed by anti-realists – namely, the attempt to establish the existence of *anti*-realist explanations of success and the attempt to show that realism doesn't follow from the premise that truth is the best explanation of success. We have established the former in section 6.2. We conclude by turning to the latter.

6.5 The Circularity of the Argument from the Success of Science

Let us recap on the current status of SS. In section 6.2, we noted that Leplin's contention that there is no anti-realist explanation for scientific success (at best) fails to decide the issue. In section 6.4, we argued that Laudan's argument that neither truth nor truthlikeness can explain the success of science was inconclusive. So the anti-realist side in the debate cannot yet be declared the victor. Yet the most powerful anti-realist argument is still to come. The charge, made independently by both Laudan (1981) and Fine (1984), is that SS is circular.

Even if we grant that science is successful and that the truth of our scientific theories is the *only* viable explanation for that success, the conclusion of SS – that we have grounds for believing our theories – does not follow unless we assume additionally that the explanatory virtues of hypotheses are reasons for believing them. This, however, is an assumption that the anti-realist need not accept – indeed, anti-realists generally do not. Van Fraassen (1980), for example, distinguishes between the *epistemic* and *pragmatic* virtues of theories. There are many pragmatic virtues, an example being the property of allowing us to make quick and easy calculations. Suppose that T_1 and T_2 are empirically equivalent, but with T_1 the process of calculating empirical predictions is easier. This is an excellent reason for *using* T_1 rather than T_2 in deriving predictions. But it is not necessarily a reason for thinking that T_1 is closer to the truth. We would surely want to say that ease of calculation is (merely) a pragmatic virtue. Both types of virtue are desirable features of theories, but only the former actually bear on a theory's beliefworthiness.

According to van Fraassen, the only epistemic virtues of theories are the *empirical* virtues of getting more observable consequences right, or fewer of them wrong.

A corollary of this view is that the *explanatory* virtues of theories, since they are other than empirical, can only count as pragmatic virtues. The fact that a theory, T, provides the best (or only) explanation of a set of phenomena has no bearing on its beliefworthiness. So if anti-realists are correct on this score, then SS fails even if both of its premises are true. And whether they are right or wrong, it is question-begging to wield an argument against anti-realists that merely *presumes* that explanatoriness is a reason for belief. This is exactly what is presumed in SS, so the argument from the success of science accomplishes nothing in the realism/anti-realism debate.

Boyd (1984) responds to Laudan and Fine's circularity counter-arguments. He claims that scientists routinely decide what to believe about the observable world on the basis of which hypothesis best explains the data by a process of inference to the best explanation. This is to say that they use "abductive inference" in choosing between empirical hypotheses. But then, Boyd argues, it must be permissible for *philosophers* to use abductive inference to defend a philosophical thesis about science. Sober (1990) counters this move by claiming that the problem with SS is not that its use of abduction is question-begging but, rather, that it is a *very weak* abductive argument. We propose that, in fact, the original charge of circularity can be sustained despite Boyd's objection. His defense belongs to a recurring pattern of realist arguments designed to show internal inconsistency in the anti-realist position. Such arguments note that anti-realists refuse to give epistemic weight to some nonempirical theoretical virtue (such as simplicity or explanatoriness) when these apply to theoretical statements, but that they are willing to use the very same principle when dealing with observational claims. But the evidence does not warrant the charge of inconsistency. It is false that the appeal to nonempirical virtues in assessing the status of observational hypotheses logically commits the anti-realist to applying the same principles to theoretical hypotheses.

Suppose we use some rule, R, for giving epistemic weight to, say, the explanatory virtues of hypotheses: R tells us to give greater credence to hypotheses on the basis of how well and how much they explain. R may or may not specify circumstances under which we should elevate the epistemic status of theoretical hypotheses. In case it does, let R* be the same rule with the specification that it only applies to observational hypotheses. Anti-realists commit no logical fault in subscribing to R*. Yet R* allows for abductive inferences to observational hypotheses whilst blocking abductions to theoretical hypotheses.

It must be conceded that this account of the matter leaves the anti-realist open to the (lesser) charge of arbitrariness. But arbitrariness is not, by itself, a decisive point against a philosophical position. (Indeed, it can be a legitimate part of a philosophical position, as in the arbitrariness of prior probability assignments in personalism.) Thus, a small change in Laudan and Fine's argument insulates it against Boyd's objection. Laudan and Fine claim that the use of abduction is question-begging, since anti-realists deny the validity of abduction. Boyd counters that everyone, including the anti-realist, uses abduction. This may be so for observational hypotheses, but anti-realists can consistently disallow the use of abduction in the service of theoretical hypotheses. Realism itself is a theoretical hypothesis, for it entails that some theoretical entities exist. Thus, those who endorse SS are guilty of begging the question by engaging in abduction to *theoretical* hypotheses, when such abductions are *precisely* what anti-realists regard as illegitimate.

Our answer to the question that was formulated at the beginning of section 6.1, then, is a resolute "No": the predictive success of a theory does not give us reason to believe in the unobservable entities postulated by that theory. The argument from the success of science fails to do the job for which it was designed.

Bibliography

Boyd, R. 1984: The current status of scientific realism. In J. Leplin (1984), pp. 41–83.

Cushing, J. T. 1991: Quantum theory and explanatory discourse: Endgame for understanding? *Philosophy of Science*, 58, 337–58.

Fine, A. 1984: The natural ontological attitude. In J. Leplin (1984), pp. 83–107.

—— 1986: Unnatural attitudes: realist and instrumentalist attachments to science. *Mind*, 95, 149–79.

Hardin, C. L. and Rosenberg, A. 1982: In defence of convergent realism. *Philosophy of Science*, 49, 604–15.

Kukla, A. 1998: *Studies in Scientific Realism*. New York: Oxford University Press.

Latour, B. and Woolgar, S. 1979: *Laboratory Life: The Social Construction of Scientific Facts*. London: Sage.

Laudan, L. 1981: A confutation of convergent realism. *Philosophy of Science*, 48, 19–49.

—— 1984: *Science and Values*. Berkeley, CA: University of California Press.

Leplin, J. (ed.) 1984: *Scientific Realism*. Berkeley, CA: University of California Press.

—— 1987: Surrealism. *Mind*, 96, 519–24.

—— 1997: *A Novel Defense of Scientific Realism*. New York: Oxford University Press.

McAllister, J. W. 1993: Scientific realism and criteria for theory-choice. *Erkenntnis*, 38, 203–22.

Putnam, H. 1979: *Mathematics, Matter and Method: Philosophical Papers*, vol. 1, 2nd edn. Cambridge: Cambridge University Press.

Sober, E. 1990: Contrastive empiricism. In C. W. Savage (ed.), *Minnesota Studies in the Philosophy of Science*, vol. 14. Minneapolis: University of Minnesota Press, 392–410.

van Fraassen, B. 1980: *The Scientific Image*. Oxford: The Clarendon Press.

Further reading

Fine, A. 1996: *The Shaky Game*, 2nd edn. Chicago: The University of Chicago Press.

Kukla, A. 1998: *Studies in Scientific Realism*. New York: Oxford University Press.

Lipton, P. 1991: *Inference to the Best Explanation*. London: Routledge (see, especially, ch. 9: "Truth and explanation").

Newton-Smith, W. 1981: *The Rationality of Science*. London: Routledge (see, especially, ch. II: "Observation, theory and truth").

Suppe, F. 1989: *The Semantic Conception of Theories and Scientific Realism*. Urbana: University of Illinois Press (see, especially, ch. 11: "Scientific realism").

ARE THERE LAWS IN THE SOCIAL SCIENCES?

The field of physics includes many principles referred to as "laws": Newton's laws of motion, Snell's law, the laws of thermodynamics, and so on. It also includes many "equations" – the most important being named after Maxwell, Einstein, and Schrödinger – that function in exactly the same way that the so-called laws do. While none of these laws are universally true – they all fail within one domain or other – physics is clearly in the business of looking for universal laws, and most physicists believe that there are laws "out there" to be discovered. Taking physics as their model, some philosophers have taken the existence of laws to be the *sine qua non* of a genuinely scientific enterprise. (This view was particularly prominent in the middle of the twentieth century.) Since nothing like the laws of physics is to be found within the social sciences, it was concluded that these were not genuinely scientific fields. John Roberts defends the traditional view according to which laws are universal regularities. He goes on to argue that there can be no laws of social science, particularly criticizing attempts to classify social scientific principles as *ceteris paribus* laws. Although Roberts believes that it would be myopic to conclude that the social sciences are sciences in name only, the absence of laws does mark an important difference between physics and the social sciences. Kincaid argues that some physical laws, such as Newton's law of gravitation, do not state universal generalizations, but rather identify causal mechanisms. These laws can describe general causal tendencies, even though the tendencies may often go unobserved because of the disrupting effect of further causes. Kincaid argues that many laws of social science, such as the economic law of supply and demand, work in exactly the same way. In the process, Kincaid takes up a number of other challenges to the status of the social sciences.

CHAPTER SEVEN

There are no Laws of the Social Sciences

John T. Roberts

7.1 The Significance of the Question

What is at stake in the question of whether there are laws of the social sciences? It is tempting to suppose that what is at stake is something about the nature of human beings and human history; for example, whether our history proceeds according to an inevitable, predetermined course. But this would be a mistake. The existence of laws in a given domain is consistent with indeterminism in that domain (Earman, 1986). Moreover, the nonexistence of social-scientific laws would not guarantee indeterminism, for it could be that all human events are predetermined by laws of nature that can be articulated only in microphysical terms, and not in terms of concepts of the social sciences. Whether there are or are not laws of the social sciences, we cannot infer much about the nature of human history from this fact alone.

A different answer is that what is at stake is whether the social sciences really deserve the name "science." If the social sciences do not, or cannot, discover laws, then perhaps they should not be considered truly scientific. Why should we believe this? A tempting line of thought is the following: Science is about prediction and explanation. Phenomena are explained by subsuming them under laws of nature, and they are predicted by showing that, given the current conditions and the laws of nature, such-and-such is bound to happen (Hempel and Oppenheim, 1948). Hence, a field of study can only perform the tasks of science if it is capable of discovering laws.

The view of science presupposed by this answer has been influential in the history of the philosophy of science. But it is overly simplistic, and ought to be rejected. For one thing, it is far from clear that any good scientific explanation must cite laws. Over the past few decades, philosophers of science have proposed a variety of models of explanation according to which this is not so. Salmon (1994) proposes a model according to which explanation involves locating the event to be explained in a network of causal processes and interactions, and need not involve citing any laws; Kitcher (1981) argues that explanation involves showing how to unify a large body of phenomena by bringing a unified set of argument-patterns to bear on it, where these patterns need not appeal to laws; van Fraassen (1980) defends a pragmatic theory of explanation in which laws play no special role. Moreover, it is not at all clear that the other characteristic activities of science, such as prediction, can only be performed by a field of study that recognizes laws of nature; for arguments that this is not so, see van Fraassen (1989) and Giere (1999). It seems hard to deny that in order to make predictions, one needs reliable information about regularities, or causal mechanisms, or statistical distributions, with relevance for future occurrences. But even if this is so, it needn't follow that *laws* are required, for the concept of a law of nature involves more than this (for more on what is distinctive about laws, see section 7.2).

Why, then, does it seem so tempting (and why, historically, have so many been tempted) to suppose that a field of study cannot count as science unless it discovers laws? I suspect the answer is that there has traditionally been a strong tendency to think of physics, which appears to be strongly focused on discovering laws of nature, as the very paradigm of a science. Other fields of study get to count as sciences only to the extent that they emulate the methods, theory forms, and successes of physics. To give in to this temptation is to put oneself in the difficult position of having to say, for every putative branch of science, either that its theories, hypotheses, and results can be forced into the pigeonhole provided by the model of physics, or else that it does not really qualify as scientific. And a careful examination of contemporary biology and psychology, and much of the social sciences as well, shows how difficult it is to do the first, and how appalling it would be to do the second. It is of course very difficult to say with any precision what makes a field of study scientific, but a reasonable starting point would be the idea that a discipline can qualify as a science if it: has certain general aims, including understanding and prediction of phenomena; uses certain methodologies, including careful observation, statistical analysis, the requirement of repeatability of results, and perhaps controlled experimentation; and has certain forms of institutional organization, including distribution of intellectual authority across a large body of scientists who are (at least nominally!) not beholden to political, religious, or other extra-scientific authorities with respect to the content of their scientific opinions. By this kind of standard, much work in the social sciences deserves to be called scientific (see Kincaid, 1996). It does not immediately follow, and should not be decided in advance, that any field of study that measures up according to this kind of standard must produce hypotheses and theories that mimic the typical form of fundamental theories in physics – in other words, hypotheses and theories that posit laws of nature (among other things). Hence, it should remain an open question whether the social sciences must discover laws (or at least putative laws) in order to qualify as scientific.

One can assume a demarcation criterion according to which discovering laws is a necessary condition for being scientific, and then ask whether there are social-scientific laws in order to assess the so-called social "sciences." Or, one can assume that at least some work in the social sciences is genuinely scientific, and then ask whether there are social-scientific laws in order to assess the aforementioned demarcation criterion. The view to be defended here is that we should go the second route. There are no social-scientific laws, I shall argue, and what follows from this is not that the social sciences aren't really sciences, but that any view of science that entails that science is essentially about discovering laws must be rejected.

7.2 What is a Law?

Before we can come to grips with the question of whether there are laws of the social sciences, we need to have some idea of what a law is. Unfortunately, no answer is possible that is simultaneously informative and uncontroversial. The concept of a law of nature has proven terribly difficult to explicate, and the literature on laws is rife with disagreement (for some of the controversy, see Swoyer, 1982; Armstrong, 1983; Earman, 1986; Carroll, 1994; Lewis, 1994; Cartwright, 1999). In this section, I will present a working characterization of laws that is more or less uncontroversial, though not terribly deep. Fortunately, it will be sufficient to inform the argument below.

To begin with, I shall assume that there are such things as laws of nature, and that physics, at least, is engaged in the project of trying to discover laws. This view can be supported by noting that the succession of fundamental physical theories since Newton is a succession of theories positing laws of nature, and each of these theories enjoyed considerable empirical success for a while. Classical mechanics posited Newton's three laws of motion and various special-force laws such as the law of universal gravitation and Hooke's law. Classical electromagnetism posited Coulomb's law, the Biot–Savart law, the Lorentz force law, and Maxwell's equations, the latter being standardly interpreted as laws even though the word "law" does not occur in their names. More recently, it has been less common to use the word "law" in names of fundamental physical principles, but Einstein's field equations are commonly described as the laws of general relativity, and the Schrödinger equation as the basic law of (nonrelativistic) quantum mechanics.

To be sure, there are philosophers of science who have argued that this is a misinterpretation, that neither physics nor any other science should be understood as a search for the laws of nature, and that there are probably not any such things as laws of nature anyway (van Fraassen, 1989; Giere, 1999). I think these arguments fail, but for present purposes, it isn't necessary to refute them. For if there are no laws, then it follows trivially that there are no laws of the social sciences. There is an interesting question here that is specifically about the social sciences only if the physical sciences, at least, *can* discover laws.

So, what is a law of nature? To begin with, laws of nature are closely related to regularities. Some philosophers maintain that laws are regularities of a certain sort (Lewis, 1994; Earman, 1986, ch. 6), while others deny that they *are* regularities but insist that they *entail* regularities (Dretske, 1977; Tooley, 1977; Swoyer, 1982; Armstrong, 1983). We can distinguish three kinds of regularities. First, there are strict regularities: universally quantified conditionals, holding throughout the universe – an example is "Anything made of copper conducts electricity." Secondly, there are statistical or probabilistic regularities, characterizing an unrestricted domain; for example, "Any atom of uranium-238 has a probability of 0.5 of decaying within any time-interval of 4.5 billion years."[1] Thirdly, there are what might be called "hedged regularities," which are sometimes called "*ceteris paribus* regularities."[2] These are (statistical or nonstatistical) regularities that are qualified by admitting that they have exceptions in various circumstances, and these circumstances are not made explicit but are given the label "disturbances" or "interferences."[3] Most philosophers who have written on the topic of laws of nature allow that laws may be (or may entail) either strict or statistical regularities. Some allow that laws may also be (or entail) hedged regularities (Armstrong, 1983; Fodor, 1991; Pietroski and Rey, 1995; Lipton, 1999; Morreau, 1999). Others have argued that *all* laws of nature are (or entail), at best, hedged regularities (Cartwright, 1983; Lange, 1993).

Almost everyone agrees, however, that just being a regularity is not sufficient for being a law of nature (but for the contrary view, see Swartz 1985). It seems that there are more regularities than there are laws. For example, consider the proposition that every sphere of solid gold has a diameter of less than one kilometer. This has the form of a strict regularity, and it is quite plausible that it is true. (If it turns out not to be, then presumably, we can just substitute some larger diameter and make it true.) But it is implausible to suppose that this proposition states (or is entailed by) a law of nature. No law that we have any inkling of rules out gargantuan spheres of gold; it just happens to be the case that there are no such spheres. More generally, a regularity (strict, statistical, or (perhaps) even hedged) can be true "just by accident," because of the way the world happens to be as a matter of brute fact. So there is more to being a law than just being a regularity.

Exactly what it takes to be a law, beyond being (or entailing) a regularity, is a matter about which there is wide disagreement. But a couple of things are uncontroversial. First, laws are logically and mathematically contingent, and can only be known *a posteriori*.[4] Secondly, laws have what John Carroll (1994) calls a "modal character." They "govern" the course of events, in that they constrain the scope of what is physically or naturally possible. The cash-value of this idea includes at least the fact that laws have a certain explanatory power and a certain counterfactual robustness. The former means that a law is a suitable general principle to appeal to in order to explain particular phenomena.[5] The latter means that in reasoning about

1 Examples of statistical regularities with restricted domains include "Three-quarters of the socks I now own are black" and "30 percent of American adults are cigarette smokers." (I do not know whether either of these is true.) I am excluding such regularities from consideration here, because they do not seem to have the kind of generality required by a law of nature; rather, they are descriptions of local conditions that are effects of a number of contingent causes.

2 As many writers have noted, the literal meaning of "*ceteris paribus*" – other things being equal – is not exactly appropriate here; "there being no interfering or disturbing conditions" would be better. For this reason, I will stick to the term "hedged." The topic of hedged regularities will be explored further below.

3 If these conditions were made explicit, then we would really have a strict or statistical law. For example, if we are told that all F's are G's, unless there is an interference, and every interference is of type H, then we have the strict regularity: all F's not in the presence of H are G's.

4 This is distinct from the somewhat more controversial claim that laws are *metaphysically* contingent. See Swoyer (1982) for an argument that laws are metaphysically necessary though knowable only *a posteriori*.

5 Of course, it is important to distinguish this claim from the very strong claim that scientific explanation always must appeal to a law.

hypothetical, counterfactual situations, we tend to hold constant the laws of the actual world, and it is reasonable for us to do so.

How to account for this "modal character" is a matter of dispute. In the early modern period, natural philosophers such as Boyle and Newton maintained that the laws represented free legislative decrees of God, which could not be violated because they were backed by the divine will. Much more recently, many philosophers have argued that the modal character of laws can only be accounted for by supposing laws to be facts of a fundamentally different kind than facts about the actual antics of particular natural objects, and regularities and patterns in these antics. For the latter concern only what happens to be the case in the actual world, and as such have no implications for what would have happened had circumstances been different, or for the ultimate reasons why things are as they are. Since laws are expected to have such implications, they must be a fundamentally different kind of feature of the world – such as relations among universals (Dretske, 1977; Tooley, 1977; Armstrong, 1983), the essences of natural properties (Swoyer, 1982), the natures and capacities of kinds of systems (Cartwright, 1983, 1999), or irreducible, *sui generis* modal principles (Carroll, 1994). Philosophers of a more Humean bent, on the other hand, have argued that the "modal character" of laws can be adequately accounted for by supposing them to be regularities or patterns in the actual course of events that are among the world's deepest or most pervasive structural features. (There is more than one way to work out the details here; Skyrms (1980) describes one, and Lewis (1994) another.) Such regularities have explanatory and counterfactual import, according to this view, not because they have metaphysical natures that confer such import on them but, rather, because our practices of explanation and counterfactual reasoning place great importance on such pervasive structural aspects of the world. All parties to the dispute, however, agree that any successful account of the nature of laws must provide some explanation of what it is that gives laws their peculiar significance for explanation and counterfactual reasoning.

It might be thought that the question of whether there are social-scientific laws should be settled by consulting the most successful contemporary theories from the social sciences, and seeing whether or not they say that anything is a law. It is important to see why this is not a promising way to settle the issue. Being *called* a law by scientists, or being given a name with the word "law" in it (e.g., "the law of universal gravitation"), is neither necessary nor sufficient for being a law. Schrödinger's equation is not called "Schrödinger's law," and textbooks on quantum mechanics do not always explicitly call it a law. However, it plays a role within quantum mechanics that is analogous to the role played in classical mechanics by Newton's laws of motion, and for this reason it seems to deserve to be thought of as a law. On the other hand, many things commonly called "laws" by scientists are actually mathematical truths and so, unlike laws of nature, they are necessary and *a priori*. (One example is the Hardy–Weinberg law in population genetics.) Thus scientists do not use the word "law" uniformly to mark the distinction that we are interested in here. This is no cause for blame; the purposes of scientists typically do not require the use of a standard term for marking this distinction. The upshot is that deciding what, according to a given scientific theory, is a law and what is not is a task that requires some philosophical interpretation.

The interpretation required can be subtle. It is tempting to suppose that any general principle or regularity that is true, logically contingent, and plays a role in explanation and counterfactual reasoning counts as a law of nature. Surely (one might think) the social sciences discover such general principles; for social scientists engage in explanation and their explanations often make appeal to general regularities. So how could anyone deny that there are laws of the social sciences? The problem with this quick argument is that while it is plausible that every law is a logically contingent general principle that plays an important role in explanation and counterfactual reasoning, it does not follow that every such principle is a law, and there are reasons to doubt that this is so. Consider the following regularity: "All seawater is salty." It supports counterfactuals: If the glass from which I am now drinking had been filled with water taken straight out of the Atlantic, I would be making an awful face right now. It can play a role in explanation; for example, in explaining why the oceans are populated with organisms with certain physiological characteristics. But it would be very odd to call this regularity a law of nature (or even a law of geology); surely it is just a contingent fact resulting from what the initial conditions of the earth just happened to be, and these conditions could have been otherwise without there being any violation of the laws of nature.

This general principle about seawater is not a law because it is too fragile. It would be upset by differences in the contingent facts about our world that need not imply any violation of the laws of nature. The question of whether there are social-scientific laws depends on whether there exist, within the realm of the social sciences, principles that are robust enough not to count as nonlaws for the same reason that our regularity concerning seawater does, and at the same time logically contingent. Before attacking this question, though, it still needs further clarification.[6]

7.3 Distinguishing some Questions

The question of whether there are any laws of the social sciences is ambiguous, and needs to be sharpened up. To see this, consider the analogous question: "Are there

---

6 The preceding discussion might suggest that what it takes for a proposition to be a law, over and above it's being a logically contingent generalization and playing an important role in explanation and counterfactual reasoning, is *nonfragility* – that is, the property of being such that it would still be true even if circumstances differed in some nomologically possible way. This is not a view that I would endorse, however. For one thing, it would be viciously circular if proposed as a definition of "law of nature," since it uses the phrase "nomologically contingent," which just means "consistent with, but not entailed by, the laws of nature." For another, this kind of fragility is no doubt something that comes in degrees, so that if we adopted this view, we would be committed to the view that lawhood comes in degrees. Many philosophers would be happy with this result (see, e.g., Lange, 2000). But it seems to me that the laws that have lawhood "to the highest degree" – the fundamental laws of physics – play roles within the practice of physics that are so special and so important (I have in mind the roles played by laws in characterizing symmetry principles and in constraining the probability measures used in statistical mechanics) that it makes sense to reserve the term "law" for them. This is admittedly a terminological preference, and the arguments to follow do not depend on it: Even if we agree to use the term "law" in a broad sense, in which it is correct to speak of degrees of lawhood, the arguments of sections 4–6 show that there cannot be laws (of any degree) in the social sciences.

laws of physics?" A quick and obvious answer is: "Yes, of course: There are Newton's second law of motion, Coulomb's law, Boyle's law, and so on." But now consider a follow-up question: "Were there any laws of physics one million years ago?" None of the aforementioned laws had been thought up one million years ago, which tempts one to answer negatively. But a negative answer implies that laws of physics have come into being because of the evolution and activities of human beings – which seems outrageous! But there is an easy resolution: The question "Are there laws of physics?" is ambiguous between:

(a) "Have physicists discovered any laws?"

and

(b) "Are there, really, any laws that are within the subject matter of physics?"

These questions are in the present tense, but they can be put into the past tense. Our temptation to say that there were no laws of physics one million years ago is explained by our taking the question to be the past-tense version of (a), and the intuition that this could not be right comes from thinking of the question as the past-tense version of (b). Once the question is disambiguated, there is no problem.

That easy resolution is not quite good enough, though, as we can see by noting that all of the "laws" of physics given as examples above are no longer thought to be (universally) true. So citing them does not suffice for an affirmative answer to (a). It would be nice if we could justify an affirmative answer to (a) by citing more up-to-date laws, but alas, things are not so nice. Which up-to-date laws would we cite? Those of quantum field theory? Those of general relativity? Each has a good claim, but they could not both be right, since quantum field theory and general relativity are incompatible with each other. Physicists and physics fans hope that one day, we will be able to answer (a) affirmatively by citing "real" laws of physics. But we are not there yet, and it is difficult to give any reason – other than pious optimism – for believing that we ever will be. This makes it look as if we are not really justified in giving affirmative answers to either (a) or (b). Yet, a glance at the historical development of modern physics makes it hard to deny that physics has something important to do with laws. To do justice to this consideration, we can distinguish a third sense of the question "Are there laws of physics?":

(c)   Do the successful theories of physics posit laws?

Here, "successful" means counting as successful by the evidential standards of science; a theory need not be true to be successful, and a theory that is successful can nonetheless be overturned by a later theory if the latter is even more successful. The answer to (c) is clearly affirmative. A sensible conclusion, then, would be that if we want to know whether there are laws of physics, we need to get more precise about what we mean. If we mean (a), then the answer is "Probably not yet, but maybe some day." If we mean (b), then the answer is "We do not really know yet." If we mean (c), then the answer is "Yes, certainly." Which answer we get thus depends crucially on the way in which we disambiguate the question.

From these considerations, we should learn that what answer we give to the question "Are there laws of the social sciences?" depends on how we disambiguate that question. Again, we have three options:

(A) Have the social sciences discovered any laws?
(B) Are there really any laws within the subject matter of the social sciences?
(C) Do the successful theories of the social sciences posit laws?

To anticipate, I will argue for the following: We have pretty good reasons for thinking that the answer to (B) is negative. The empirically successful theories of the social sciences can, generally, be plausibly interpreted as not positing any laws, and since we have good reason to believe that the answer to (B) is "No," it is more charitable to interpret social-scientific theories as not positing laws. Hence, we should so interpret them, and so we should answer (C) in the negative. (The reason why interpretation is needed to answer (C) was explained above: to find out whether a scientist has posited a law, it is not sufficient to look and see whether and how she uses the word "law.") It follows straightforwardly that the answer to (A) is negative as well.

Question (B) concerns laws that are "within the subject matter of the social sciences" – What does that mean, exactly? It is hard to give a satisfying positive answer, but it is not hard to give examples of things that would clearly not count. For example, suppose that it is a law of nature that every social class is such that the bodies of all of its members move in such a way that their collective center of mass has an acceleration proportional to the sum of all forces acting on the bodies of its members. This law concerns social classes, which fall within the subject matter of sociology, but clearly the law itself is not a law of sociology: It is a special case of a law of physics, restricted to physical objects that are picked out using a social-scientific concept. Suppose, for the sake of argument, that it is a law of biology that the process of respiration always includes the Krebs cycle. Then it is arguably a derivative law that all heads of state undergo respiration that includes the Krebs cycle. But this is just a rather arbitrarily restricted special case of a biological law, rather than a law of political science, even though it is formulated using the term "head of state." In each case, the putative "law" is about some kind of thing that falls within the subject mater of a social science, but the law itself holds in virtue of some fact in the subject matter of the natural sciences. Speaking roughly and intuitively, it is not because something is a social class that it obeys the first law mentioned above; rather, it is because it is a set of objects with physical properties. It is not because someone is a head of state that he or she obeys the second law just mentioned; it is because he or she is a biological organism. For a law to be a law within the subject matter of the social sciences, it must be a law that applies to the kinds of things studied by the social sciences *because of* their properties or natures *qua* social entities. This is a crudely stated principle, and it could use a more careful and detailed explication. But it will serve well enough for the purposes of the discussion to follow.

7.4 The Case against Social Laws

Let us introduce the term "social laws," to refer to laws wholly within the subject matter of the social sciences. Then the negative answer to question (B) can be expressed by denying that there are any social laws. This denial, I claim, is supported by the following argument:

(i) If there are any social laws, then they are hedged laws.
(ii) There are no hedged laws.

Therefore, there are no social laws.

By "hedged law," I mean a law that is, or entails, only a hedged regularity (in the sense described in section 7.2), rather than a strict or statistical regularity. The argument is sound if both of its premises are true. In section 7.5, I will argue that premise (i) is true, and in section 7.6, I will argue that premise (ii) is true.

7.5 Why Social Laws must be Hedged

It will help to start by considering an example. One familiar putative example of a social-scientific law comes from economics – the law of supply and demand:

If the supply of a commodity increases (decreases) while the demand for it stays the same, the price decreases (increases); if the demand for a commodity increases (decreases) while the supply remains the same, the price increases (decreases).

(There are more careful ways of formulating this putative law – see Harold Kincaid's companion chapter (chapter 8, this volume) for more details – but the remarks to follow would apply equally well to a more careful formulation.) This claim is not a bad rule of thumb for predicting and explaining various economic phenomena. But there are numerous kinds of cases in which it will be false. These include cases in which the government imposes price controls, in which either vendors or consumers are ignorant about changes in supply or demand, in which there is widespread irrationality on the part of either vendors or consumers, in which humanitarian feelings on the part of many vendors motivate them not to raise prices on goods, such as medical supplies, for which there is a pressing need, and so on. So the law cannot be a strict law. One might hold out hope of reformulating it as a statistical law, but in order to do this, one would need to find probabilities of all the kinds of phenomena that can lead to violations. Since these phenomena are so diverse and can have such diverse causes, this seems hopeless. One would need to find the probability of, for example, an outbreak of a terrible disease; one would then need to find the conditional probability that humanitarian concerns would motivate vendors of medical supplies not to increase prices, were such an outbreak to occur; and so on. It seems that, at best, the law of supply and demand is a hedged law. The regularity it describes holds, unless for some reason it does not, and we cannot specify all of the possible conditions under which it will not.

It might be objected that the last claim is false: We *can* specify the class of conditions under which the law of supply and demand will hold, and the class of conditions under which it will not. Since we can do this, we can reformulate the law as a strict law, simply by appending "As long as conditions C obtain . . ." to the front of the statement of the law, where C stands for the set of conditions under which it will not be violated. What would conditions C be? They would have to be sufficient to rule out *all* factors that can interfere with the working of the market in such a way as to disturb the regularities posited by the law of supply and demand. The list of disturbing factors offered in the preceding paragraph was only a start. Who knows how many others there may be? Consider that the regularities described by the putative law are regularities in the mass behavior of groups of human beings; psychological quirks, external pressures of a noneconomic variety, failures of communication, natural disasters, cultural norms, and cultural conflicts can all potentially have a disturbing influence on the regularities in this behavior. But in many cases, such factors will not disturb these regularities. The problem is to characterize the class of possible conditions that will disturb the regularities, so that we can state the law in a form that applies only to the others. We could do this by defining conditions C as those conditions in which there are no factors of any kind that result in violations of the regularities described in the law of supply and demand. But if we do this, then we render this putative law a tautology: "These regularities will obtain unless for some reason they do not."[7] But short of doing this, there seems to be no way of adequately characterizing the conditions C; the range of possible interfering factors is indefinitely large and indefinitely varied. It seems that the law of supply and demand can only be a hedged law.

The issues that we have encountered in this examination of a single example illustrate a more general characteristic of social phenomena, which strongly suggests that there are no nonhedged laws to be found in the social realm. The kinds of system that are studied by the social sciences – markets, states, social classes, political movements, and so on – are *multiply realizable*. Every social system is constituted by a large and complex physical system. This is because a social system is made of a group of humans and their environment, which is in turn constituted ultimately out of physical parts, perhaps elementary particles. (There is an interesting question about what exactly the relation of constitution is here. Is a social system *identical* with the system of physical particles out of which it is constituted? Or is it a distinct thing that supervenes on the latter? Or what? Fortunately, we need not settle this issue here. I will assume, however, that a social system *supervenes* on the large and complex physical system out of which it is composed, in the sense that there could not be a difference in the social system without a physical difference somewhere in the physical system that constitutes it.) The multiple realizability of a kind of system studied by the social sciences is the fact that there is a large and heterogeneous class of kinds of physical systems that could constitute a social system of a given kind. An example due to John Searle (1984) is *money*; there is a huge class of kinds of physical objects that

---

7 You might think this is unfair: "Unless for some reason they do not" is not obviously equivalent to "unless they do not." But this point doesn't help; see the discussion of the proposal of Pietroski and Rey in the following section.

could count as money. *A fortiori*, there is a tremendous variety of kinds of physical system that could constitute a *market*. Examples are easily multiplied.

Suppose that we have a kind of social system studied by the social sciences; call it kind F. Further suppose that we are interested in the behavior of F-systems under circumstances C. Consider the hypothesis that it is a law that in circumstances C, F-systems will exhibit behavior G. Now, suppose that C and G, like F, are kinds (of circumstances, of behavior) that belong to the classificatory schemes of the social sciences. So, C might be the circumstance of currency inflation, but it will not be the circumstance of having a collective center of mass that moves uniformly. This is necessary to guarantee that the regularity proposed by our hypothesis is "within the subject matter of the social sciences" in the sense described above. Since all these kinds are multiply realizable, there is a great plurality of kinds of physical system that could be a system of kind F in circumstances C. How such a system will evolve is sensitive to the details of how the underlying physical system will evolve. Since all we know about this physical system is that it belongs to one of a certain enormous and heterogeneous class of kinds of physical system, there is a great variety of different ways in which it might evolve. It is very implausible that there is any single answer to the question of how an F-system will behave in circumstances C, beyond that it will continue to conform to the truths of logic and the laws of physics. But if G specifies only that the system continues to conform to the truths of logic, then our hypothesis is not contingent, so it cannot be a law. And if it specifies only that the system will continue to conform to the laws of physics, then our regularity is not within the subject matter of the social sciences. So our hypothesis will admit of exceptions; it will not state a strict law.

This argument is extremely abstract and it is very quick. But consideration of a concrete example helps to drive the point home. Suppose that F stands for currency markets, and C stands for the condition of inflation. One kind of physical system that can constitute an F-system in circumstances C is a population on a planet in the direct path of an enormous comet that will arrive within a few hours. (Other examples include populations about to be stricken by a new and terrible virus, populations in the paths of hurricanes, and so on.) Such a physical system is bound to evolve in such a way that, in a short amount of time, it will belong to no recognizably economic kind at all.

This is a rather brutal example (both in terms of its content and the rhetorical use to which I am putting it), but it is easy to imagine more subtle ones. For example, imagine a population in which the molecules in the brains of all individuals are coincidentally arranged in such a way that very soon, they will undergo psychological changes that completely revise their behavior patterns, so that they all eschew their former acquisitive ways and become ascetic humanitarians. This scenario is a far-fetched one, but improbable as it is, there is no obvious reason why it should be nomologically impossible, and it does supply an example of a kind of physical system that would constitute a system of kind F in circumstances C, whose future evolution is likely to be quite different, economically speaking, from what we would expect. Any kind of behavior G, belonging to a social-scientific classificatory scheme (rather than, say, a physical one), which would apply correctly to the future state of such a system as well as to the future state of a more "normal" currency market in conditions of inflation, would have to be extremely descriptively weak. It would have to cover cases in which everybody becomes extremely altruistic, as well as cases in which everybody gets obliterated, and cases in which things proceed as we normally expect them to do. If G is so broad, then our hypothesis – that all F-systems in circumstances C exhibit behavior G – would have to be completely uninformative. Again, the case described is an extreme one. But if it is so easy to imagine circumstances in which the antecedent of our hypothesized social-scientific regularity is fulfilled but bizarre phenomena (the obliteration of all life; the sudden widespread imitation of Mahatma Gandhi) result, how much more plausible it is that there exist extremely complex physical circumstances in which this antecedent is fulfilled and what occurs is something more ordinary, yet still not what would be predicted by any given, reasonably informative, social-scientific hypothesis?

The basic point here is that no matter which social-scientific kinds F, C, and G are, there are likely to be kinds of physical system that constitute social systems that instantiate F and C but, under normal physical evolution, lead to bizarre outcomes, which will not be covered by G (unless G is so weak that our hypothesis is a truth of logic or a law of physics). One natural response is to point out that just because there are such kinds of physical system, it does not follow that any real social system will ever actually be constituted by one. So, there might not actually be any exceptions to our hypothesis. The problem is that even if this is so, still our hypothesis will state a regularity that does not seem to count as a law: Even though it is true, it would have been violated if the actual circumstances had been different in a way that there is no reason to regard as impossible. The regularity will be fragile, in a sense that seems incompatible with being a law. Another natural response is to revise the hypothesis, by building into the circumstances C the requirement that the F-system not be constituted by any of the kinds of physical system that will lead to bizarre behavior. But since *any* social kind is multiply realizable, this is not likely to succeed unless we define the circumstances C in physical terms – any definition of C that is couched solely in terms of social-scientific kinds is multiply realizable, so the same problem will arise again. And if we define C in physical terms, then we no longer have a regularity that could be a *social* law. In order to rescue our hypothetical social regularity, then, it seems that we must hedge it: "F-systems in circumstances C will exhibit behavior G, *unless something goes wrong.*"

7.6 The Case against Hedged Laws

It has been argued that hedged regularities cannot be discovered by science, because they are not empirically testable. The hedge functions as an "escape clause" that allows any hypothesized hedged regularity to escape empirical refutation: Whenever you discover a counterexample, claim that there has been an interference of some kind, so that the case is outside the scope of the hypothesis and does not falsify it (Earman and Roberts, 1999). Although I think that there is something important to this line of thought, I will not pursue it here. I will focus not on the empirical testing of hedged laws, but on the supposed hedged laws themselves. There are no hedged laws, because any hedged law would be (or would entail) a hedged regularity, and there is no coherent concept of a hedged regularity that could be (or could be entailed by) a social law.

A hedged regularity takes the following form:

Whenever A happens, B happens, unless there is an interference.

What does the "unless" clause add to the content of the statement? This depends on how "interference" should be understood. There are a number of possibilities:

1 "Interference" means any event or circumstance that we can identify as a cause of B failing to happen, even in the face of A.
2 "Interference" refers to an event or circumstance in a certain, definite class *I* of events or circumstances:
   (a) where we can identify the class *I* independently of saying that it is the range of cases in which there is an exception to the A–B regularity;
   (b) where we cannot so identify the class *I*, but nonetheless we understand what "interference" means in this context;
   (c) where we have no idea how to identify the class *I* or any implicit understanding of the range of cases that would count as an interference.
3 "Interference" just means any case in which A happens but B fails to happen.

I can offer no proof that this list of possibilities is exhaustive, but it is hard to imagine what a further alternative would be. None of these possibilities provides us with a coherent concept of a hedged regularity that would allow us to make sense of a hedged social law.

I will consider the easy cases first. In case 3, our hedged regularity is a tautology: "Whenever A happens, B happens, unless B does not happen." So it could not be a law. In case 2(c), the regularity need not be a mere tautology, but in stating it, we have no idea what we are stating. No particular fact is stated by the hedged-regularity statement. But that just means that no fact is stated by it. In case 2(a), what we really have is a strict regularity: "Whenever A happens and no event in the class *I* occurs, B happens." We have just stated this strict regularity in an abbreviated form.

Case 1 is more complicated. It echoes a proposal of Pietroski and Rey (1995) concerning how we should understand hedged laws. The problem with it is that it trivializes the notion of a hedged regularity. Consider an example. Every material object either exhibits ferromagnetism, or fails to do so. It is plausible that for every object, there is some feature of its molecular constitution and structure that explains why it does or does not. So, for every spherical object S, it is true either that S exhibits ferromagnetism, or else that there is some property of it (some feature of its molecular constitution and structure) that explains why it does not. Hence, if we understand "interference" in the sense of case 1, then we can say that every spherical object is ferromagnetic unless something interferes. But it would be absurd to suppose that this fact is a "hedged regularity"; at any rate, if we do, then we are going to find hedged regularities everywhere. Whenever it is true that any object that fails to have property G does so because of some factor that explains why it fails to have property G, we will have a hedged regularity to the effect that every F is a G, unless something interferes, *no matter what F is*, and even if F is completely irrelevant to G (cf., Earman and Roberts, 1999, pp. 452–4).

What remains is case 2(b). It is, in essence, the possibility favored by Marc Lange (1993) in his general discussion of hedged laws. For Lange, the content of a hedged law is, for example, that every F is a G unless there is an interference, and we understand the meaning of "interference" even if we cannot give a precise, illuminating definition of it. There is nothing particularly troubling or unusual about this; we understand the meanings of plenty of words, even though we cannot give informative definitions of them all in a noncircular way. As Wittgenstein argues, language is a rule-following activity in which a person can engage even if she cannot state every rule in a complete form that leaves nothing to interpretation. Why should we demand explicit definability of "interference" in other terms, when we cannot consistently make this demand for every word that we use? Why does "interference" deserve to be picked on more than, say, "force" or "distance"?

This is an interesting and subtle proposal. I will not try to refute it here (but for an objection, see Earman and Roberts, 1999, pp. 449–51). However, even if the general point is a good one, it does not seem to be of any help with respect to the kinds of hedged regularities that would be expressed by social-scientific laws. Lange illustrates his proposal with the example of a law from phenomenological physics, the law of thermal expansion. This law says that when a metal bar is heated or cooled, it undergoes a change in length that is proportional to the change in temperature. Exceptions can occur when some kind of external stress force is being exerted on the bar; for example, by someone pounding on the ends of the bar with a sledge hammer. So the law must be hedged. (This example might be called "Lange's hedge slammer.") According to Lange, a physicist can have an implicit understanding of what sort of thing would count as an interference covered by the hedge, even if she is unable to sit down and write out a complete list of those things. That is, she can tell in advance whether any particular specified condition is covered by the hedging clause, without having to check experimentally to see whether the law is violated in that condition (Lange, 1993, p. 233).

Be that as it may, things are different with the case of hedged social laws. As we saw above, social systems are constituted by physical systems of enormous complexity, and social-scientific kinds are multiply realizable, so the kinds of situations in which any social regularity might be violated include extremely complex states of the underlying physical system which cannot be characterized in terms of social kinds. It is implausible that a social scientist proposing a hedged social law could have the kind of implicit grasp of all possible such conditions that Lange's physicist has, and still less plausible that it is possible to tell in advance whether or not a specified physical state of the underlying physical system is covered by the hedging clause. Lange's proposal thus does not help with the case of hedged social-scientific laws.

Hence, there appears to be no way of understanding what a "hedged regularity" is which will allow us to make sense of a hedged social law. If we want to find a coherent and charitable interpretation of the contemporary social sciences, then, we had better find a way of interpreting their hypotheses and theories as doing something other than positing hedged laws.

7.7 Why Social Science need not Posit Laws

Consider one example of highly successful social-science research: Jeffrey Paige's (1975) study of revolutions in agrarian societies. (Paige's study is analyzed by Kincaid (1996, pp. 70–80), and here I rely on Kincaid's analysis.) Paige identified a variety of factors that tend to influence the degree and kind of collective political activity among cultivators. For example, Paige finds that commercial hacienda systems tend to lead to agrarian revolt, and plantation systems tend to lead to labor reforms instead (Kincaid, 1996, p. 77). These tendencies are not uniform; for example, the occurrence of a given type of political action is made more likely if the same kind of event has occurred in the same society in the recent past ("contagion effects"; Kincaid, 1996, p. 78). Paige's results were arrived at by a painstaking classification and statistical analysis of a large body of carefully collected data. Kincaid argues persuasively that Paige's methodology is scientifically sound and that we are justified in believing his conclusions.

The question before us now is how we should interpret Paige's results. One interpretation is that Paige has discovered a number of social laws governing agrarian societies. The structure of his inference is this: (1) He gathered data on the statistical distributions of various kind of political events across agrarian societies, and identified certain complex statistical relations; for example, that there is a positive correlation between hacienda systems and agrarian revolt, and that this correlation is affected in certain ways when one conditionalizes on the presence or absence of various kinds of political actions having occurred in the recent past. (2) From this statistical data, Paige inferred that it is a law that, for example, hacienda systems lead to agrarian revolt, although this is a hedged law rather than a strict one. (3) From this law, in turn, we can now derive predictions and explanations of particular political events.

If the above arguments are sound, then this interpretation of Paige's results cannot be right. But what alternative is available? Well, why not take the results at face value, as very precise and informative statistical information about the characteristics of the actual agrarian societies that Paige studied? As such, they may well be projectible to other agrarian societies, including future societies. This projectibility would amount to its being rationally justified to expect similar statistical patterns to prevail elsewhere. This would be useful for predictive purposes, and on many models of explanation, it could be explanatory as well (e.g., Salmon's (1971) statistical-relevance model and Kitcher's (1981) unification model). None of this requires that anything be considered a law, if law is understood in the general way sketched in section 7.2. On this interpretation, the idea that Paige's results are put to predictive and explanatory work by inferring, from his statistical conclusions, *first* to a hedged law, and *then* to particular predictions and explanations, is just an inferential detour. Moreover, it is an ill-advised detour, since it takes us through a hedged law-statement, which could not be an accurate description of anything in the world, since there is no such thing as a hedged law. It is thus possible to interpret Paige's results in a way that does not deny them predictive or explanatory import, but does refrain from positing any social-scientific laws.

Prediction and explanation require reliable sources of information about the world, in the form of strict or statistical regularities. Laws of nature are regularities that have certain features: they are global or universal, and robust, in the sense that they do not depend on contingent details of particular systems of objects, and they would not be upset by changes in the actual circumstances that are physically possible. In order to have explanatory and predictive value, though, a regularity (strict or statistical) need not have these special features. Hence, not only does social science have no laws; it needs no laws.

Bibliography

Armstrong, D. M. 1983: *What is a Law of Nature?* Cambridge: Cambridge University Press.

Carroll, J. 1994: *Laws of Nature.* Cambridge: Cambridge University Press.

Cartwright, N. 1983: *How the Laws of Physics Lie.* Oxford: Oxford University Press.

— 1999: *The Dappled World: A Study of the Boundaries of Science.* Cambridge: Cambridge University Press.

Dretske, F. I. 1977: Laws of nature. *Philosophy of Science*, 44, 248–68.

Earman, J. 1986: *A Primer on Determinism.* Dordrecht: D. Reidel.

— and Roberts, J. 1999: *Ceteris paribus*, there is no problem of provisos. *Synthese*, 118, 439–78.

Fodor, J. 1991: You can fool some of the people all the time, everything else being equal: hedged laws and psychological explanations. *Mind*, 100, 19–34.

Giere, R. 1999: *Science without Laws.* Chicago: The University of Chicago Press.

Hempel, C. G. and Oppenheim, P. 1948: Studies in the logic of explanation. *Philosophy of Science*, 15, 135–75.

Kincaid, H. 1996: *Philosophical Foundations of the Social Sciences: Analyzing Controversies in Social Research.* Cambridge: Cambridge University Press.

Kitcher, P. 1981: Explanatory unification. *Philosophy of Science*, 48, 507–31.

Lange, M. 1993: Natural laws and the problem of provisos. *Erkenntnis*, 38, 233–48.

— 2000: *Natural Laws in Scientific Practice.* Oxford: Oxford University Press.

Lewis, D. 1994: Humean supervenience debugged. *Mind*, 103, 473–90.

Lipton, P. 1999: All else being equal. *Philosophy*, 74, 155–68.

Morreau, M. 1999: Other things being equal. *Philosophical Studies*, 96, 163–82.

Paige, J. 1975: *Agrarian Revolutions.* New York: The Free Press.

Pietroski, P. and Rey, G. 1995: When other things aren't equal: saving *ceteris paribus* laws from vacuity. *The British Journal for the Philosophy of Science*, 46, 81–110.

Salmon, W. C. 1971: Explanation and relevance: comments on James G. Greeno's "Theoretical entities in statistical explanation." In R. C. Buck and R. S. Cohen (eds.), *Proceedings of the 1970 Biennial Meeting of the Philosophy of Science Association.* Dordrecht: D. Reidel, 27–39.

— 1994: Causality without counterfactuals. *Philosophy of Science*, 61, 297–312.

Searle, J. 1984: *Minds, Brains and Behavior.* Cambridge, MA: Harvard University Press.

Skyrms, B. 1980: *Causal Necessity.* New Haven, CT: Yale University Press.

Swartz, N. 1985: *The Concept of Physical Law.* Cambridge: Cambridge University Press.

Swoyer, C. 1982: The nature of natural laws. *Australasian Journal of Philosophy*, 60, 203–23.

Tooley, M. 1977: The nature of laws. *Canadian Journal of Philosophy*, 7, 667–98.

van Fraassen, B. C. 1980: *The Scientific Image.* Oxford: The Clarendon Press.

— 1989: *Laws and Symmetry.* Oxford: The Clarendon Press.

Further reading

Cartwright, N. 1995: *Ceteris paribus* laws and socio-economic machines. *Monist*, 78, 276–94.

Hempel, C. G. 1988: Provisoes: a problem concerning the inferential function of scientific theories. *Erkenntnis*, 28, 147–64.

There are no Laws of the Social Sciences

CHAPTER EIGHT

There are Laws in the Social Sciences

Harold Kincaid

In 1973 OPEC drastically reduced the amount of oil that it supplied to the world, and in short order there were long lines at the gas pumps in the USA and a much higher price for gasoline. When the gasoline prices rose, sales of large gas-guzzling automobiles declined. At the same time, exploration of new oil sources increased and eventually the known world oil reserves remained steady, despite earlier widespread predictions that oil reserves would soon be depleted.

Why did these things happen? Economists claim to know why; the law of supply and demand governs market-oriented economies and thus we would expect just the kinds of cause and effect relations between supply, demand, and price that occurred after the OPEC boycott.

So it seems that the social sciences can and do describe laws. This chapter argues that this appearance is correct. Of course, this sketchy example will not convince skeptical philosophers, who will argue that the appearances are deceptive for various reasons. To turn this example into a compelling argument will require first a discussion of what laws of nature are and do, then a demonstration that the law of supply and demand has similar traits and functions, and finally a response to the many alleged obstacles to laws in the social sciences.

Clarification of the nature and role of laws is the task of the first two sections below. Section 8.3 will develop in detail an argument that there are laws in the social sciences, using the law of supply and demand as a paradigm case. Section 8.4 will answer various objections, both to the reasoning of section 8.3 and to very idea of social laws.

8.1 Initial Issues

Do the social sciences study natural laws? The answer seems obviously "Yes," because of the following argument: physics describes natural laws governing physical entities, human beings are physical entities and thus so are societies, and hence there must be laws that describe societies.

Unfortunately, the critic won't be convinced. She will respond by saying that maybe there are laws describing the behavior of social entities as physical processes, but that does not mean that physics can state those laws, and even if it could, that is far from having laws in the social sciences, disciplines that describe the world in an entirely different vocabulary than physics.

That's fine, but here is another reason to think that there are obviously laws in the social sciences: actual examples of such laws are easy to find. Here are two: (1) No nation–state survives intact for longer than 1,000 years. (2) Every human society is composed at least in part of humans (in part, because buildings, factories, and so on might be part as well).

Case closed? No, because, as in every philosophical debate, answering one philosophical question requires answering a host of others. In this case, two of these other questions concern what is a natural law and what role such laws play in science. Those questions arise because critics will not "like" our alleged laws – they will deny that a tautology such as (2) can be a natural law (since it is a logical truth) or that a tautology can be used to explain anything. They will deny that (1) is really a law because it holds by chance.

The moral is that before we can debate laws in the social sciences, we have to first discuss what we think laws are. This is the task of this and the next section.

The question "What is a law of nature?" is not one question but several. To see why, consider the following possible different answers to the question:

> Temperature is a function of pressure and volume.
> A universal statement that expresses a necessity.
> An exceptionless regularity between events.

The first answer provides us with an *example* of a law statement. The second answer tells us that a law is a piece of language – a statement. The final answer claims that a law is something in the world, something that presumably would exist if language had never been invented. Each answer seems to be independent from the others in that someone could give any one of the answers without being logically committed to the others: I might agree with the example of a law, but object to the other two answers (as many have) that try to explain what a law is. Likewise, I might accept that laws are universal statements but deny that they are about universal regularities. For example, "the electron is a fundamental particle" looks like a law, but it is not about a relation between events. So we have three different questions to answer about laws: What kind of thing in the world is a law? What are the kinds of statements that pick out laws? Which particular alleged statements of laws actually do so? Philosophical discussions of laws have not always kept these questions separate, perhaps in part out of confusion and in part out of the legitimate hope that there might be a systematic answer to all three questions.

These distinctions point out that the question "Are there laws of the social sciences?" raises at least two distinct issues. As things in the world, there might be laws that govern social phenomena. That claim could be true independently of social science's ability to find and state such laws.

There are likewise several issues at stake behind this second question as to whether social science can state laws. We can imagine two law skeptics, the moderate and the radical. The moderate claims that current social science has identified no laws. The radical claims that it never could.

So we have a series of progressively more substantial or logically stronger theses that a defender of social laws might propound, namely that:

(1) Laws of social phenomena exist.
(2) Social science can provide statements that pick out laws.
(3) There are specific parts of current social science that pick out laws.

In what follows, I defend laws in the social sciences by defending all three theses.

8.2 What is a Law?

Philosophers and scientists have given different answers to this question, even when they were aware of the different issues at stake. Hence any debate over laws in the social sciences is at the same time a debate over what laws are. In this section I discuss what laws do and do not involve.

A quintessential example of a law of nature is Newton's law of universal gravitation: between any two bodies there is an attractive force proportional to their respective masses and inversely proportional to the inverse of the square of the distance between them. What kind of thing in the world does Newton's law pick out? The most natural answer is that it identifies a force. What is a force? It is a causal factor. A force is *causal* in that it influences something. It is a *factor* in that it need not be the only influence present. Modern physics, for example, identifies electromagnetic and nuclear forces that can be present at the same time as gravity. So a paradigm case of a law is a force or causal factor. (Smith (2002) is a good general defense of these ideas.)

Which statements then pick out laws? An obvious answer is that at least those statements that pick out causal forces are laws. The law of universal gravitation picks out gravitation forces, Maxwell's equations pick out electromagnetic force, the Darwinian law of the survival of the fittest picks out a force – fitness – that plays a major role in the biological realm, and so on. How then do we go about deciding if there are laws in the social sciences? The short answer is again by whatever means we can use to decide if there are causal factors influencing social phenomena.

It is important to note that I am not claiming that every law in science identifies a force, only that some of the clearest and most important cases of laws do. If we show that the social sciences can do something similar, we will have shown that there are laws in the social sciences.

There are obvious objections to this proposal. Answering those objections will further explore the view of laws that I favor, so I turn next to consider the following criticisms:

A law states a universal regularity, but statements picking out causal factors do not. Diet, for example, is a causal factor in health, but there is no exceptionless regularity that I can cite between diet and health outcomes.

Laws are different from accidental generalizations. (Lange (2000) is one of the more elaborate attempts to spell out this difference.) "All the coins in my pocket are copper" might be a true generalization, but it holds only by chance. It is not a fundamental fact about the world. However, statements that pick out causal factors might not be fundamental truths – they can be true because of the way the world happens to be. "The coins in my pocket interacted with sulfuric acid to produce copper sulfate" might be a true causal statement if I poured acid into my pants, but it would be accidentally true – true only due to the chance fact that all the coins in my pocket were copper.

Laws tell us reliably what would happen if things were different than they are (they "support counterfactuals" in philosophy lingo) *in a way that statements picking out causal factors do not.* I cannot infer that if I had poured acid into my pants yesterday, then copper sulfate would have been produced.

Laws allow us to reliably predict what unobserved events will look like, but statements citing causal factors may not. If I pour acid in my pants tomorrow, I cannot expect to produce copper sulfate.

Laws are universal in that they do not refer to particular entities. The law of gravitation does not mention any specific body, but the generalization about the coins in my pocket does. It is thus not universal in the way needed for a law. (Earman (1978) explores the issues in this criterion of lawfulness.)

Laws must state precise quantitative relationships; statements picking out causal factors need not.

There are laws in the natural sciences that do not refer to causes at all. Snell's law, for example, tells us the relation between the angle of incidence and the angle of reflection for a wave. Causes are not mentioned, only functional relationships.

Laws of nature describe fundamental causal forces that are not the byproduct of some deeper causal forces, but statements citing partial causal factors may be such byproducts.

Finally, identifying laws with statements that pick out causal factors doesn't explain laws. Any account of causation will need the notion of a law to explain it, and even if it doesn't, the concept of causation is just as obscure as that of a law.

I think that these claims are misguided in various ways. My responses to these objections are as follows:

Statements about causal factors need not entail universal regularities. Laws have no automatic connection to universal regularities. (Cartwright (1983, 1999) is a dedicated defender of this claim.) The inverse square law identifies the force due to gravity. However, it makes no mention of other forces. The actual behavior of objects will be a resultant of all those forces present and thus we do not expect them to move as the law of gravity by itself would predict. The law at most tells us what regularities we would see if gravity were the only force; it does not tell us when or if that is the case and thus it does not generally entail any specific regularities on its own. It nonetheless allows us to explain and predict reliably when conjoined with other knowledge – and that is what we want laws to do.

The concept of cause is no clearer than the concept of law and/or presupposes it. Laws in the natural sciences need not be causal. I place these two objections together because they rest on a particular view of what a philosophical account ought to do, one that I reject. Traditionally, philosophers have hoped to define the concept "law" by giving a set of jointly sufficient and individually necessary conditions for being a law. That definition was to be tested by two means: (1) it was to count as laws as much as possible all and only those things that we now call or would call laws; and (2) the defining features of a law should not commit us to any philosophically questionable assumptions or concepts. For example, a tradition going back to Hume wanted an account that attributed no necessities to nature (on the grounds that all we have evidence for are regularities that happen, not that must happen) and did not invoke the notion of causation (because again our experience only presents evidence of regularities).

I reject this picture of what philosophers should be doing in giving an account of laws for several reasons:

1 Few, if any, of our concepts are amenable to definition by necessary and sufficient conditions. (Stephen Stich's work (1990, ch. 4) is an interesting exploration of this claim and its implications for conceptual analysis is philosophy.) Instead, we work with paradigm cases and make rough judgments about how close specific cases are to that paradigm. There is no reason to think that the concept of law is any different.

2 Finding a definition that fit philosophers' or even scientist's intuitions about what we call laws need not tell us much about the practice of science. What gets called a law no doubt rests on the vagaries of convention and historical contingencies. The point of a philosophical account should instead be to shed light on the practice of science – in this case, on what role laws play in science. So the motivation for explaining laws as partial causal factors is not to provide a definition of the concept "law." That is a project of dubious merit. Rather, our project should be to get clear enough on how laws function in science to ask the question as to whether the social sciences can function that way as well. Then what role do laws play in science? Perhaps many, but above all, science produces laws to explain and reliably predict the phenomena. That is precisely what identifying causal factors should allow us to do. A claim to know a causal factor is dubious to the extent that it does not allow us to explain and predict. This is why we should care if there are laws in the social sciences. If the social sciences cannot explain and predict, then it is not clear that they are sciences at all.

3 The kinds of metaphysical constraints that an account of laws must meet cannot be determined *a priori* and independently of an account of how science itself works. If a philosopher's metaphysical ideals conflicts with the successful practice of science, I would give up the former, not the latter.

Perhaps some day a completely universal account of laws will be produced. If it is, I assume that it must count those parts of science that identify forces as laws. So if I can show that the social sciences can identify forces, then I will have shown that on any account of laws the social sciences produce laws. While I am dubious that we can provide a universal account in terms of necessary and sufficient conditions that actually illuminates science, my claim is nonetheless that identifying causal forces is a paradigm of law, not that all laws must do so:

Citing causal factors does not preclude referring to particulars, but laws cannot. Whether something does or does not refer to a particular entity depends on the language that we use. Darwin's laws of natural selection apparently refer to organisms on this planet. We can eliminate that reference by defining a "Darwinian system" as one with differential reproduction and trait inheritance. Then Darwin's laws stop referring to particulars. So the issue isn't referring to particulars.

Causes can be cited without identifying precise quantitative relationships. There are two responses to this worry: laws in the natural sciences don't always do so either and specifying precise quantitative relations is not essential for the role that laws play in science. Darwinian processes involving fitness specify no quantitative relationship – relative fitnesses have to be plugged in by hand and do not follow from any general theoretical claims made by Darwin. The central tenet of modern molecular genetics is that DNA produces RNA, which produces proteins. This tenet certainly explains much and has allowed for many successful predictions. Yet it is not quantitative either.

Claims citing causal factors may be picking out accidental truths that do not support counterfactuals and cannot predict unobserved phenomena. Behind this objection is a certain vision about the place of laws in nature and science that I reject. That vision sees a sharp division between laws and other parts of science. Counterposed to this vision, I urge the view that there is no fundamental divide between the laws of science and the other causal claims that it makes. The ability to predict unobserved phenomena and to support counterfactuals accrues to all causal claims, but to varying degrees; all causal claims hold to some extent by necessity; or, in other words, being an accidental generalization is a relative matter.

Consider first the ability to say what might have happened. Any time that we have good evidence for a causal claim, we have evidence for some claims about what might have happened. If an organism's fitness is a positive causal influence on the genes found in the next generation, then various counterfactuals are supported – for example, that if A had been less fit than B (contrary to fact), then A's genes would have been less well represented than B's. Whenever there is a causal relation, we know that if the cause had not been present and everything else had been the same, then the effect would not have occurred.

Think next about whether laws are necessary. Let's assume that a relation is necessary when it holds across different possible arrangements of things. Then the claim that fitness is a causal factor in inheritance is also necessary to a degree. In any world in which there is differential survival and inheritance of traits, fitness is a causal factor – even if different species exist or even if those organisms have a different physical basis for inheritance than DNA/RNA. Yet the laws of evolution by natural selection are accidental in that they hold only of systems with the right inheritance and competitive characteristics. They need not describe other life forms that do not meet these criteria. Even the claim that "all the coins in my pocket are copper" looks less accidental if there is some causal mechanism that excludes noncopper coins – maybe my pocket has holes in it and only pennies do not fall out. And the basic laws of physics are accidental in that the values of the fundamental constants are apparently the result of chance events in the big bang that need not have happened.

Similar arguments can be made about the ability to predict unobserved events. To the extent that we think we have picked out a cause, we are committed to thinking that future instances of the cause in similar circumstances will produce similar effects. If there is a causal mechanism that explains why my pocket only has copper coins, then we can expect that future coins examined will be copper as well. Changes in fitness will predict changes in the distribution of traits of organisms. This is true despite the fact that not all pockets and not all possible living systems allow for such predictions:

Laws of nature describe fundamental causal forces that are not the byproduct of some deeper causal forces, but statements that cite partial causal factors may be such byproducts. The request that a law of nature be fundamental in this way eliminates many apparent laws in the natural sciences, and is at odds with the role that laws play as well. The gas laws tell us that a change in the pressure in a gas will cause a change in temperature if volume is held constant. This is just one of a number of thermodynamic laws that apply to aggregations of molecules. These laws describe causal processes that are the byproduct of processes at the molecular level. Other compelling examples of such laws come from chemistry, where laws of association were explained in terms of valence, a chemical force, as it were, that is the byproduct of more fundamental forces of quantum mechanics. The no-byproduct criterion asks too much.

Moreover, the citation of causes that are the byproduct of more fundamental causes can nonetheless explain and predict. The gas laws are a case in point. They identify causes and predict new changes – and they did so long before the underlying molecular details were explained in terms of Newton's laws.

The moral I then take from these objections and the answers to them is this: laws exist in one important sense where we can cite causal factors that explain and allow us to successfully predict. But not all statements that cite causal factors are alike – they vary in how broadly they apply, in how wide an array of changing circumstances they apply to, and in how they tell us about how the world might be or will be. In general, our confidence that causal claims are true is a function of how widely they explain and predict. The key question thus is whether the social sciences provide causal claims that provide relatively extensive explanations and predictions.

8.3 Problems and Prospects for Generalizable Causal Knowledge

So far, we have argued that we can show there are laws in the social sciences if we can show that they pick out causal processes in a way that allows for significant explanation and prediction. This section first discusses how science in general goes about finding such processes, specific obstacles to implementing these strategies in the social sciences, and the ways in which the social sciences can deal with those obstacles. I then argue that empirical research on the force of supply and demand shows that the social sciences sometimes succeed in overcoming these difficulties.

What does it take to establish explanatory causal knowledge that allows successful prediction? In rough terms, here is how the sciences do so: they take background knowledge about causes, observe various changes in factors of interest, and infer what causes what. So, in an ideal experiment, all possible causes are known and all are held fixed but for one, which is varied in a specific way. The relevant changes in effects are observed and the cause inferred.

Such knowledge is strengthened and deepened by showing that similar effects are observed in repetitions of this setup, and that the causal knowledge generated can be combined with knowledge about other causes to produce successful predictions about different setups. In this way, explanations are broadened and new phenomena predicted.

So the question is whether the social sciences can meet this requirement for establishing causal claims. We should note immediately that one standard complaint about the social sciences – that they are nonexperimental and qualitative – cuts no ice once we see what causal claims require. Recall that we need to observe one factor varying while others remain constant. We can achieve this goal without performing experiments and without stating how much one factor influences another. In other words, we can observe so-called natural experiments – situations in which the factor of interest varies and everything else stays the same. Whatever consequences result we can attribute to the causal influence of the varying factor, even though we did not directly manipulate it or directly hold other things fixed and even though we do not measure the consequences in quantitative terms. Thinking about the nonexperimental natural sciences helps to show that this must be the case. Astronomy, cosmology, geology, ecology, and evolutionary biology all rely in large part on nonexperimental evidence, evidence that is often not quantitatively measured either. Of course, controlled experiments generally are a more reliable and efficient way of identifying fundamental causes, but they are not the only way or even always the best way to gain knowledge.

A second obstacle comes from the fact that social science research often relies on assumptions that are literally false. We can divide such assumptions into two rough categories: idealizations and abstractions. Both involve assumptions that are literally false, but in different ways. Here is an example that involves both. The gas laws tell us that the temperature of a gas is a function of pressure and volume. This causal relation can be explained by Newton's laws of motion, by applying them to the motions of the particles in a gas and equating temperature with mean kinetic energy. Such an explanation assumes that particles are solid bodies and that gravitational forces are the only ones present. However, atoms are only approximately perfectly hard bodies and gravitational forces are not the only ones present. Our claim about atoms is an idealization in that atoms do approximately have the property in question. It is not even approximately true that only gravitational forces are present – we are in this case abstracting from such forces in that we are considering how particles *would* behave if gravity were the only force present. But in both cases we are relying on false assumptions. The social sciences rely heavily on both idealizations and abstractions.

How can we have knowledge of causes by using false assumptions? If our models only approximate the real world, should we believe them? If our explanations describe how things would be if the world were different, do they explain the actual world at all?

We have good reason to think these are not insurmountable obstacles. After all, as the example of the gas laws illustrates, the natural sciences rely widely on idealizations and abstractions. Perhaps the most famous example concerns explanations in mechanics where more than three bodies are interacting. We do not have the mathematics to solve the equations that describe such situations, and we are forced to use approximations. Nonetheless, we can use Newtonian mechanics to land space probes on Mars. Obviously, not all false assumptions are an obstacle to causal knowledge.

So the interesting questions are: How do the natural sciences manage to produce knowledge of basic causal processes while using idealizations and abstractions? and Can the social science do something similar? The key issue is this: when a theory or model involving idealizations or abstractions seems to successfully predict and explain, how do we know that its success is not due to the falsity of its assumptions rather than to the truth of its causal claims? How do we know that results are not spurious?

The answer is that the natural sciences use a variety of techniques to show that the postulated causes are responsible for the data despite the false assumptions involved in obtaining them. For example, we can show that as approximations are reduced, the predictive accuracy of our theory improves, or we can show that the causal factors that we left out would not change the predictions that we derived from a theory that abstracts. So, for example, in the explanation of the gas laws, the walls of the container holding the gas are assumed to be perfectly rigid. If we allowed for walls that absorbed some energy and our resulting predictions improved, we would have evidence that our results were not spurious. Similarly, we might use our knowledge of atoms and of the gas in question to argue that the nongravitational forces present – that is, electromagnetic forces – had a constant effect regardless of temperature or pressure, and thus that changes in temperature were caused by changes in pressure.

So the question is now whether the social sciences can in principle and in practice make the same kind of arguments. We know that the nonexperimental natural

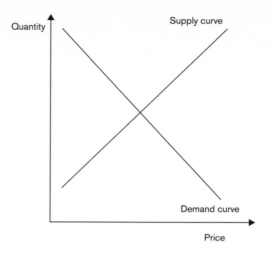

Figure 8.1 Curves representing the amount supplied and demanded for each specific price.

sciences can, so that gives some reason to think the social sciences can as well. Better evidence would come from showing real cases in the practice of social research. Our next task is to provide such an example.

Perhaps the most successful piece of social research around comes from the body of work describing the "law of supply and demand." The law of supply and demand tells us that the aggregate demand for a good, the aggregate supply for that good, and the good's price are causally interdependent. More precisely, it tells us that:

Changes in price cause changes in quantity demanded and quantity supplied.
Changes in supply and demand curves cause changes in price.

A demand or supply curve graphs how much individuals are willing to produce or buy at any given price (see figure 8.1). When there is a shift in price, that causes corresponding changes in the amount produced and purchased. A shift in the supply or demand curve is a second causal process – when it gets cheaper to produce some commodity, for example, the amount supplied for each given price may increase, so that the entire supply curve shifts from S_1 to S_2 (figure 8.2). A shift in the supply curve causes a change in price toward the point at which supply and demand match. In reality, both processes interact, leading to the causal interdependence between changes in demand, supply, and price.

This "law of supply and demand" does just what we argued earlier that a law does – it picks out a force or causal factor. It does not describe an actually existing regularity that relates events, since supply and demand is not the only causal factor. For example, if people experience a drastic drop in income at the same time that the price of a commodity drops, the quantity purchased might decrease instead of increasing, as the law of supply and demand on its own would predict. What the law of supply and demand instead asserts is that there exists a causal force in certain circumstances

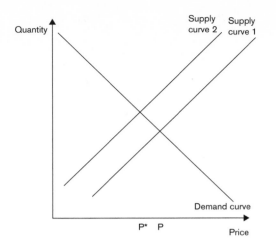

Figure 8.2 Change in price from P to P* due to a shift in the supply curve.

– roughly, where there is a market. Supply and demand influences what happens even if it need not be the only influence.

What is the evidence for the law of supply and demand? Of course we have lots of common-sense experience that seems to support it. Economists have gone far beyond that, however. They provide evidence of two sorts: deductions of what must happen from intuitively obvious assumptions about individual behavior, and detailed observational studies of price, supply, and demand in particular markets. The deductive evidence shows that to the extent that individuals maximize what they have or get, then we can expect price, supply, and demand to interact as the law of supply and demand describes. The observational evidence comes from many studies of diverse commodities – ranging from agricultural goods to education to managerial reputations – in different countries over the past 75 years. Changes in price, demand, and supply are followed over time. Study after study finds the proposed connections. (Deaton and Muellbauer (1980) is a good general survey. Early systematic work was undertaken by Stone (1954), and Pollack and Wales (1992) present some of the more sophisticated recent research.)

Such studies do not, however, work simply by observing that an increase in price results in decrease in quantity demanded, and so on. The law of supply and demand does not claim to identify all forces that influence price and we do know that other forces can interact with it. So to confirm that the supply and demand force is at work in any given situation requires taking those other factors into consideration. Observational studies of specific commodities do precisely that, and have done so with increasing sophistication as research has progressed. Changes in income, in the price of substitutes and complements (Pepsi is a substitute for Coke, chips are a complement of both), in tastes and technology, and in a host of other factors have been observed along with data about price, supply, and demand. By including these factors, economists help to rule out confounding factors and more strongly confirm the existence of the supply and demand force in economic affairs.

These practices are examples of the methods described earlier for dealing with idealizations and abstractions. Economists have refined their evidence in part by moving from studies of highly aggregative to disaggregated data – for example, from studies of the demand for food to studies of specific foods. The former studies assume that food is homogeneous with a single price; the latter drop that idealization for a more realistic picture. Abstractions are, of course, being dealt with as economists factor in other causal factors, such as income, in applying the law of supply and demand.

8.4 Objections

I want to finish my case for laws in the social sciences by considering a number of objections that have not been addressed explicitly already. Some are objections to the reasoning given above; others are directed toward the conclusion that there are laws in social science. Answering them will give my main thesis further credibility, and simultaneously tie the debates about laws to other debates about the social sciences.

A common objection to laws in the social sciences is that the laws that the social sciences produce are bound to have exceptions, exceptions that social scientists cannot identify in advance. In philosopher's lingo, any law claim in the social sciences must be qualified with the phrase "*ceteris paribus*" – other things being equal. This alleged fact raises doubts of varying sorts – that all such claims are "vacuous" or "meaningless; that they are trivially true or tautologies ("A causes B unless it doesn't"); and that they cannot be confirmed or disconfirmed because of the open-ended escape clause. (Many of these worries are discussed in Earman and Roberts (1999), as well as in Roberts' contribution to this volume – chapter 7.)

The first thing to note is that these charges are not mutually consistent, even though they are sometimes made by the same authors. If a *ceteris paribus* law cannot be confirmed, then it cannot be a tautology and vice versa; if it is trivially true, then it cannot be meaningless. We should also note that this problem of exceptions should not be unique to the social sciences. Some have argued that even the most fundamental laws in physics are qualified *ceteris paribus*. However, we don't need that radical claim to make the point. Outside of fundamental physics, most natural science deals with complex phenomena in which it is hard and sometimes practically impossible to control, or even know, all of the possible interacting causes. Yet they produce reliable causal knowledge nonetheless.

Various defenders of laws have tried to provide an account of the *meaning* of *ceteris paribus* laws in order to defer objections to them. (Hausman (1981) is an early systematic attempt; more recent accounts are surveyed in Earman and Roberts (1999).) That is a project that is perhaps best left unpursued, for it assumes there is a unitary meaning to be found, and that the scientific problem of idealizations and abstractions could be solved by analyzing the relevant concepts. The problem of analyzing *ceteris paribus* laws can be ignored, because we need not think of laws in the social sciences as qualified *ceteris paribus* in the first place. Recall the picture of laws described in section 8.2. A law picks out a causal force or factor. The law of universal gravitation, for example, asserts that there exists the force gravity. It does not describe a regularity, for gravity is not the only force and the law of gravity is silent on how other forces might combine with it. Since it does not claim to cite a regularity without exceptions, there is no reason to think of the law as qualified *ceteris paribus*. The law is true if there is indeed a gravitational force. Laws in the social sciences work in the same way: they claim to identify causal factors and make no commitment by themselves to what other causal factors there might be and how they might combine.

Of course, like the natural sciences, the social sciences do have to worry in each particular situation that there is sufficient evidence that the causal factor in question is operative – in other words, they have to worry about confounding causes. Those worries are dealt with by the various means that we described earlier for handling idealizations and abstractions. But those are methods for applying the relevant laws, not *ceteris paribus* clauses attached to the laws themselves.

Another long-standing objection to laws in the social science complains that a "social physics" misses what is uniquely human about us. On the view defended here, the social sciences are very much like physics in the sense that both describe causal forces. But humans are not inert objects or automatons – they make choices and actively interpret the world. Human behavior is free and meaningful, and thus not amenable to natural science style explanations. Explanations in the social sciences are interpretations, not the citing of causes. Arguments in this vein come from defenders of "hermeneutic" or "interpretivist" social scientists and like-minded philosophers (Geertz, 1973; Taylor, 1971, 1980).

These worries, even if they were entirely convincing (they are not), might not undermine the claims of this chapter. We have defended the claim that the social sciences produce some laws, not the claim that the social sciences only explain via laws. It might be that some social phenomena – for example, aggregate economic activity – are best explained by identifying causal factors and that other phenomena – for example, symbolic rituals – are best explained in terms of meanings. However, interpretivists usually defend the more radical view that the meaningful and/or the free nature of human behavior undermines any noninterpretive social science. Moreover, if large parts of social behavior were not amenable to explanation in terms of causes, then the conclusions of this chapter would be less significant. So we must take these arguments seriously.

Let's begin with the worry that human free will makes laws in the social sciences impossible. Causal laws should allow for prediction of future events. But if human actions are free, the argument runs, then they are not fixed in advance. This argument has two serious flaws: it assumes a particular and controversial notion of human freedom and it presupposes that the social sciences are essentially about individual behavior.

Consider first the notion of freedom needed for this argument to work. There are two competing notions of freedom discussed in debates over freedom and determinism, the libertarian and the compatibilist. The libertarian holds that human actions are in some way uncaused – at least, uncaused by anything outside the agent. The compatibilist holds that free will and causal influences on behavior can coexist. They can do so because being free requires that your choices make a difference – that if you had chosen otherwise, it would have influenced your behavior. So changes in price cause changes in consumption, but my choice is free in that if I had chosen to

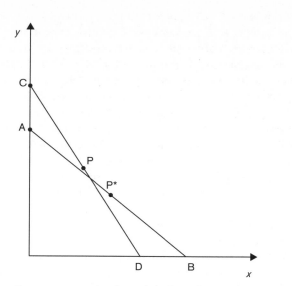

Figure 8.3 P and P* are the average quantity demanded of *x* and *y* as *x* becomes more expensive when consumers choose at random.

spend my money differently I could have. Obviously, the argument against laws only has force if we assume the libertarian conception of free will.

However, the libertarian conception is quite problematic. The main problem raised in debates over freedom and determinism is to specify a sense of freedom that anyone could actually have and that would really be free will. If our actions are uncaused in the sense of random, then it is hard to see that these are free acts. But if our actions are free in that we are the full cause of our actions, then it seems we must have created ourselves – if something else made us the way we are, then we are determined. But the notion of self-creation is of dubious sense and is not something that is reasonably attributable to humans.

The argument from freedom against laws also fails because it presupposes that the social sciences are only about individual behavior. That presupposition is needed for the argument because it is possible that each individual act might be uncaused and yet that there be causes of the aggregate behavior of individuals. Why? Because, as Marx succinctly put it, "men make their own history, but they do not make it as they please." We can't avoid death and taxes, as the saying goes. Individual decisions must be between the real possibilities; the facts about those possibilities may allow us to identify causal patterns in the combined behavior of individuals.

The law of supply and demand discussed earlier provides a nice illustration of how this can happen. That law implies, among other things, that an increase in price means a decrease in the quantity demanded – a move along the demand curve. There is good reason to think that this relation would hold for aggregate markets even if individual choices are free in the libertarian sense, because consumers are still constrained by relative prices (having free will does not entail the ability to determine what things cost!). (So far as I know, this argument occurs first in Becker (1976).) Figure 8.3 illustrates the relative prices of two goods x and y – it gives the relative amounts of x and y that can be bought with the same fixed budget. On curve CD, x is more expensive relative to y than on curve AB (you have to give up more x to get the same amount of y). Assume that the libertarian notion of freedom implies that the choices of how much individuals will consume are unpredictable – that any choice along the line is equally probable. Then the average of the choices when the price of x to y is represented by AB will fall in the middle at P*. When x increases relative to y, the new price is represented by CD and the average choice will become P. But at P less of x is consumed than at P*. Random choices facing a fixed budget and changing prices will demand less when the price increases.

Is there some reason to think that the social sciences can only be about individual behavior? There is a doctrine – sometimes called methodological individualism – that claims that this is so. (The classic statement is Watkins (1973) and a more recent defense is Elster (1985).) Yet the social sciences seem often to be about things other than individual behavior: the nature of institutions, the causes of unemployment and economic growth, the class structure of society, and so on. It takes a long and widely criticized story to get around these apparently obvious facts (see Ruben, 1985; Kincaid, 1996, 1997).

Let's turn now to the second version of the interpretivist objection. The second objection to laws stems from the fact that human behavior is meaningful. These are complicated issues, in part simply because the notion of "meaning" has proven hard to clarify. Thus, below is a sketch of how the defender of laws can respond, not a definitive treatment.

The root idea in these objections is that human behavior essentially involves interpretation in two senses:

1 Subjects interpret the world through their own categories, unlike inert objects.
2 What subjects do and say has to be interpreted. Unlike with inert objects, categorizing behavior involves determining the meaning of the behavior.

Given these two senses in which meaning is involved in the social sciences, the question is how these facts preclude finding confirmed and generalizable claims about causal factors.

One influential argument (Taylor, 1980) asserts that interpreting behavior always must be done with the aid of some previous categories on the part of the investigator. Unlike the natural sciences, there are no "brute data" in the social sciences. So the social sciences cannot be about the objective confirmation of causal claims.

A first problem with this version of the argument (and with all others) is that it seems to rule out causal explanations in some natural sciences as well. For example, when applied to the higher mammals, ethology and behavioral ecology seem to be at risk. Mammals seemingly interpret the world too and in categorizing their behavior – its point – we seem to use prior notions about what the animal is trying to do, how it reads stimuli, and so on.

In fact, there is a deeper objection lurking here; namely, that all science approaches the data in terms of previously assumed categories. In short, there are no brute data in natural science either, because data do not come precategorized and individuated.

Taylor's argument attributes to the natural sciences a very sharp distinction between theory and data that is now widely rejected.

Another common argument from the meaningful nature of human behavior claims that interpreting a subject's behavior must be done in the subject's own categories and thus makes the social sciences interpretive, not causal. We need to ask about this argument, first, whether it is true that we must use the subject's own categories and, secondly, just how that implies that causal explanations are inappropriate.

Mandating that we always explain in the subject's own categories implies that individuals never misunderstand their own behavior. That is a drastic assumption which much evidence belies – there is good evidence, for example, that individuals' conceptions of their own attitudes and beliefs are only weakly associated with their actual social behavior (Liska, 1975).

We might nonetheless grant that the subject's categories are at least important evidence, even if not infallible. But the question is then how this implies that ordinary causal explanation is impossible. There are good arguments to show that reasons and beliefs can be causes of behavior (Davidson, 1980; Henderson, 1993). A standard view on the opposite side holds that it takes a special kind of empathetic insight to grasp the subject's take on the world, and that this precludes the objective confirmation needed to establish causal claims. Perhaps intuitive insight is important evidence, but it certainly is not infallible and it must be checked against other kinds of data – by how it fits with what we know about human psychology in general, with various actions that we want to explain, and so on. Taken this way, it is just one species of evidence about causes, not a mystical fusion of observer and observed that would make the social sciences beyond the pale of generalizable causal knowledge.

8.5 Conclusion

Those parts of science that pick out forces are a paradigm case of laws in the natural sciences. Forces are partial causal factors. Some causal factors are more basic than others, but laws in the natural sciences are not restricted to only the most fundamental causes; identifying causes at all levels does what we want laws to do, namely explain and predict. The social sciences do not always tell us all relevant causes when they pick out a specific force, but then neither do the natural sciences – witness the law of gravity. The key question is whether other complicating causes can be handled in way that allows reliable causal knowledge. Arguably the social sciences can do so, as the law of supply and demand illustrates. There are various arguments that allege that humans are special and could never be explained by identifying the causal forces governing their social behavior. However, those arguments are unconvincing. While the complex nature of social phenomena certainly presents difficulties, there is good reason to think that laws in the social sciences are possible both in principle and in practice.

Acknowledgments

I am grateful to Marc Lange for helpful comments on parts of this chapter.

Bibliography

Becker, G. 1976: *The Economic Approach to Human Behavior*. Chicago: The University of Chicago Press.

Cartwright, N. 1983: *How the Laws of Physics Lie*. New York: Oxford University Press.

Cartwright, N. 1999: *The Dappled World*. Cambridge: Cambridge University Press.

Davidson, D. 1980: Actions, reasons and causes. In *Essays on Actions and Events*. Oxford: Oxford University Press.

Deaton, A. and Muellbauer, J. 1980: *Economics and Consumer Behavior*. Cambridge: Cambridge University Press.

Earman, J. 1978: The universality of laws. *Philosophy of Science*, 45, 173–81.

Earman, J. and Roberts, J. 1999: *Ceteris paribus*, there is no problem of provisos. *Synthese*, 118, 439–78.

Elster, J. 1985: *Making Sense of Marx*. Cambridge: Cambridge University Press.

Geertz, C. 1973: *The Interpretation of Cultures*. New York: Basic Books.

Hausman, D. 1981: *Capital, Profits and Prices*. New York: Columbia University Press.

Henderson, D. 1993: *Interpretation and Explanation in the Human Sciences*. Albany, NY: SUNY Press.

Kincaid, H. 1996: *Philosophical Foundations of the Social Sciences*. Cambridge: Cambridge University Press.

Kincaid, H. 1997: *Individualism and the Unity of Science: Essays on Reduction, Explanation, and the Special Sciences*. Lanham, MD: Rowman and Littlefield.

Lange, M. 2000: *Natural Laws in Scientific Practice*. New York: Oxford University Press.

Liska, A. 1975: *The Consistency Controversy: Readings on the Effect of Attitudes on Behavior*. New York: John Wiley.

Pollak, R. and Wales, T. 1992: *Demand System Specification and Estimation*. New York: Oxford University Press.

Ruben, D. H. 1985: *The Metaphysics of the Social World*. London: Routledge and Kegan Paul.

Smith, S. 2002: Violated laws, *ceteris paribus* clauses, and capacities. *Synthese*, 130, 235–64.

Stich, S. 1990: *The Fragmentation of Reason*. Cambridge, MA: The MIT Press.

Taylor, C. 1971: Interpretation and the sciences of man. *Monist*, 25, 3–51.

Taylor, C. 1980: Understanding in the human sciences. *Review of Metaphysics*, 34, 3–23.

Watkins, J. N. 1973: Methodological individualism: a reply. In J. O'Neil (ed.), *Modes of Individualism and Collectivism*. London: Heinemann, 179–85.

Further reading

Giere, R. 1999: *Science without Laws*. Chicago: The University of Chicago Press.

Hempel, C. 1994: The function of general laws in history. In Martin and McIntyre, op. cit., pp. 43–55.

Follesdal, D. 1994: Hermeneutics and the hypothetical-deductive method. In Martin and McIntyre, op. cit., pp. 233–47.

Little, D. 1991: *Varieties of Social Explanation*. Boulder: Westview Press.

Martin, M. and McIntyre, L. (eds.), *Readings in the Philosophy of Social Science*. Cambridge, MA: The MIT Press.

Salmon, M. 1994: On the possibility of lawful explanation in archaeology. In Martin and McIntyre, op. cit., pp. 733–47.

Scriven, M. 1994: A possible distinction between traditional scientific disciplines and the study of human behavior. In Martin and McIntyre, op. cit., pp. 71–9.

Stone, J. R. 1954: *The Measurement of Consumer Expenditure and Behavior in the U.K., 1920–38*. Cambridge, UK: Cambridge University Press.

ARE CAUSES PHYSICALLY CONNECTED TO THEIR EFFECTS?

It is common (at least among philosophers) to talk of causes being "connected" to their effects. This talk is usually metaphorical, but according to some ways of thinking about causation, causes are *literally* connected to their effects. The connection consists of a *causal process*, a certain kind of physical process that has been characterized in different ways by different authors. According to Phil Dowe, a causal process involves an object that possesses a conserved quantity. Such process theories of causation seem to run into trouble with *negative* causation: cases in which one event *prevents* another from occurring, or in which the *failure* of some event to occur brings about some outcome, or in which there is some other combination involving the nonoccurrence of events. Jonathan Schaffer presents this problem for process theories in his chapter. Phil Dowe claims that negative causation is not genuine causation: he calls it "quasi-causation." Although some cases of quasi-causation may look like genuine cases of causation, it is important to maintain a theoretical distinction. Schaffer, by contrast, argues that negative causation gives us everything we could want from causation, and criticizes Dowe's attempt to draw a dividing line between the different cases. The debate is interesting, in part, because of the nature of the examples discussed. Many mechanisms (whether causal or quasi-causal) turn out to work in a manner quite different from what common sense might assume.

Causes are Physically Connected to their Effects: Why Preventers and Omissions are not Causes

Phil Dowe

9.1 Introduction

This morning I didn't throw a rock through the window across the road. Did I cause the window not to break? According to some theories of causation I didn't. The Salmon–Dowe theory of causation says that for there to be causation, there must be a set of causal processes and interactions, understood in terms of conserved quantities linking a cause with its effect.

To give more detail: the conserved quantity (CQ) theory of causation (Salmon, 1998; Dowe, 2000) can be expressed in two propositions:

CQ1. A *causal process* is a world-line of an object that possesses a conserved quantity.
CQ2. A *causal interaction* is an intersection of world-lines that involves exchange of a conserved quantity.

For example, suppose I had thrown the rock through the window. Then my throwing the rock caused the window to break precisely because there is a causal process, the trajectory of the rock, possessing momentum, which links my throw to the window's breaking. And the window's breaking involves an exchange of momentum. But quite clearly there is no such set of causal processes and interactions linking my not throwing the rock and the window (see also the theories of Fair, Aronson, Ellis, Bigelow and Pargetter, and Ehring).

According to other theories of causation, I did cause the window to break. The counterfactual theory says that it is causation since it is true that had I thrown the rock the window would have broken (the *sine qua non* condition). The agency theory says that it is causation because not throwing the rock is a good way of bringing about that the window does not break (Price, Gasking). The regularity theory says that it is causation because there is a regular pattern – a "constant conjunction" – between not throwing rocks and windows not breaking (Hume), while the probabilistic theory says it is causation because not throwing the rock makes it more likely that the window won't break (Reichenbach, Suppes, Humphreys). The same can be said for non-Humean versions of these last two, such as those in terms of natural necessitation, and non-Humean notions of chance (Mellor).

However, there are also versions of these latter accounts according to which such cases are not causation. According to Lewis (1986), the most influential version of the counterfactual theory, any case that involves negative "events" exhibits no *relation* of counterfactual dependence (had event *A* obtained, event *B* would not have), a relation that is required for causation. (Later, Lewis argues that such cases count as causation on account of the relevant counterfactual *truth* – see Lewis, forthcoming). And David Armstrong's necessitation theory (1999) rules them out since the fact that I didn't throw the rock is not a first-order state of affairs.

Our case involves negative events (or facts): *not* throwing the rock and the window *not* breaking. This can be described as a case of prevention by omission. A similar divide arises for other cases involving negatives, in which only the alleged cause is a negative (causation by omission – for example, not taking the medication caused his death), in which only the alleged effect is a negative (prevention – for example, taking the medication "caused" him not to die), and other cases in which some intermediate events in an alleged causal chain is a negative (for example, she hid his medicine, thereby "causing" his death). Let's call the range of theories that count "causation" by negatives as causation *negative-friendly*, and those that do not *negative-excluding*.

I wish to defend the claim that cases involving negatives in these ways are not, strictly speaking, cases of causation. However, more should be said about this class of cases. I will use the label *quasi-causation* for prevention, "causation" by omission, and the other cases just mentioned, and in section 9.3 I will explain what this amounts to, given that it is not causation. In section 9.4 I will show why it is that we sometimes are tempted to think that cases of quasi-causation really are causation. But first we need to clear the ground by thinking about some of the intuitions that lead people to take the view that these cases either are or are not cases of causation.

9.2 Arguments from Intuition

Arguments for and against the view that I am defending often appeal to "clear intuitions" that we supposedly have about this case or some other; for example, "I clearly didn't cause the death of those starving children in *x* because I'm thousands of miles away, and had nothing whatsoever to do with them," or "Clearly your failure to regularly clean your teeth is the cause of your tooth decay."

However, arguments that appeal simply to allegedly clear intuitions should be treated with suspicion. In the days before the Copernican theory was widely accepted, someone might have said "Our clear intuition is that the sun rises."

Of course, the Copernican theory had to give a clear account of what was going on, an account that explained why, loosely speaking – if not literally – it could be taken as true to say the sun rises in the morning. (Sections 9.3 and 9.4 will do this for quasi-causation.) So common-sense intuitions about individual cases cannot be taken as sacrosanct.

Consider the following claims:

(A) I caused her death by holding her head under water for five minutes.
(B) I can see the darkness outside.
(C) I caused the terrorist attack in London by failing to report information that I had about it.
(D) The hospital administration caused the death of an elderly patient by refusing to release funds to ship expensive equipment from the USA and thereby allowing her to die by "natural causes."
(E) I caused the terrorist attack in London by failing to be in a pub on a certain night, where I would have overhead about the plot, and by failing to travel to the UK and blow up the terrorists' van before they could do any damage.
(F) I caused the death of some penguins by failing to hire a plane and travel to the Antarctic to intervene in a shark attack.
(G) A man is engrossed in the view at a lookout, and doesn't see a small girl (who he doesn't know) playing nearby. He sees her just as she is about to slip off the cliff, and runs as fast as he can to the edge, hurls himself headlong across the rocks, gets just a finger to the child's shirt as she slips off the edge, but is unable to prevent her from hurtling to her death. He caused her death by omission – had he not been so engrossed, he would have seen her earlier and been able to save her.
(H) My not throwing the rock caused the window not to break.

If we focus on what intuitions we might have about each of these cases, and ask "Is this a case of causation?" we get different answers for different cases. I submit that we would naturally suppose the earlier ones as causation and the later ones as not being causation. However, all the theories of causation mentioned above treat them uniformly. The negative-friendly theories say that these are *all* cases of causation, while the negative-excluding theories say that none of them are.

If this analysis is correct, it spells the failure of any argument that appeals simply to "folk intuitions" about some particular case to establish (1) that indeed it is a case of causation, and (2) that a theory of causation that gives the opposite result is therefore wrong.

So, any account that wishes in some sense to respect intuitions will need to explain why our intuitions vary across cases that are theoretically similar. (I don't myself place high importance on respecting intuitions (Dowe, 2000), but since many theorists do, let's grant this desideratum.)

There is a well-established tradition – probably originating with Mill – which appeals to so-called "pragmatic considerations" to explain why there are counterintuitive cases of causation. On Mill's theory, the scientific "total" cause of an event is sufficient for its effect, but we can consider any part of the total cause that is necessary for the effect a *partial cause*, and we could call partial causes "causes" if we wish. Two types of counterintuitive cases then arise. Intuitively, some partial causes count as causes and others do not (striking the match rather than the presence of oxygen is considered the cause of the explosion). And, intuitively, some very remote sufficient or partial causes are also counterintuitive (the big bang is not, intuitively, the cause of the explosion, even if it were a sufficient condition). But "pragmatic considerations" explain this. Our intuitions arise in a context in which we focus on aspects of human interest. Some partial causes are of more interest to us for various reasons, and more or less proximate causes are of more interest to us than remote causes; for example, in establishing legal responsibility.

This approach can explain why, for example, we might have different intuitions about (C) and (E). (C) involves additional knowledge, and arguably negligence, features that don't obtain in (E). Or it can explain the difference between (A) and (G) in terms of the agent's intention.

So, in general terms, this solution takes all cases involving negatives as causation and appeals to "pragmatic considerations" to explain why some are counterintuitive. There are two problems with this answer. First, at best it works for those negative-friendly theories of causation that take negatives to be causes or effects, but it won't work for negative-excluding theories. Secondly, it's not clear how this solution works in detail. Take (H), the case of my not throwing the rock causing the window to not break – intuitively not a case of causation, compared with the case in which my throwing the rock causes the window to break. Considerations of human interest – for example, involving partial causes, remoteness, negligence, or value – don't appear to distinguish these cases. This does not mean that no such account could be given, but this remains an unanswered challenge.

If we could give an alternative account that (1) is available to all theories of causation, and (2) that does deliver an explanation for the inconsistency of our intuitions, that account would therefore have to be the preferred answer. We now turn to such an alternative account.

9.3 The Counterfactual Theory of Quasi-Causation

The general outline of this account is to assert that negatives cannot be linked by genuine causation, but we can give a principled account of the near relative to causation, quasi-causation. So what is quasi-causation? (For more details, see Dowe (2000, ch. 6) and Dowe (2001), which the following summarizes.)

> Prevention by omission: not-A quasi-caused not-B if neither A nor B occurred, and
> (1) if A had occurred, A would have caused B.
> where A and B name positive events or facts.

This is the case in which an omission prevents something simply by not causing it. So, for example, "My not throwing the rock caused the window not to break", is true in the circumstances if the counterfactual "If I had thrown the rock it would have caused the window to break" is true; a counterfactual about genuine causation. So quasi-causation is essentially the mere possibility of genuine causation.

The counterfactual account of standard prevention is as follows:

> Prevention: A prevented B if A occurred and B did not, and there occurred an x such that
> (P1) there was a causal interaction between A and the process due to x, and

(P2) if A had not occurred, x would have caused B.
where A and B name positive events or facts, and x is a variable ranging over events and/or facts.

The reason this is expressed only as a sufficient condition is that there are other kinds of preventions (see Dowe, 2001).

For example, "I prevented the terrorist attack in London by blowing up the terrorists' van before they could do any damage" is true in virtue of the fact that I blew up the terrorists' van, the attack didn't occur, and there was a genuine causal process – the terrorists beginning to carry out their plan – where (P1) I interacted with that process by blowing up the van, and where (P2) had I not blown up the van, the terrorists beginning to carry out their plan would have led the attack, via a genuine causal process.

Again, quasi-causation is essentially the mere possibility of genuine causation (P2). Such prevention does involve some actual genuine causation (P1), but this is not a relation between A and B, nor A and not-B.

In the case of omissions, we have quasi-causation by omission whenever not-A quasi-causes B, where A and B are positive events or facts, and not-A is an "act of omission":

Omission: not-A quasi-caused B if B occurred and A did not, and there occurred an x such that
(O1) x caused B, and
(O2) if A had occurred then A would have prevented B by interacting with x where A and B name positive events/facts and x is a variable ranging over facts or events, and where prevention is analyzed as above.

To illustrate, "I caused the terrorist attack in London by failing to travel to the UK and blow up the terrorists' van before they could do any damage" is true in virtue of the fact that I didn't blow up the terrorists' van, the attack did occur, and there was a genuine causal process – the terrorists carrying through their plan – and the truth of the counterfactual "Had I interacted with that process by blowing up the van, that would have prevented the attack."

More complex cases, which we will pass over here, include cases of prevention by omission where had A occurred, it would have quasi-caused B (one reason why the definition above is just a sufficient condition), prevention by prevention (where both relata are positive but there are some quasi-causal links), and the preventing of such prevention-by-prevention, where A prevents B, but in more complex fashion than that described above (this is why the definition of prevention is given just as a sufficient condition) (see Dowe, 2000, ch. 6).

The counterfactual account of quasi-causation presumes nothing about causation itself. Notice that causation appears in the definition – in other words, is primitive in the theory – so quasi-causation is defined in terms of genuine causation. For this reason, the account is compatible with any theory of causation – one can plug any theory into the definition. This shows us why the counterfactual account of quasi-causation is compatible with any theory of causation. If causation is x, we merely stipulate that negatives may not enter into causation, and analyze quasi-causation as counterfactuals about x. This will work even for the counterfactual theory: moreover, it solves the problem that Lewis (1986) has with negative events.

But the account does nothing yet to explain why we have a variety of intuitions about cases of quasi-causation. It takes all our cases to be cases of quasi-causation, not causation. We now turn to this question.

9.4 Epistemic Blur and Practical Equivalence

First, although the theoretical distinction is clear enough, in practice it may not always be clear to us whether an event or fact in question is positive or negative. Events that we think of as negative may turn out really to be positive. In those cases, apparent omissions and preventions turn out to be cases of ordinary genuine causation. Alternatively, apparently positive events may turn out to be negative events and, consequently, cases of apparently genuine causation may turn out to be omissions or preventions. The latter is especially convincing. Take the case of "causing" drowning. Actually, this is quasi-causation, since holding her head under water prevents her from getting oxygen. Thus there is an epistemic blur between quasi-causation and causation.

So, preventions and omissions may be very much more commonplace than we commonly recognize. "Smoking causes heart disease," but perhaps the actual effect of smoke is to prevent normal processes from impacting certain cells in a certain way, so that, in the absence of those processes, diseased cells prosper (causation by omission).

These considerations are of merely epistemic concern. We may not know whether a given case is a prevention or genuine causation, but the conceptual distinction between genuine causation and omissions/preventions is clear enough.

But this widespread uncertainty shows why it is useful practically to treat causation and quasi-causation as if they were the same thing. And the epistemic blur may explain why we might not immediately see the distinction.

Secondly, causation and quasi-causation play very similar practical roles. Negatives (negative facts or events), when they figure in quasi-causation, can be ends and means, and can raise chances. As well as serving as means and ends, since they raise chances they can be evidence for their quasi-effects and quasi-causes, and they can also feature in explanation. Arguably, quasi-causation may also, subject to "pragmatic" considerations, track moral responsibility in just the way that causation does. This is why it does not matter that for practical purposes we don't bother to, or can't, distinguish quasi-causation from causation. The distinction only becomes important theoretically, in metaphysics.

The counterfactual theory can also explain why there is a practical equivalence, given a significant theoretical difference. The unity of causation and quasi-causation lies in the fact that, in essence, quasi-causation is possible causation. That a causes b is not the same thing as the nonactual possibility that a causes b. But if a might have caused b, then a's absence explains and is evidence for b's absence, and bringing about a's absence was a good strategy for ensuring the absence of b.

9.5 Explaining Intuitions

It remains to draw together the argument that *negative-excluding* theories explain the range of intuitions better than *negative-friendly* theories. First, pragmatic considerations appealing to knowledge, intention, duties of care, and remoteness are available to *negative-excluding* theories as much as to *negative-friendly* theories. Any difference of this sort that can be used to explain the cases that are counterintuitively causation in virtue of the obtaining of such factors can, on the other side of the coin, also be used to explain cases that counterintuitively are not causation in virtue of the lack of the same factors.

Secondly, the two facts outlined in the previous section – the epistemic blur and the practical equivalence – furnish the answer to cases such as (A) "I caused her death by holding her head under water for five minutes." The answer is that the epistemic problem and the practical equivalence together suggest that we take our undoubted intuition of causation as an intuition that the case is either causation or quasi-causation, rather than as an intuition that it is causation not quasi-causation. It's hard to see how folk could intuit the latter given the deep epistemic problem, and it's hard to see why folk would have such as intuition as the latter given the practical equivalence.

Thirdly, there is a straightforward explanation for the difficult case of *negative-friendly* theories; that is, as to why we think that cases such as (H), my not throwing the rock causing the window not to break, intuitively are not causation. The explanation is simply that this is not causation, as directly entailed by *negative-excluding* theories. The details will depend on which theory of causation we are trying to defend, but will involve features such as the fact that nothing happened and that there is no process between the two "events." I submit that these factors place the *negative-excluding* theories ahead of *negative-friendly* theories in explaining the range of differing intuitions that we have about quasi-causation.

Bibliography

Armstrong, D. 1997: *A World of States of Affairs*. Cambridge: Cambridge University Press.
—— 1999: The open door. In H. Sankey (ed.), *Causation and Laws of Nature*. Dordrecht: Kluwer, 175–85.
Aronson, J. 1971: On the grammar of "cause." *Synthese*, 22, 414–30.
Beebee, H. unpublished: Causes, omissions and conditions.
Bennett, J. 1995: *The Act Itself*. New York: The Clarendon Press.
Bigelow, J. and Pargetter, R. 1990: *Science and Necessity*. Cambridge: Cambridge University Press.
Collingwood, R. 1974: Three senses of the word "cause." In T. Beauchamp (ed.), *Philosophical Problems of Causation*. Encino, CA: Dickenson, 118–26.
Dowe, P. 2000: *Physical Causation*. New York: Cambridge University Press.
—— 2001: A counterfactual theory of prevention and "causation" by omission. *Australasian Journal of Philosophy* 79(2), 216–26.
Ehring, D. 1998: *Causation and Persistence*. Oxford: Oxford University Press.

Fair, D. 1979: Causation and the flow of energy. *Erkenntnis*, 14, 219–50.

Gasking, D. 1996: *Language, Logic and Causation*. Melbourne: Melbourne University Press.

Glover, J. 1977: *Causing Death and Saving Lives*. Harmondsworth: Penguin.

Hart, H. and Honoré, T. 1985: *Causation in the Law*. Oxford: The Clarendon Press.

Hausman, D. 1998: *Causal Asymmetries*. New York: Cambridge University Press.

Lewis, D. 1986: *Philosophical Papers*, vol. II. New York: Oxford University Press.

—— forthcoming: Void and object. In J. Collins, N. Hall, and L. A. Paul (eds.), *Causation and Counterfactuals*. Cambridge, MA: The MIT Press.

Mellor, D. 1995: *The Facts of Causation*. London: Routledge.

Salmon, W. 1997: Causality and explanation: a reply to two critiques. *Philosophy of Science*, 64, 461–77.

—— 1998: *Causality and Explanation*. New York: Oxford University Press.

Suppes, P. 1970: *A Probabilistic Theory of Causality*. Amsterdam: North Holland.

von Wright, G. 1971: *Explanation and Understanding*. Ithaca, NY: Cornell University Press.

Further reading

Dowe, P., Causal processes. In E. Zalta (ed.), *Stanford Encyclopedia of Philosophy*. Stanford University: http://plato.stanford.edu/entries/causation-process/causation-process.html

Hart, H. and Honoré, T. 1985: *Causation in the Law*. Oxford: The Clarendon Press.

Salmon, W. 1998: *Causality and Explanation*. New York: Oxford University Press.

Causes need not be Physically Connected to their Effects: The Case for Negative Causation

Jonathan Schaffer

Negative causation occurs when an *absence* serves as cause, effect, or causal intermediary. Negative causation is genuine causation, or so I shall argue. It involves no physical connection between cause and effect. Thus causes need not be physically connected to their effects.

10.1 Negative Causation

The terrorist presses the detonator button, and the bomb explodes. Here is *causation*.

But I have not said *which way* the detonator is wired. Perhaps pressing the button generates an electrical current that connects to the bomb and triggers the explosion:[1]

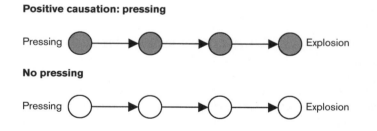

---

1 What follows are *neuron diagrams*, which are a useful way to represent causal structures. The conventions are as follows. Filled circles doubly represent firing neurons and occurring events. Unfilled circles doubly represent nonfiring neurons and absences. Arrows doubly represent stimulatory synapses and physical connections. Lines headed with black dots doubly represent inhibitory synapses and preventions (which are *not* physical connections in the relevant sense: section 10.2). If two neurons are connected by a stimulatory connection and the first fires, then the second will fire unless some other neuron inhibits it.

Or perhaps pressing the button causes the absence of an inhibiting shield that had been preventing the source current from triggering the explosion:

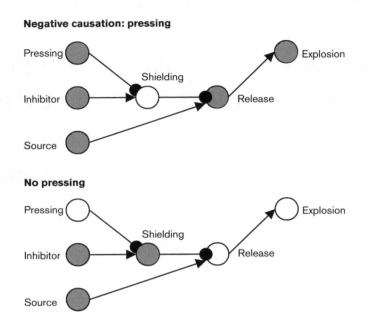

Negative causation: pressing

Pressing

Shielding

Inhibitor

Release

Source

Explosion

No pressing

Pressing

Shielding

Inhibitor

Release

Source

Explosion

If the detonator is wired as per the second diagram, then the case is one of negative causation, where an absence (the unfilled circle) serves as causal intermediary. Does it really matter, causally, which way the detonator is wired?

Whichever way the detonator is wired, the explosion is *counterfactually dependent* on the pressing. That is, the following counterfactual conditional is true: if the terrorist had not pressed the button, then the bomb would not have exploded. Such counterfactual dependence is sometimes thought to be constitutive of causation, and serves as the standard test for causation used in the legal system.

Whichever way the detonator is wired, the pressing is of a type that is *statistically relevant* to explosions in such circumstances. That is, the following probabilistic inequality holds: the probability of an explosion given the circumstances and a button pressing is greater than the probability of an explosion given the circumstances and no button pressing. Such statistical relevance is sometimes thought to be constitutive of causation, and serves as the basis by which statisticians infer causal relationships.

Whichever way the detonator is wired, the pressing serves as an *agential means* to achieve the explosion. That is, someone (for instance, a terrorist) whose end is to explode the bomb may manipulate the button to that end. Such agential means is sometimes thought to be constitutive of causation, and provides the original wellspring of our causal concept.

Whichever way the detonator is wired, the pressing may provide *predictive evidence* that the bomb will explode. That is, if one knows that the terrorist has pressed the button, then one is in position to know that the bomb will explode. (Likewise, the explosion may provide *retrodictive evidence* that the terrorist did press the button.) The scientist would be able to use the button pressing to predict an explosion. Such a prediction is a causal prediction, and thus would not be possible unless the pressing and explosion were causally related.

Whichever way the detonator is wired, the pressing may help to *explain why* the explosion occurred. That is, if one asks "Why did the bomb explode?" then someone may informatively answer "Because the terrorist pressed the button." The government would certainly cite the button pressing to explain the explosion. Such an explanation is a causal explanation, and thus would not be possible unless the pressing caused the explosion.

Whichever way the detonator is wired, the terrorist is *morally responsible* for the explosion. That is, pressing the button exposes someone to blame (or praise) for the consequences of the explosion. The judge would hold the terrorist morally responsible, especially if anyone was hurt. Such responsibility is causal responsibility, and thus would not be possible unless the pressing did cause the explosion. In summary, whichever way the detonator is wired, all the central *conceptual connotations* of causation are satisfied, including counterfactual dependence, statistical relevance, agential means, inferential evidence, explanatory grounds, and moral responsibility.

Still not convinced? The pattern of negative causation features in even the most *paradigmatically causal* cases. Suppose that the sniper feels murderous, pulls the trigger, fires a bullet through the victim's heart, and the victim dies. Here is a paradigmatic causal sequence, *every step of which* is negative causation.

Workings backwards, *surely* the firing of the bullet through the victim's heart causes the victim to die. But heart damage only causes death by negative causation: heart damage (*c*) causes an absence of oxygenated blood flow to the brain (~*d*), which causes the cells to starve (*e*). The mechanism of death can thus be represented as:

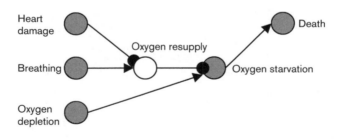

The Center for Disease Control (fully aware of the wiring) identifies heart disease as "the leading cause of death" in the United States.

At the next step backwards, *surely* the pulling of the trigger causes the bullet to fire. But trigger pullings only cause bullet firings by negative causation: pulling the trigger (*c*) causes the removal of the sear from the path of the spring (~*d*), which causes the spring to uncoil, thereby compressing the gunpowder and causing an explosion, which causes the bullet to fire (*e*). This may be seen in the following blueprint for a gun:

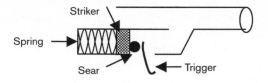

The mechanism by which bullets are fired can thus be represented as:

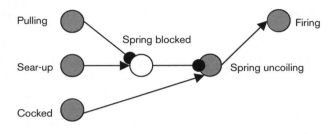

Even the National Rifle Association, which insists that "guns don't kill people, people kill people," concedes thereby that people who fire guns can cause death.

At the third and final step backwards, *surely* the sniper's feeling murderous causes him to pull the trigger. But nerve signals only cause muscle contractions (such as that of the sniper's trigger finger) by negative causation: the firing of the nerve (*c*) causes a calcium cascade through the muscle fiber, which causes calcium–troponin binding, which causes the removal of tropomyosin from the binding sites on the actin (~*d*), which causes myosin–actin binding, and thereby causes the actin to be pulled in and the muscle to contract (*e*). This may be seen in the following blueprint for a muscle fiber:

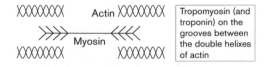

The mechanism by which muscles contract can thus be represented as:

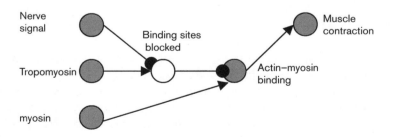

Since all voluntary human actions are due to muscle contractions, it follows that *all voluntary human actions* (perhaps the most paradigmatic of all causes) involve negative causation.

Other paradigm cases of causation involving negative causation are cases of negligence and breech of contract, which are *explicitly recognized as causal by the law*. In this vein, H. L. A. Hart and Tony Honoré note:

> There are frequent contexts when the *failure* to initiate or interrupt some physical process; the failure to provide reasons or draw attention to reasons which might influence the conduct of others; and the failure to provide others with opportunities for doing certain things or actively depriving them of such opportunities are thought of in causal terms. (1985, pp. 2–3)

Here is one of the many examples that Hart and Honoré then cite: "So a failure to deliver to a manufacturer on time a piece of machinery which he has ordered, may, like the destruction of the existing machinery, be held the cause of the loss of the profits which would have been made by its use" (1985, pp. 59–60).

Other paradigm cases of causation involving negative causation are cases that fall under the *causative verbs of ordinary language*. These include cases of removing, releasing, unveiling, unblocking, untying, unplugging, severing, disconnecting, letting go, cutting loose, switching off, and so on. In summary, negative causation features in the most paradigmatic cases of causation, as is supported by our intuitions, the law, and ordinary language.

Still not convinced? Negative causation is required by the most useful *theoretical applications* of causation. Saul Kripke (1972) proposes that *reference* is a causal notion: the reference of names is transmitted via causal chains. Kripke's idea is that the reference of a name is not a matter of what description a speaker would associate with the name but, rather, of the causal chain by which the name was produced, transmitted, and ultimately entered into the speaker's lexicon. It should be obvious that this transmission process is indifferent to positive versus negative causation. If a name is printed in a book, then its reference may be transmitted thereby, whichever way the printing press is wired.

Brian Skyrms (1980) and David Lewis (1981) argue that *rational decision-making* is a causal affair: one needs to calculate the expected effects of candidate actions. Skyrms's and Lewis's idea is that rational decision-making is not a matter of calculating the expectations *per se* but, rather, a matter of calculating the expected effects. It should be obvious that this calculation is indifferent to positive versus negative causation. If one's overriding end is to explode the bomb, then one will act rationally by pressing the button, whichever way the detonator is wired.

Alvin Goldman (1977) maintains that *perception* is a causal affair: perception is a species of causal relation by which the world acts on the mind. Goldman explicitly notes that one can perceive black holes without any energy coming from them (pp. 281–2). In fact, black holes cause an absence of light that would otherwise be visible. Michael Tye then adds that the point extends to any perfectly black object: "This difficulty is not peculiar to astronomical contexts. It seems to me that perfectly black objects which are not too small or too distant may be seen with the naked eye provided that they are located against light backgrounds" (1982, p. 324). Thus the fact that one can see both white (positive) and black (negative) shows that perception is indifferent to positive versus negative causation. In summary, negative causation is required by the most useful theoretical applications of causation, including applications to the theories of reference, decision, and perception.

Still not convinced? Negative causation is routinely recognized in *scientific practice*. Psychologists routinely invoke negative causation, such as in hormonal theories of sexuality:

> The theory is that androgen causes masculine behavior and its absence causes feminine behavior . . . [M]ale rats were deprived of androgens by castration or by treatment with anti-androgenic drugs, which was seen to result in the later manifestation of the female pattern of lordosis . . . Thus a simple causal chain was established between the sexual behavior of animals such as guinea-pigs and rats and their hormonally sexually differentiated bodies. (Roberts, 2000, p. 2)

Biologists routinely invoke negative causation, such as in explaining disease. What causes scurvy is an absence of vitamin C, what causes rickets is an absence of vitamin D, what causes diabetes milletus is an absence of insulin, and what causes dwarfism is an absence of growth hormone, and so on. The way in which HIV causes death is by disconnecting the immune system – an absence of functioning CD4+ T-cells serve as causal intermediary, allowing opportunistic infections and cancers to spread unchecked. One finds explicit discussion of the chains of negative causation involved, for instance, in Duchenne's muscular dystrophy:

> Muscular dystrophy in the *mdx* mouse has been described as a mutation in a colony of C57B1/10ScSn mice, which results in the absence of the 427 kDa membrane-associated protein dystrophin . . . A deletion on the human X-chromosome causes the absence of an analogous protein and leads to Duchenne muscular dystrophy (DMD) . . . The absence of dystrophin leads to the destabilization of [the transmembrane glycoprotein complex], yielding weaker muscle fibers that undergo progressive degeneration followed by massive necrosis. Ultimately, premature death of DMD patients occurs . . . (Caceres et al., 2000, p. 173)

Nor is the invocation of negative causation limited to psychology and biology. Chemists routinely invoke negative causation, such as in acid–base reactions. The addition of NH_3 (base) to H_2O (acid) causes the formation of OH^- (together with NH_4^+) because the H_2O loses a proton. Likewise, in oxidation–reduction reactions, reduction is understood as "a process in which the oxidation state of some element decreases," as for instance when it loses O atoms. Physicists routinely invoke negative causation, such as in characterizing the process of "electron–hole pair generation":

> When an electron (which is a negative charge carrier) is freed from the atom, it leaves behind a *hole*, or the absence of an electron (which acts as a positive charge carrier). Free carriers are generated when electrons have gained enough energy to escape their bonds to the atom and move from the *valence band* to the *conduction band*. This process is called "electron–hole pair generation". Electron–hole pairs can be created by any mechanism which delivers sufficient energy to an electron, including absorbing energy from light (as in a photo diode) and thermal excitation (absorbing heat energy). (Mason, 2000, p. 4)

Jonathan Schaffer

Such electron–hole pair generation is routinely understood as causal: "The electron absence created by this process is called a hole." And: "These positively-charged holes can cause a catastrophic negative shift in the threshold voltage of the device" (Wall and Macdonald, 1993). In summary, negative causation is routinely recognized in scientific practice, including throughout the practice of psychology, biology, chemistry, and physics.

Bringing the results of this section together, here is the full case for negative causation. First, negative causation is supported by all the central conceptual connotations of causation, including counterfactual, statistical, agential, evidential, explanatory, and moral connotations. Secondly, negative causation features in paradigm cases of causation including heart failure, gun firings, and all voluntary human actions, and is considered causal by the law and by ordinary language. Thirdly, negative causation is required by the most useful theoretical applications of causation, including applications to the theories of reference, decision, and perception. And, fourthly, negative causation is routinely recognized in scientific practice, including the practice of psychologists, biologists, chemists, and physicists. What stronger case can be made that *anything* is causal?

10.2 Physical Connection

David Hume (1748) glossed our naïve conception of causation as that of *necessary connection*. While Hume thought the connection not in the objects but, rather, projected by the mind, a number of subsequent philosophers have sought a causal connection in the objects via *physical connection* such as energy flow. These philosophers have maintained that causes need to be physically connected to their effects.

Negative causation refutes this program. Causation may well be objective, but it does not require physical connection.

There are three thematically related versions of the physical connections view. First, there is the idea that causation requires *transference*, of a property, or more specifically of energy-momentum. This idea has been developed by such philosophers as Jerrold Aronson (1971), David Fair (1979), and Hector-Neri Castaneda (1984).

Secondly, there is the idea that causation requires *processes*. This idea traces back to Bertrand Russell (1948). It was developed by Wesley Salmon (1984), who characterizes a causal process as a continuous qualitative persistence that is capable of transmitting a mark, of propagating structure. This idea was further developed by Phil Dowe (1992, 1995, 2000; see also Salmon, 1994, 1998), who characterizes a causal process as the world-line of an enduring conserved-quantity-bearing object.

Thirdly, there is the idea that causation requires an intrinsic *tie*. This idea has been developed by J. L. Mackie (1974). Douglas Ehring (1997) specifies this tie as the persistence line of a trope, and Max Kistler (1998, 2001) further develops this thought, while bringing it closer to Dowe's view, by restricting the persisting tropes to those of conserved quantities.

These three approaches owe their distinctive aspects as much to historical pedigree as to philosophical difference. All understand physical connections as *lines of persistence*. They differ only in what is said to persist: unspecified for Russell and Mackie, properties for Aronson, tropes for Ehring and Kistler, energy for Fair and Castaneda, structure for Salmon, and objects (those instantiating conserved quantities) for Dowe.

Thus the question of whether causes need to be physically connected to their effects becomes: Must there be lines of persistence from cause to effect? That is, must there be anything like energy flowing from cause to effect?

It is obvious that these research programs have no room for negative causation, since negative causation involves *no persistence line* between cause and effect. As Fair explains, "Omissions are non-occurrences. They obviously cannot be [physically connected] because, being only possible, non-occurrences cannot be the sources or the sinks of actual energy-momentum" (1979, p. 246). Thus when heart damage causes death, there will be no energy flow or other persistence line between the stopped heart and the starving brain. Rather, what is causally salient here is the *absence* of a physical connection. If only there had been a physical connection between heart and brain in the form of a flow of oxygenated blood, then the victim would have lived.

The physical connection view of causation may seem plausible if one concentrates on colliding billiard balls, or other cases of connection. But negative causation reveals the view to be a *hasty generalization*. Not all cases of causation involve physical connection. There is more than one way to wire a causal mechanism.

At this point, the road diverges. The physical connections theorist may *accept* the lesson of negative causation, grant that causes need not be physically connected to their effects, and hope that the notion of physical connection may still prove useful in some more complex account of causation. Fair takes this road, viewing the genuineness of negative causation as more or less beyond dispute:

> I think a flagpole causes its shadow as an omission; the failure of the flagpole to trans-mit incident light causes the failure of the light to reach the shadow region. There is little question about the truth of this statement simply because it is very plausible that if the flagpole were removed, the light would reach the shadow region. Similarly, the ice on the road caused the auto accident because the road failed to transmit its usual fric-tional force to the tires. (1979, p. 248)

Fair then offers a more complex account of causation in terms of counterfactuals about physical connection. Causes, on this account, need no longer be physically connected to their effects. It is enough, to take the case in which both cause and effect are absences, that if the cause had occurred, then it *would have been* physically connected to an occurrence of the effect.

The other road open to the physical connections theorist is to *reject* the lesson of negative causation, insist that causes need to be physically connected to their effects, and swallow the consequences. Aronson takes this road, dismissing negative causa-tion outright:

> Consider a weight that is attached to a stretched spring. At a certain time, the catch that holds the spring taut is released, and the weight begins immediately to accelerate. One might be tempted to say that the release of the catch was the cause of the weight's accel-eration. If so, then what did the release of the catch transfer to the weight? Nothing, of course. (1971, p. 425; see also Armstrong, 1999; Dowe, 2000, 2001; Kistler, 2001)

Such a response might have been tolerable were the consequences limited to the isolated case of launching a weight by a spring. But the consequences are not so limited. As shown in section 10.1, to dismiss negative causation is to swallow the following: counterfactual, statistical, agential, evidential, explanatory, and moral implications are not marks of causation; the folk are wrong that voluntary human action is causal, the law is wrong that negligence is causal, ordinary language is wrong that "remove," "release," "disconnect," and so on are causal; philosophers are wrong that reference, decision, and perception are causal; and scientists are wrong that electron–hole pair generation and other negative processes are causal. I submit that no theory so dismissive deserves to be considered a theory of *causation*.

10.3 Meaning and Method

" 'When *I* use a word,' Humpty Dumpty said in a rather scornful tone, 'it means just what I choose it to mean – neither more nor less.' " If Humpty Dumpty were a physical connections theorist, he might have added "When *I* use the word 'causes' it means just physical connection – neither more nor less." The sequel would of course be: " 'When I make a word do a lot of work like that,' said Humpty Dumpty, 'I always pay it extra.' "

So far, I have argued that negative causation proves that causes need not be physically connected to their effects. The argument that negative causation is genuine causation involved considering the conceptual connotations, paradigm cases, theoretical applications, and scientific practices concerning causation (section 10.1). But why should these considerations *matter*? That is, suppose that some philosopher simply *refused* to countenance negative causation, and insisted that causes must be physically connected to their effects, maintaining this thesis against even the most damning counterexamples, swallowing whatever absurdities might arise. "When I use the word 'causes,' " this philosopher might say "it means just physical connection."

Or suppose some truly crazed philosopher were to say "When I use the word 'causes,' it means just being over a mile apart." What might one say to such a philosopher? One could point out that none of the connotations of causation – counterfactual, statistical, agential, and so on – require being over a mile apart. One could wheel out paradigm cases of causation in which cause and effect are not over a mile apart. One could point out that none of the theoretical applications of causation to reference, decision, and perception require being over a mile apart. One could mention that scientific practice in attributing causation does not require being over a mile apart. But if this philosopher remained *unmoved*, what more could one say?

Perhaps the crazed philosopher who defined "causes" in terms of being over a mile apart was merely engaged in *stipulation*. Such a stipulation may seem pointless and confusing, but it is still legitimate as such (especially so long as the word *is* paid extra). It is also of no further interest: mere *wordplay* cannot reveal the real structure of the world.

If this crazed philosopher were engaged in *description* of our actual causal concept, though, then he would have missed the mark by miles. There is an excellent method for explicating the meanings of our actual concepts. It is the method of *functional definition* (Lewis, 1970). Roughly speaking, a functional definition of "causes" may be obtained by conjoining our most central platitudes involving the concept, replacing the term "causes" by the variable R, and uniquely existentially quantifying over the conjunction. The best satisfier of this definition then deserves to be considered the best candidate to be the meaning of our actual concept.

The central platitudes involving causation are to be drawn from the conceptual connotations, paradigm cases, and theoretical applications of the concept. Thus the functional definition of causation will look something like:

> There exists a unique relation R such that: R is associated with counterfactual dependence & R is associated with statistical relevance & R is associated with agential means & R is necessary for inferential evidence & R is necessary for explanation & R is necessary for moral responsibility & . . . & R holds between heart damage and death & R holds between trigger pullings and gun firings & R holds between volitions and actions & . . . & R secures the reference of names & R is involved in rational decision & R is the genera of perception & . . .)

This is why the definition of "causes" in terms of being over a mile apart has no claim to descriptive adequacy. It violates virtually all the platitudes. And for *exactly* this reason, the definition of "causes" in terms of physical connection has no claim to descriptive adequacy, either. (Indeed, both definitions look to be about *equally* amiss.)

It may turn out, in the end, that no account of causation is perfect. Perhaps all violate some platitudes. Functional definitions allow for this possibility, by merely asking for *best* satisfiers. Negative causation is not merely a counterexample to the thesis that causes must be physically connected to their effects. There is a deeper point; namely, that any account of causation that requires physical connection between causes and effects is so far off the mark that it is *not even in the running* for the meaning of "causes."

Not all philosophers will accept functional definitions as explicating the meaning of "causes," though. The main alternative is to think of "causes" as a *natural kind term*, whose meaning (or essence or whatnot) is fixed, as a matter of *a posteriori* necessity, by the actual nature of causal relatedness. Many physical connections theorists hold this view, including Fair: "The hypothesized relationship between causation and energy-momentum flows is expected to have the logical status of an empirically discovered identity, namely that the causal relation is identical with a certain physically specifiable relation" (1979, p. 231; see also Aronson, 1982; Castaneda, 1984, p. 23). So just as chemists are said to have discovered that water is H_2O, so physicists are said to have discovered that causation is energy flow or some other species of physical connection.

To begin with, I believe that the natural kind term view of "causes" is just semantically wrong – "causes" is *not* a natural kind term in the way that "water" is. The mark of a natural kind term is the way in which we use it to describe counterfactual situations: given that the actual nature of natural kind k is x, we refuse to call counterfactual things that are k-like but not x by the k term. For instance, we refuse to use "water" to describe a hypothetical water-like liquid made of XYZ rather than H_2O (Putnam, 1975; see also Kripke, 1972). "Causes" does not work this way. We do not hesitate to use "causes" to describe a hypothetical spell that works by magic rather than by energy. Indeed, the concept of causation belongs to the *nomic family* of concepts, alongside lawhood, explanation, disposition, and so on. *None* of the terms for nomic concepts have the characteristic counterfactual usage profile of natural kind terms.

But suppose for the sake of argument that "causes" *is* a natural kind term. And suppose that some philosopher insisted that the nature of causation is energy flow (or some other physical connection), dismissing even the most damning counterexamples as merely *intuitive*, ignoring whatever absurdities may arise as merely *conceptual*. "The nature of causation is energy flow," this philosopher might intone "and no mere human intuitions or concepts can affect this."

Or suppose that some truly crazed philosopher were to say "The nature of causation is being over a mile apart, and no mere human intuitions or concepts can affect this." What might one say to such a philosopher? One could point out that none of the connotations of causation – counterfactual, statistical, agential, and so on – require being over a mile apart. One could wheel out paradigm cases of causation in which cause and effect are under a mile apart. One could point out that none of the theoretical applications of causation to reference, decision, and perception require being over a mile apart. One could mention that scientific practice in attributing causation does not require being over a mile apart. But if this philosopher remained *unmoved*, what more could one say?

Supposing, for the sake of argument, that causation is a natural kind, the question arises: *Which*? The crazed philosopher who identifies causation with being over a mile apart has obviously shot wide. Why so? There are three main views as to *which* natural kind a given term targets. The first view, Kripkean in spirit, is that a natural kind term refers to the dominant kind it was targeted at when the term was *born*. The Kripkean view can explain why the relation of being over a mile apart has no claim to be the actual nature of causation – the dominant kind that "causes" was originally targeted at includes events under a mile apart. And for *exactly* this reason, the Kripkean view rules that the relation of physical connection has no claim to be the actual nature of causation either. The dominant kind that "causes" was originally targeted at includes voluntary human actions without physical connection. Indeed, our term "cause" is the etymological descendant of the Latin "*causa*" and the Greek "$\alpha i \tau i \alpha$," whose aboriginal use (which still survives in the law) is to identify grounds for moral complaint.

The second main view as to which natural kind a given term targets is the view, Putnamian in spirit, that a natural kind term refers to what the *community experts* now use it to cover. The Putnamian view can explain why the relation of being over a mile apart has no claim to be the actual nature of causation – the expert scientists in our community do not use "causes" in this way. And for *exactly* this reason, the Putnamian view rules that the relation of physical connection has no claim to be the actual nature of causation either. Psychologists, biologists, chemists, and physicists routinely use "causes" to cover negative causation, as when biologists speak of the absence of vitamin D as the cause of rickets, or when physicists speak of the causes and effects of electron holes in semiconductors.

The third main view as to which natural kind a given term targets is the view, Kaplanesque in spirit, that a natural kind term *rigidly* refers to the actual extension of its functional definition. The Kaplanesque view can explain why the relation of being over a mile apart has no claim to be the actual nature of causation – the functional definition does not actually have this extension. And for *exactly* this reason, the Kaplanesque view rules that the relation of physical connection has no claim to be the actual nature of causation either. The actual extension of the functional definition of "causes" includes such cases of heart failures, gun firings, and voluntary actions, where there is no physical connection.

Of course, both the physical connections theorist and the crazed miles-away theorist might reject both functional definition and natural kind views. But the onus is on them, at this point, to say what possible conception of meaning could justify their theories. I conclude that there is no known and decent conception of meaning that could possibly *justify* the physical connections view (or even accord it *any* higher standing than the crazed miles-away view).

So far, I have spoken about meaning. But the situation would be essentially the same if one abjured the search for meaning, and simply proposed an *empirical specification* of actual-world causation in terms of actual physical connection, or in terms of actual distances of over a mile. The question would arise: Why *that*?

Dowe, for instance, has proposed to empirically specify actual-world causation as physical connection (understood in terms of the world-lines of objects that instantiate conserved quantities). But why *that*? Dowe is laudably explicit about his method. Empirical specification is to draw on science: "The empirical analyst can reply that there are procedures for investigating such an entity [as objective causation], namely, the methods of science, which is in the business of investigating language-independent objects. Empirical philosophy can draw on the results of science, . . ." (2000, p. 7). In particular, empirical specification is to draw on scientific judgments about the use of a term, so that it may be viewed as "a conceptual analysis of a concept inherent in scientific theories" (2000, p. 11).

It follows that Dowe's method is similar to the Putnamian view mentioned above, in that it looks to the usage of the experts (but without the assumption that "causes" is a natural kind term). It follows for this reason that the relation of being over a mile apart has no claim to be the empirical specification of causation – the expert scientists in our community do not use "causes" in this way. And for *exactly* this reason, Dowe's method rules that the relation of physical connection has no claim to be the empirical specification of causation either. Expert psychologists, biologists, chemists, and physicists routinely use "causes" to cover cases of negative causation.

To understand Dowe's method better, it might help to understand why Dowe himself rejects alternative views such as statistical relevance. Here Dowe provides a *counterexample* involving a certain radioactive decay process (2000, pp. 33–40). Dowe's counterexample is an extension of an *intuitive* counterexample involving a squirrel kicking a golf ball, but the radioactive decay version "is an idealization of a real physical nuclear decay scheme" (2000, p. 33), which the *physicist* Enge has glossed with the causal-sounding phrase "production process" (2000, p. 38). Well and good, I say. I only ask that the empirical philosopher not ignore intuitively paradigmatic cases of negative causation, and not ignore the testimony of psychologists, biologists when discussing the causes of disease, chemists, and physicists who speak of "electron–hole pair generation."

10.4 Dowe's Arguments

In light of negative causation, why would anyone maintain that causes must be physically connected to their effects? That is, what possible considerations could convince someone that heart failures cannot cause death, that trigger pullings cannot cause bullet firings, that desires cannot cause actions, and so on? Likewise, what possible considerations could convince someone that decapitating someone (thereby stopping an influx of oxygenated blood) cannot cause that person to die?

Dowe (2000, 2001, and see chapter 9 of this volume) is perhaps the leading apologist for the physical connections view. Dowe deserves praise for being the first physical connections theorist to offer explicit *arguments* against negative causation, rather than merely biting so deadly a bullet. Dowe's overall strategy is to define a relation of "quasi-causation" in terms of merely possible physical connection, and then claim that negative causation is merely quasi-causal.

Dowe offers exactly two arguments for downgrading negative causation to the status of quasi-causation, the first of which is that he feels an *intuition of difference* between presence and absence cases:

> There are cases where we have the intuition that an instance of quasi-causation is just that, "quasi", and not strictly speaking a genuine case of causation. "The father caused the accident by failing to guard the child." It's natural enough to use the word "cause" here, but when we consider the fact that the child's running onto the road was clearly the cause of the accident, but that the father did nothing to the child, in particular that the father did not cause the child to run onto the road, then one has the feeling that this is not a real, literal case of causation. So I claim that we do recognize, on reflection, that certain cases of prevention or omission are not genuine cases of causation. I call this the "intuition of difference". (2001, pp. 217–8)

Dowe recognizes that there are also cases, such as when chopping off someone's head *seems* to cause them to die, in which we have strong intuitions of genuine causation. But he criticizes the view that negative causation is genuinely causal as running afoul of the intuition of difference:

> According to the genuinists – who say "quasi"-causation is genuine causation – the first intuition [the intuition of difference] can be dismissed but the second [the intuition of genuine causation] is important. Indeed, it would be fair to say that the main argument of the generalists (so far) is simply to cite intuitions of the second kind. However, without further explanation, this is an inadequate response to the existence of the intuition of difference, at least for those genuinists who purport to respect folk intuitions in their philosophical theorizing. Some account of the intuition of difference is required, and articulation and reiteration of the genuinist intuition does nothing to answer the point. (2001, p. 218)

In summary, Dowe maintains that our intuitions are conflicted here, and that only the quasi-causation view can explain this.

I offer three responses to Dowe's argument from the intuition of difference, the first of which is that Dowe has *misdescribed his own intuitions*. Look closely at the first Dowe quote from the previous paragraph. Dowe begins with a case of parental negligence. He then immediately *acknowledges* that "It's natural enough to use the word 'cause' here."[2] But this is just to *acknowledge* that the case is intuitively *causal*. So much for intuitions.

What happens next in the passage under consideration is that Dowe presents a miniature *theoretical argument* to try to *overturn* our natural intuition of causality. This argument is not only beside the point as far as intuitions, but it is a cheat for two reasons. The first reason is that this argument implicitly assumes the physical connection view in sliding from "the father did nothing to the child" to "this is not a real, literal case of causation." The corrective would be to note that by doing nothing the father thereby negligently *abandoned* the child, and that if he hadn't the child would have been safe, which is what makes this a real case of causation. The second reason why Dowe's argument is a cheat is that it turns on the misleading expression "the cause," in "the child's running onto the road was clearly the cause of the accident." It is well known that this sort of locution invites people to make invidious distinctions between genuine causes.[3] The corrective would be to acknowledge that the child's running onto the road was *a cause* of the accident, as was the father's negligence. Thus it comes as no surprise that Dowe reverses his natural intuitions at this point, driven by his theory and the invidious terminology of "the" cause. In summary, a close reading of Dowe's own presentation reveals *no intuition of difference whatsoever*; rather, it reveals an intuition of genuine causality attacked by a theory-laden and misleading argument.

The second response that I offer to Dowe's argument from the intuition of difference is that he has *understated the case for genuinism*, in two respects. First, genuinists do more than "simply to cite intuitions of the second kind." The arguments of section 10.1 invoke not just paradigm case intuitions but also conceptual connotations, theoretical applications, and scientific practice. Moreover, as emerges in section 10.3, these are the considerations that determine the very meaning of "causes," and that are supposed to guide Dowe's own method of empirical specification. To be fair, Dowe explicitly limits his claim to "the genuinists (so far)." But in any case it should be clear that the genuinist can and should do far more than toss off a handful of intuitions.

The second respect in which Dowe's argument from the intuition of difference understates the case for genuinism is that it treats all intuitions as if they were equal. But really intuitions come in *degrees*: some intuitions are paradigmatic near-certainties; others are mere borderline tugs. Forget the genuinist arguments from conceptual connotations, theoretical applications, and scientific practice, and consider only the genuinist argument from intuitions about paradigm cases, such as decapitation. Weigh this against whatever intuitions one might have about negligence, and

2   On this claim, Dowe and the law are in agreement. As Hart and Honoré note, "The use of this notion in the law is an extension of the general idea, common in non-legal thought, that the neglect of a precaution ordinarily taken against harm is the cause of that harm when it comes about" (1985, p. 195).
3   The misleading nature of "the cause" was first remarked on by J. S. Mill: "Nothing can better show the absence of any scientific ground for the distinction between the cause of a phenomenon and its conditions, than the capricious manner in which we select from among the conditions that which we choose to denominate the cause" (1846, p. 198).

here assume that one really does intuit that negligence is not genuinely causal (*contra* to what Dowe himself really intuits, and *contra* to what the law acknowledges). Would this be a case of "conflicting intuitions"? *Hardly*. This would be a case in which a paradigmatic near-certain judgment was being weighed against a tentative borderline tug. So even waving all the genuinist arguments but the paradigm case argument, and even granting Dowe's alleged intuition of difference, this would at most provide a *negligible counterweight* to genuinism.

The third response that I offer to Dowe's argument from the intuition of difference is that *genuinism does acknowledge a difference*. Positive and negative causation are different: the first involves physical connection and the second doesn't. These are different ways of wiring a causal mechanism. All the genuinist denies is that this difference in the type of causal mechanism is a difference between genuine and fake *causation*. The physical connections theory has some initial plausibility (witness the fact that excellent philosophers such as Dowe defend it), although it turns out to be a hasty generalization. It is no surprise if we intuit that positive and negative causation are different: they are. It is no surprise that we are sometimes hesitant about whether this difference is a causal difference, because we are sometimes tempted to hasty generalization. (It is especially not a surprise if those who are most tempted to this hasty generalization are those who feel the most hesitancy.)

In summary, what Dowe has done in his first argument is to compare two cases of negative causation, one of which (decapitation) is a paradigm of genuine causation, and the other of which (parental negligence) is more borderline but still intuitively causal. Should one really conclude that it is *genuinism* that is in intuitive trouble here?

Dowe's second argument for downgrading negative causation to the status of quasi-causation is that he sees negative causation as posing a *universal problem* for theories of causation:

> So now we can outline a second argument against genuinism. At least for a number of significant theories of causation, allowing quasi-causation to count as causation leads to considerable theoretical difficulty. So there is considerable advantage to the claim that quasi-causation is not genuine causation – *viz.*, this solves those problems. (2001, p. 220)

Dowe then develops a detailed theory of quasi-causation in terms of counterfactuals concerning genuine causation, on which, for instance, the claim that the absence of *c* quasi-causes the absence of *e* is analyzed as: if *c* had occurred, then *c* would have caused *e*. Finally, Dowe uses his theory of quasi-causation to explain away genuinist intuitions. The explanation is that, because causation and quasi-causation are difficult to distinguish epistemically, and because they play the same roles conceptually, we get confused:

> Further, our two results – the epistemic blur and the practical equivalence – furnish the answer to the objection that there are cases of quasi-causation which according to strong intuitions are definitely cases of causation, such as chopping off someone's head causing death being a prevention since chopping off the head prevents processes which would have caused the person to continue living (the genuinist intuition). The answer is that the

epistemic problem and the practical equivalence together suggests that we take the intuition of similarity as an intuition that the case is either causation or quasi-causation rather than as an intuition that it is causation not quasi-causation. It's hard to see how folk could intuit the latter given the deep epistemic problem, and it's hard to see why folk would have such an intuition as the latter given the practical equivalence. (2001, p. 225)

In summary, Dowe maintains that quasi-causation solves theoretical difficulties while serving to explain away intuitions for genuinism.

I offer three responses to Dowe's argument from a universal problem, the first of which is that the problem simply is *not* universal. The only theories that Dowe cites as having a problem are Salmon's, Armstrong's, and Lewis's, the first two of which are just physical connections theories. Moreover, Dowe admits that "There are accounts of causation according to which cases of omission and prevention come out as clear cases of causation" (2001, p. 220). Here he cites the statistical relevance and agential manipulation theories. So the situation is that negative causation poses a problem for only some theories. What should one conclude from that? I would have thought that the logical conclusion would be that negative causation provides an argument *in favor of the theories that can handle it*. Should one really conclude, from the fact that negative causation is a counterexample to a few theories, that it is *negative causation* that is in trouble?

Moreover, I think a Lewis-style counterfactual theory has more resources to handle negative causation than has been recognized. Lewis considers three strategies, one of which (Lewis's third strategy: 1986, pp. 192–3) is to regard the absence description as a way of referring to a present event, so that "the father's negligence" is just a way of referring to his actual nap, or whatever he actually did. Lewis worries, though, that this does not fit his general counterfactual approach, because the effect might not depend counterfactually on whether or not the father napped. The worry is that, if the father hadn't napped, perhaps he would have merely read a book instead, and so merely have been negligent by another means. I think the solution is near at hand though, since we may take the use of the absence description to guide the interpretation of the supposition that the nap does not occur. In such a context, it is to be interpreted as the supposition of a watchful father, which delivers counterfactual dependence. If so, then the only theories that face difficulties here are Salmon's and Armstrong's, both of which are versions of the physical connections theory. If so, then Dowe's argument effectively reduces to the following: if negative causation is genuine causation, then it refutes the physical connections theory.

The second response that I offer is that Dowe's theory of quasi-causation is *unsuccessful*. It goes wrong in more complex cases in which negative causation is combined with *overdetermination*, in that there are multiple failures, each of which suffices for the effect. Suppose that the mason and the carpenter both fail to work on the building, as a result of which the building fails to be completed. Then the mason's failing to work ($\sim c_1$) and the carpenter's failing to work ($\sim c_2$) should each count as causes of the building's failing to be completed ($\sim e$).[4] But neither $\sim c_1$ nor $\sim c_2$ even

---

4   This case is borrowed from Hart and Honoré:

> [A] similar rule should apply as regards concurrent omissions, each sufficient to produce the harm. If a mason and a carpenter both fail to do their part of the work on a building, so that it cannot be completed

212

Jonathan Schaffer

count as quasi-causes, on Dowe's treatment. For even if $c1$ had occurred, e would still not have occurred, due to $\sim c2$. (Likewise, even if $c2$ had occurred, e would still not have occurred, due to $\sim c1$.)

Dowe himself recognizes that his account of quasi-causation is troubled by a variety of complex cases, and in fact explicitly restricts his account to giving a *sufficient* condition. But what sort of solution fails to provide a necessary condition? And in what sense have genuinist intuitions been accommodated, by an account that only matches them in simple cases?

The third response that I offer is that Dowe's theory of quasi-causation is *semantically unstable*. Either Dowe's theory is the best solution to the allegedly universal problem of negative causation, or not. If not, then of course it proves nothing. But if so, then it deserves to be considered *the true theory of genuine causation*. For at that point Dowe will have offered us two relations, one (that of physical connection) which he labels "causation," and the other (that of counterfactuals involving the first) which he labels "quasi-causation." But don't be fooled by the labels: there is nothing "quasi" about the second relation.

For the sake of neutrality, I shall relabel Dowe's two relations "R1" (physical connection) and "R2" (counterfactuals about connection). Now consider: Which of R1 and R2 is more deserving of the label "causation"? This is a semantic question. There are methods for answering such questions (section 10.3). And R2 is the runaway winner by every measure. For it is R2 that squares with the central conceptual connotations of causation, R2 that covers all the paradigm cases of causation, R2 that underwrites the must useful theoretical applications of causation, and R2 that does justice to actual scientific practice.[5]

The "epistemic blur" and "practical equivalence" that Dowe trumpets are, if anything, even further evidence of genuineness. Suppose that Dowe's R2-theory is the best solution to the "universal problem" of negative causation. Then Dowe would have explicated one relation, R1, with none of the epistemic, practical, conceptual, intuitive, theoretical, or scientific markings of being the causal relation, and a second relation, R2, with all of the epistemic, practical, conceptual, intuitive, theoretical, and scientific markings of being the causal relation. In short, it would be R2 that paddles, waddles, and quacks like causation.

In summary, what Dowe has done in his second argument is to start with a problem that is not universal, partially re-solve it, and then stick the label "quasi" onto his solution. Should one really conclude that is it *genuinism* that is in theoretical difficulty here?

by the due date, and the building-owner loses an expected profit, each would be liable for the lost profit provided it was in their contemplation at the time of contracting. So, when a supplier of webbing delayed in delivering it to an overall manufacturer, and the latter incurred contractual penalties on the main contract, the jury was entitled to hold the webbing supplier liable to indemnify the overall manufacturer for the penalties paid, the failure to deliver the webbing being "a substantial factor" in causing the loss and "sufficient in itself to have delayed" the fulfillment of the main contract, despite the existence of other "contributory" causes of the default on the main contract. (1985, p. 236)

5   The attentive reader may recognize R2 as, in essence, Fair's modified theory of *genuine causation* (from the conclusion of section 10.2).

By way of conclusion, I submit that Dowe and other physical connections theorists are in a *dialectically impossible* situation. I mean, what would it take to convince you that decapitating someone cannot actually cause them to die? Let the physical connections theorist make what arguments they may. At the end of the day, their arguments will still need to be weighed against the fact that, by their theory, chopping off someone's head cannot cause them to die.

10.5 Misconnections and Differences

What is causation? While here is not the place to detail a positive theory, I would like to sketch one moral of negative causation. But first, it will help to have in mind a well-known counterexample to the *sufficiency* of physical connection for causation, namely the problem of *misconnection* (Hitchcock, 1995; Dowe, 2000; Schaffer, 2001). For instance, suppose that while Pam throws a rock at the window, Red sprays red paint in the air, so that the rock turns red on route to breaking the window. Then there is a physical connection from Red's hand to the rock over to the window, namely the line of red paint. But there is obviously no causation – Red's spraying red paint in the air did not cause the window to shatter. Or suppose that while Pam throws a rock at the window, Tim the innocent bystander gapes in horror. Then there is a physical connection from Tim's gaping to the window shattering, namely the lines traced by innumerable photons and other microparticles emitted from Tim and absorbed by the window. But there is obviously no causation – Tim's gaping did not cause the window to shatter.

So what is causation? What is it that positive and negative causation shares, and that misconnection lacks? The moral I would draw is that causation involves at least some aspect of *difference making.* In both positive and negative causations, whether or not the cause occurs *makes the difference* as to whether or not the effect will occur. For instance, the pulling of the trigger makes the difference as to whether or not the gun will fire, and the absence of blood flow makes the difference as to whether or not the victim will live. Misconnection is not causation, because it *makes no difference* as to whether or not the effect will occur. For instance, Red's spraying makes no difference as to whether or not the window will shatter, and Tim's gaping makes no difference there either.

Physical connections are often difference makers. *But not always*: some connections are of the wrong sort (Red's paint) or of the wrong magnitude (Tim's photon) to make a difference to whether the window shatters. *And not exclusively*: absences (victim's lack of blood flow) also work as difference makers. This is why the physical connections theory is plausible and often enough right, but no more.

The moral I would draw, then, is that causation has a *counterfactual* aspect, involving a comparative notion of difference making. I leave open whether causation is a purely counterfactual affair, or whether it involves some hybrid of counterfactuals and physical connections, as Fair suggests. Either way, negative causation proves that causes need not be physically connected to their effects.

Acknowledgments

I am grateful to Russ Colton, Phil Dowe, and Chris Hitchcock; and also to *Philosophy of Science* and the University of Chicago Press for allowing me to reprint material from my "Causation by disconnection" (Schaffer, 2000; © 2000 by the Philosophy of Science Association. All rights reserved).

Bibliography

Armstrong, D. M. 1999: The open door: counterfactual versus singularist theories of causation. In H. Sankey (ed.), *Causation and Laws of Nature*. Dordrecht: Kluwer, 175–85.

Aronson, J. L. 1971: On the grammar of "cause." *Synthese*, 22, 414–30.

— 1982: Untangling ontology from epistemology in causation. *Erkenntnis*, 18, 293–305.

Caceres, S., Cuellar, C., Casar, J. C., Garrido, J., Schaefer, L., Kresse, H., and Brandan, E. 2000: Synthesis of proteoglycans is augmented in dystrophic *Mdx* mouse skeletal muscle. *European Journal of Cell Biology*, 79, 173–81.

Castaneda, H.-N. 1984: Causes, causity, and energy. In P. French, T. Uehling, Jr., and H. Wettstein (eds.), *Midwest Studies in Philosophy IX*. Minneapolis: University of Minnesota Press, 17–27.

Dowe, P. 1992: Wesley Salmon's process theory of causality and the conserved quantity theory. *Philosophy of Science*, 59, 195–216.

— 1995: Causality and conserved quantities: a reply to Salmon. *Philosophy of Science*, 62, 321–33.

— 2000: *Physical Causation*. Cambridge: Cambridge University Press.

— 2001: A counterfactual theory of prevention and "causation" by omission. *Australasian Journal of Philosophy*, 79, 216–26.

Ehring, D. 1997: *Causation and Persistence*. Oxford: Oxford University Press.

Fair, D. 1979: Causation and the flow of energy. *Erkenntnis*, 14, 219–50.

Goldman, A. 1977: Perceptual objects. *Synthese*, 35, 257–84.

Hart, H. L. A. and Honoré, T. 1985: *Causation in the Law*. Oxford: The Clarendon Press.

Hitchcock, C. R. 1995: Salmon on explanatory relevance. *Philosophy of Science*, 62, 304–20.

Hume, D. 1748: *An Enquiry Concerning Human Understanding*. New York: Doubleday, 1974.

Kaplan, D. 1989: Demonstratives: an essay on the semantics, logic, metaphysics, and epistemology of demonstratives and other indexicals. In J. Almog, J. Perry, and H. Wettstein (eds.), *Themes from Kaplan*. Oxford: Oxford University Press, 481–563.

Kistler, M. 1998: Reducing causality to transmission. *Erkenntnis*, 48, 1–24.

— 2001: Causation as transference and responsibility. In W. Spohn, M. Ledwig, and F. Siebelt (eds.), *Current Issues in Causation*. Paderborn, Germany: Mentis, 115–33.

Kripke, S. 1972: *Naming and Necessity*. Cambridge, MA: Harvard University Press.

Lewis, D. K. 1970: How to define theoretical terms. *Journal of Philosophy*, 67, 427–46.

— 1981: Causal decision theory. *Australasian Journal of Philosophy*, 59, 5–30.

— 1986: Causation. In *Philosophical Papers II*. Oxford: Oxford University Press, 159–213.

Mackie, J. L. 1974: *The Cement of the Universe*. Oxford: Oxford University Press.

Mason, A. 2000: Introduction to Solid State Physics. Found online October 12, 2002, at http://www.engr.uky.edu/~ee562/562H01-physics.PDF

Mill, J. S. 1846: *A System of Logic, Ratiocinative and Inductive; Being a Connected View of the Principles of Evidence and the Methods of Scientific Investigation.* New York: Harper & Brothers.

Putnam, H. 1975: The meaning of "meaning." In K. Gunderson (ed.), *Language, Mind, and Knowledge.* Minneapolis: University of Minnesota Press, 131–93.

Roberts, C. 2000: Biological behavior? Hormones, psychology, and sex. *NWSA Journal*, 12, 1–20.

Russell, B. 1948: *Human Knowledge: Its Scope and Limits.* New York: Simon and Schuster.

Salmon, W. 1984: *Scientific Explanation and the Causal Structure of the World.* Princeton, NJ: Princeton University Press.

— 1994: Causality without counterfactuals. *Philosophy of Science*, 61, 297–312.

— 1998: *Causality and Explanation.* Oxford: Oxford University Press.

Schaffer, J. 2000: Causation by disconnection. *Philosophy of Science*, 67, 285–300.

— 2001: Causes as probability-raisers of processes. *Journal of Philosophy*, 98, 75–92.

Skyrms, B. 1980: *Causal Necessity.* New Haven, CT: Yale University Press.

Tye, M. 1982: A causal analysis of seeing. *Philosophy and Phenomenological Research*, 42, 311–25.

Wall, J. and Macdonald, A. 1993: *The NASA ASIC Guide: Assuring ASICs for Space.* Pasadena, CA: California Institute of Technology Jet Propulsion Laboratory.

Further reading

Anscombe, G. E. M. 1975: Causality and determination. In E. Sosa (ed.), *Causation and Conditionals.* Oxford: Oxford University Press, 63–81.

Davidson, D. 1967: Causal relations. *Journal of Philosophy*, 64, 691–703.

Hall, N. forthcoming: Two concepts of causation. In J. Collins, N. Hall, and L. A. Paul (eds.), *Causation and Counterfactuals.* Cambridge, MA: The MIT Press.

Hitchcock, C. R. 2001: A tale of two effects. *Philosophical Review*, 110, 361–96.

Lewis, D. K. 2000: Causation as influence. *Journal of Philosophy*, 97, 182–97.

— forthcoming: Void and object. In J. Collins, N. Hall, and L. A. Paul (eds.), *Causation and Counterfactuals.* Cambridge, MA: The MIT Press.

Mackie, J. L. 1965: Causes and conditions. *American Philosophical Quarterly*, 2, 245–64.

Paul, L. A. 2000: Aspect causation. *Journal of Philosophy*, 97, 235–56.

Schaffer, J. 2003: The metaphysics of causation. *Stanford Encyclopedia of Philosophy*, online at: http://plato.stanford.edu/entries/causation-metaphysics/

Tooley, M. 1987: *Causation: A Realist Approach.* Oxford: The Clarendon Press.

Jonathan Schaffer

IS THERE A PUZZLE ABOUT THE LOW-ENTROPY PAST?

The second law of thermodynamics tells us that the entropy of a closed physical system can increase, but that it never decreases. This is puzzling, because the underlying laws of physics that determine the behavior of the parts of a system do not distinguish between the future and the past: anything that can happen in one direction of time can happen in the other. Thus, if entropy can increase, then it should be able to decrease just as easily. There have been many attempts to resolve this paradox. According to Huw Price, the issue is not really about why entropy increases, but about how it ever got to be so low in the first place. Recent evidence suggests that shortly after the big bang, the world was in fact in a state of very low entropy. Price argues, however, that we shouldn't stop there: we should seek an explanation of why the early universe was in such a state. We may not succeed, but we shouldn't give up before we have even tried. Craig Callender, drawing on an analogy from debates in natural theology, argues that we cannot and should not explain the boundary conditions of the universe as a whole. This debate brings together issues from two quite different branches of physics: thermodynamics and cosmology. And, as Callender's theological analogy shows, the question "What is in need of explanation?" arises within other areas of philosophy as well.

On the Origins of the Arrow of Time: Why there is Still a Puzzle about the Low-Entropy Past

Huw Price

11.1 The Most Underrated Discovery in the History of Physics?

Late in the nineteenth century, physics noticed a puzzling conflict between the laws of physics and what actually happens. The laws make no distinction between past and future – if they allow a process to happen one way, they allow it in reverse.[1] But many familiar processes are in practice "irreversible," common in one orientation but unknown "backwards." Air leaks out of a punctured tire, for example, but never leaks back in. Hot drinks cool down to room temperature, but never spontaneously heat up. Once we start looking, these examples are all around us – that's why films shown in reverse often look odd. Hence the puzzle: What could be the source of this widespread temporal bias in the world, if the underlying laws are so even-handed?

Call this the *Puzzle of Temporal Bias,* or *PTB* for short. It's an oft-told tale how other puzzles of the late nineteenth century soon led to the two most famous achievements of twentieth-century physics, relativity and quantum mechanics. Progress on PTB was much slower, but late in the twentieth century cosmology provided a spectacular answer, or partial answer, to this deep puzzle. Because the phenomena at the heart of PTB are so familiar, so ubiquitous, and so crucial to our own existence, the achievement is one of the most important in the entire history of physics. Yet it is little known and underrated, at least compared to the other twentieth-century solutions to nineteenth-century puzzles.

Why is it underrated? It is partly because people underestimate the original puzzle, or misunderstand it, and so don't see what a big part of it is addressed by the new cosmology. And it is partly for a deeper, more philosophical reason, connected with

---

1   In some rare cases discovered in the mid-twentieth century, the reversal also needs to replace matter with antimatter, but this makes no significant difference to anything discussed below.

the view that we don't need to explain initial conditions. This has two effects. First, people undervalue the job done so far by cosmology, in telling us something very surprising about the early history of the universe – something that goes a long way toward explaining the old puzzle. And, secondly, they don't see the importance of the remaining issues – the new issues thrown up by this story, about *why* the early universe is the way that modern cosmology reveals it to be.

I'm going to argue that the old philosophical view is mistaken. We should be interested in the project of explaining initial conditions, at least in this case. As a result, we should give due credit – a huge amount – to modern cosmology for what it has already achieved. And we should be interested in pushing further, in asking *why* the early universe is the way modern cosmology has revealed it to be.

To understand why these issues matter, we need to understand how what cosmology tells us about the early history of the universe turns out to be relevant to the puzzle of temporal bias. Let's begin in the nineteenth century, where the puzzle first comes to light.

11.2 PTB in the Age of Steam

In one sense, the temporal bias discovered by physics in the nineteenth century had never been hidden. It was always in full view, waiting for its significance to be noticed. Everybody had always known that hot things cooled down in a cooler environment, that moving objects tended to slow down rather than speed up, and so on. But two things changed in the nineteenth century. First, examples of this kind came to seen as instances of a single general tendency or law – roughly, a tendency for concentrations of energy to become more dissipated, less ordered, and less available to do work. (The impetus for a lot of the relevant physics was the goal of extracting as much work as possible from a given lump of coal.) A measure of this disorder came to be called "entropy," and the general principle then said that the entropy of a closed system never decreases. This is the famous *second law of thermodynamics*.

After formulating this general principle in the mid-nineteenth century, physicists set about trying to explain it. Early work looked at the case of gases. Other properties of gases had already been successfully explained by assuming that gases were huge swarms of tiny particles, and by applying statistical techniques to study the average behavior of these swarms. Building on earlier work by James Clerk Maxwell (1831–79), Ludwig Boltzmann (1844–1906) argued in the 1870s that the effect of collisions between randomly moving gas molecules was to ensure that the entropy of a gas would always increase, until it reached its maximum possible value. This is Boltzmann's *H-theorem*.

This connection between thermodynamics and mechanics provides the crucial second ingredient needed to bring PTB into view. We've seen that PTB turns on the apparent conflict between two facts: the lack of temporal bias in the underlying laws, and the huge bias in what we can now call thermodynamic phenomena. In 1876, Boltzmann's colleague Josef Loschmidt (1821–95) called attention to the first of these facts, noting that the Newtonian mechanics assumed to be guiding gas molecules is time-symmetric, in the sense that every process it allows to happen in one direction is also allowed in reverse. (Mathematically, all we need to do to change a description of one to a description of the other is to change the sign of all the velocities at a given instant.) Loschmidt's immediate point was that if the second law rests on mechanics, it can't be exceptionless. There are *possible* motions such that entropy decreases.

In response to Loschmidt, Boltzmann suggested a completely new way of thinking about the second law. In place of the idea that collisions *cause* entropy increase, he offered us a new idea. Corresponding to any description of the gas in terms of its "macroscopic" observable properties – temperature, pressure, and so on – there are many possible "microstates" – many possible configurations of molecules that all give the same macrostate. Boltzmann's insight is that for nonequilibrium macrostates, the vast majority of these microstates are ones such that entropy increases in the future. It doesn't have to do so, but usually it will. (It isn't a trivial matter how to carve up the space of possible microstates, to get the measure right. An important part of Boltzmann's contribution was to find the right way to do this.)

Still, Loschmidt's argument turned on the realization that thanks to the underlying time-symmetry of Newtonian mechanics, microstates come in pairs. For every possible microstate in which some process is occurring in one temporal direction, there's another microstate in which the same process is occurring in the opposite temporal direction. So where does the asymmetry come in, on Boltzmann's new picture? To ask this question is to be struck by PTB.

This issue evidently occurred to Boltzmann at this time. His response to Loschmidt includes the following note:

> I will mention here a peculiar consequence of Loschmidt's theorem, namely that when we follow the state of the world into the infinitely distant past, we are actually just as correct in taking it to be very probable that we would reach a state in which all temperature differences have disappeared, as we would be in following the state of the world into the distant future. (Boltzmann, 1877, at p. 193 in translation in Brush, 1966)

Thus Boltzmann seems to suggest that, on the large scale, there is no temporal bias (and hence no PTB). But then why do we observe such striking asymmetry in our own region of space and time, if it doesn't exist on the large scale? And why is entropy so low now, given that according to Boltzmann's own way of counting possibilities, this is such an unlikely way to be?

Boltzmann doesn't seem to have asked these questions in the 1870s, and for the next twenty years PTB dropped back out of sight. It surfaced again in the 1890s, in a debate about the H-theorem initiated by E. P. Culverwell, of Trinity College, Dublin. As one contemporary commentator puts it, Culverwell's contribution was to ask in print "the question which so many [had] asked in secret, . . . *'What is the H-theorem and what does it prove?'*" (Hall, 1899, p. 685). This debate clarified some important issues, as we'll see in a moment. All the same, no one involved – not even Boltzmann – seems to have seen how much his new approach, formulated twenty years earlier in response to Loschmidt, had actually superseded the H-theorem.

This is an early manifestation of a confusion which has persisted in the subject ever since. To avoid it, we need to distinguish two different approaches to explaining the temporal bias of thermodynamics. As we'll see, both approaches face the question we're really interested in – Why is entropy so low, now and in the past? – but they have different conceptions of what else an explanation of the second law requires. On one conception, the most interesting issue about time asymmetry is somewhere else. This is one source of the tendency to undervalue the new contribution from cosmology, so it is important to draw distinctions carefully at this point.

11.3 How many Asymmetries do we Need?

What would it take to explain the temporal bias of thermodynamic phenomena? Since what needs to be explained is a time asymmetry, it's a safe bet that an adequate explanation is going to contain some time-asymmetric ingredient. (Symmetry in, symmetry out, after all.) But there are two very different views about how many asymmetries we need. On some views, we need only one; on others, we need two.

11.3.1 The two-asymmetry approach

As the name suggests, the second law of thermodynamics was originally regarded as a physical *law*. Without delving into the philosophical issue about what this means, let's say that to think of the second law in this way is to think of it as having some kind of "force" or necessity. In some sense, what the second law dictates is "bound" to happen. The discovery that the second law is probabilistic rather than exceptionless doesn't necessarily undermine this conception. It simply means we need a constraint weaker than outright necessity – some kind of real "propensity," for example.

Given this view of the second law, the task of explaining it in mechanical terms looks like the task of finding some mechanical factor that "forces" or "causes" entropy to increase (at least with high probability). The H-theorem itself is one such approach – it rests on the idea that the randomizing effect of collisions between molecules causes entropy to increase. One of the major insights of the debate about the H-theorem in the 1890s was that if entropy is to increase, this causal mechanism must be time-asymmetric. If the H-theorem worked equally well in both directions, it would show that entropy is nondecreasing in both directions – which is only possible if it is constant.

How does this asymmetry get into the H-theorem? The first person to answer this question explicitly was Samuel Burbury (1831–1911), an English barrister who had turned to mathematical physics late in middle age, as loss of hearing curtailed his legal career. Burbury saw that the source of the asymmetry in the H-theorem is an assumption, roughly, that the motions of gas molecules are independent *before* they collide. He pointed out both that the H-theorem requires this assumption, and that if entropy is to increase, the assumption cannot hold *after* collisions (see Burbury, 1894, 1895). Burbury's argument is widely misinterpreted as showing that collisions *cause* correlations. In fact, it shows no such thing. The correlations are simply those required by the assumption that entropy decreases toward the past, and are quite independent of whether the molecules collide at all (see Price, 2002b).

There are other causal approaches to the second law. One, called interventionism, attributes the increase in entropy to random and uncontrollable influences from a system's external environment. Another is a recent suggestion that a stochastic collapse mechanism proposed in certain extensions of quantum theory provides a randomizing influence that is sufficient to ensure that entropy increases (see Albert, 1994, 2000). Again, the point to keep in mind is that, as in all such causal approaches, the mechanism needs to be time-asymmetric, if it is not to force entropy to be nondecreasing in both directions.

These causal approaches thus need two time asymmetries altogether. Why? Because a causal mechanism that ensures that entropy will not decrease won't by itself produce what we see. Entropy also needs to start low. If a system begins in equilibrium – that is, with maximum possible entropy – such a mechanism will simply keep it there. There will be no observed *increase*. To get what we see, then, we need an asymmetric "boundary condition" which ensures that entropy is low in the past, as well as an asymmetric mechanism to make it go up. What wasn't seen clearly in the 1890s, and has often been obscure since, is that this approach is thereby fundamentally different from the statistical approach suggested by Boltzmann in the 1870s, in response to Loschmidt. In Boltzmann's new approach, there is only one time asymmetry – the only asymmetry is the low-entropy boundary condition.

11.3.2 The one-asymmetry approach

Think of a large number of gas molecules, isolated in a box with elastic walls. If the motion of the molecules is governed by deterministic laws, such as Newtonian mechanics, a specification of the microstate of the system at any one time uniquely determines its entire history (or "trajectory"). This means that Boltzmann's assignment of probabilities to instantaneous microstates applies equally to whole trajectories. Accordingly, consider the set of all trajectories, with this Boltzmann measure. The key idea of Boltzmann's statistical approach is that in the overwhelming majority of possible trajectories, the system spends the overwhelming majority of the time in a high-entropy macrostate – that is, among other things, a state in which the gas is dispersed throughout the container. And there is no temporal bias in this set of possible trajectories. Each possible trajectory is matched by its time-reversed twin, just as Loschmidt had pointed out.

Asymmetry comes in when we apply a low-entropy condition at one end. For example, suppose that we throw away all the possible trajectories *except* those in which the gas is completely confined to some small region R at the initial time T_0. Restricted to the remaining trajectories, our original Boltzmann measure now provides a measure of the likelihood of the various possibilities consistent with this boundary condition – that is, consistent with the gas's being confined to R at T_0. Almost all trajectories in this remaining set will be such that the gas becomes more dispersed after T_0. The observed behavior is thus predicted by the time-symmetric Boltzmann measure, once we "conditionalize" in this way on the low-entropy condition at T_0.

On this view, then, there's no asymmetric factor that "forces" or "causes" entropy to increase. This is simply the most likely thing to happen, given the combination of the time-symmetric Boltzmann probabilities and the single low-entropy restriction in the past.

It's worth noting that the correctness of the resulting probability judgments concerning the future behavior implicitly *depends on the assumption* that that there is no corresponding low-entropy restriction in that direction. So Boltzmann's statistical approach does not enable us to predict that entropy is unlikely ever to decrease, but only to draw a much weaker conclusion: entropy is unlikely to decrease, *unless there is the kind of constraint in the future that makes entropy low in the past*. The second law holds so long as there isn't a low-entropy boundary condition in the future, but can't be used to exclude this possibility – *even probabilistically!*

As it stands, we know of no such condition in the future. The low-entropy condition of our region seems to be associated entirely with a low-entropy condition in our past. This condition is time-asymmetric, so far as we know, but this is the only time asymmetry in play, according to Boltzmann's statistical approach.

Thus we have two very different ways of trying to explain the observed temporal bias of thermodynamic phenomena. Our current interest is in what these approaches have in common, the fact that entropy is low now, and even lower in the past. On both approaches, the observed asymmetry depends on this fact – without it, we'd never see stars and hot cups of coffee cooling down, because we'd never see stars and hot coffee, full stop. The great discovery I mentioned at the beginning is a cosmological explanation of this crucial fact.

I've stressed the distinction between these approaches because although the low-entropy boundary condition plays a crucial role even in the two-asymmetry approach, it plays second fiddle there to the supposed asymmetric cause. Looking for an elusive factor that forces entropy to increase, the two-asymmetry approach often pays little heed to what seems a mere boundary condition, the fact that entropy starts low. This is one source of the tendency to discount modern cosmology's contribution to the solution of PTB.

The two-asymmetry approach faces a serious problem. To say that some asymmetric mechanism *causes* entropy to increase is to say that in the absence of that mechanism, entropy would not increase. Yet Boltzmann claims to have shown that for most possible initial microstates, entropy would increase anyway, without any such asymmetric mechanism. So friends of such mechanisms need to say that Boltzmann is wrong – that the universe (probably) starts in a microstate such that without the mechanism, entropy would not increase. It's hard to see what could justify such a claim (I develop this objection in Price, 2002a,b).

For present purposes, however, we needn't try to adjudicate between the two views. Our interest is in the low-entropy boundary condition, and in the issue as to whether it needs to be explained. So long as we keep in mind that even the two-asymmetry approach needs such a condition, we'll be in no danger of thinking that the issue is optional – a product of a questionable conception of what an understanding of PTB requires.

11.4 Did it all Happen by Accident?

In the discussion of the mid-1890s, Boltzmann himself certainly saw the importance of the question as to why entropy is now so low – much lower than its theoretical maximum. In a letter to *Nature* in 1895, he offers a tentative answer, based on a

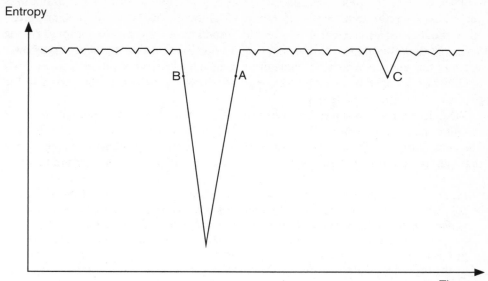

Entropy

B   A   C

Time

Figure 11.1 Boltzmann's entropy curve.

suggestion that he attributes to his assistant, Dr. Schuetz. The proposal is in two parts. First, he notes that although low-entropy states are very unlikely at any given time, they are very likely to occur eventually, given enough time. After all, if we toss a fair coin for long enough, we're very likely to get a run of a billion heads eventually – we'll just have to wait a very, very long time! So if the universe is extremely old, it's likely to have had time to produce, simply by accident, the kind of low-entropy region we find ourselves inhabiting. As Boltzmann puts it: "Assuming the universe great enough, the probability that such a small part of it as our world should be in its present state, is no longer small" (Boltzmann, 1895, p. 415).

Of course, it's one thing to explain why the universe contains regions like ours, but another to explain why we find ourselves in one. If they are so rare, isn't it much more likely that we would find ourselves somewhere else? Answering this challenge is the job of the second part of Boltzmann's proposal. Suppose that creatures like us simply couldn't exist in the vast regions of "thin cold soup" between the rare regions of low entropy. Then it's really no surprise that we find ourselves in such an unlikely spot. All intelligent creatures will find themselves similarly located. As Boltzmann himself puts it, "the . . . H curve would form a representation of what takes place in the universe. The summits of the curve would represent the worlds where visible motion and life exist" (Boltzmann, 1895, p. 415).

Figure 11.1 shows what Boltzmann here calls the H curve, except that following modern convention, this diagram shows entropy on the vertical axis, rather than Boltzmann's quantity H. Entropy is low when H is high, and vice versa, so the summits of Boltzmann's H curve become the troughs of this entropy curve. More precisely, figure 11.1 shows the entropic history of a single typical trajectory. Most of the time

225

– *vastly* more of the time than this page makes it possible to show – the universe is very close to equilibrium. Very, very rarely, a significant fluctuation occurs, an apparently random rearrangement of matter that produces a state of low entropy. As the resulting disequilibrium state returns to equilibrium, an entropy gradient is produced, such as the one on which we apparently find ourselves, at a point such as A. If intelligent life depends on the existence of an entropy gradient, it only exists in the regions of these rare fluctuations.

Why do we find ourselves on an uphill rather than a downhill gradient, as at B? In a later paper, Boltzmann offers a remarkable proposal to explain this, too. Perhaps our perception of past and future depends on the entropy gradient, in such a way that we are bound to regard the future as lying "uphill" on the entropy gradient:

> In the universe . . . one can find, here and there, relatively small regions on the scale of our stellar region . . . that during the relatively short eons are far from equilibrium. What is more, there will be as many of these in which the probability of the state is increasing as decreasing. Thus, for the universe the two directions of time are indistinguishable, just as in space there is no up or down. But just as we, at a certain point on the surface of the Earth, regard the direction to the centre of the Earth as down, a living creature that at a certain time is present in one of these isolated worlds will regard the direction of time towards the more improbable state as different from the opposite direction (calling the former the past, or beginning, and the latter the future, or end). Therefore, in these small regions that become isolated from the universe the "beginning" will always be in an improbable state. (Boltzmann, 1897; translation from Barbour, 1999, p. 342)

This is perhaps the first time that anyone had challenged the objectivity of the perceived direction of time, and this alone makes Boltzmann's hypothesis a brilliant and revolutionary idea. But the proposal also solves PTB in a beautiful way. It explains the apparent asymmetry of thermodynamics in terms of a cosmological hypothesis that is symmetric on the larger scale. So PTB simply goes away on the large scale – although without depriving us of an explanation for why we do find temporal bias locally. Boltzmann's hypothesis is the kind of idea that simply deserves to be true!

11.5 The Monkey Wrench

Unfortunately it isn't true, or at least there's a huge spanner in the works. The problem stems directly from Boltzmann's own link between entropy and probability. According to Boltzmann's famous formula, $S = k \log W$, entropy is proportional to the logarithm of probability (the latter judged by the Boltzmann measure). In figure 11.1, then, the vertical axis is a logarithmic probability scale. For every downward increment, dips in the curve of the corresponding depth become exponentially more improbable. So a dip of the depth of point A or point B is far more likely to occur in the form shown at point C – where the given depth is very close to the minimum of the fluctuation – than it is to occur in association with a much bigger dip, such as that associated with A and B. This implies that if we wish to accept that our own region is the product of "natural" evolution from a state of even lower entropy, we must accept that our region is far more improbable than it needs to be, given its present entropy.

To put this another way, if Boltzmann's probabilities are our guide, then it is much easier to produce fake records and memories than to produce the real events of which they purport to be records. Imagine, for example, that God chooses to fast-forward through a typical world-history, until he finds the complete works of Shakespeare, in all their contemporary twenty-first-century editions. It is *vastly* more likely that he will hit upon a world in which the texts occur as a spontaneous fluctuation of modern molecules than that he'll find them produced by the Bard himself. (Editions typed by monkeys are probably somewhere in between, if the monkeys themselves are products of recent fluctuations.) In Boltzmann's terms, then, it is unlikely that Shakespeare existed, 400 years ago. Someone like him exists somewhere in Boltzmann's universe, but he's very unlikely to be in *our* recent past. The same goes for the rest of what we take to be history. All our "records" and "memories" are almost certainly misleading.

There's another problem of a similar kind. Just as we should not expect the low-entropy region to extend further back *in time* than it needs to in order to produce what we see, so we should not expect it to be any more extensive *in space* than we already know it to be. (Analogy: shuffle a deck of cards, and deal a hand of 13 cards. The fact that the first six cards you turn over are spades does not give you reason to think that the rest of the hand are spades.) But we now observe vastly more of the universe than was possible in Boltzmann's day, and yet the order still extends as far as we can see.

Brilliant as it is, then, Boltzmann's hypothesis faces some devastating objections. Moreover, modern cosmology goes at least some way to providing us with an alternative. As I'll explain later, this may not mean that the hypothesis is completely dead – it might enjoy new life as part of an explanation of what modern cosmology tells us about the low-entropy past. But for the moment, our focus needs to be on that cosmological story.

11.6 Initial Smoothness

What boundary conditions, at what times, are needed to account for the time asymmetry of observed thermodynamic phenomena? There seems no reason to expect a neat answer to this question. Low entropy just requires concentrations of energy, in useable forms. There are countless ways in which such stores of useable energy could exist, at some point in our past. Remarkably, however, it seems that a single simply characterizable condition does the trick. All the observed order seems attributable to a single characteristic of the universe soon after the big bang.

The crucial thing is that matter in the universe is distributed extremely smoothly, about 100,000 years after the big bang. It may seem puzzling that this should be a low-entropy state. Isn't a homogeneous, widely dispersed arrangement of matter a disordered, high-entropy arrangement? But it all depends on what forces are in charge. A normal gas tends to spread out, but that's because the dominant force – pressure – is repulsive. In a system dominated by an attractive force, such as gravity, a uniform distribution of matter is highly unstable. The natural behavior of such matter is to clump together. Think of the behavior of water on the surface of a waxy leaf, where the dominant force is surface tension – or of a huge collection of sticky polystyrene foam pellets, whose natural tendency is to stick together in large clusters.

To get a sense of how extraordinary it is that matter should be distributed uniformly near the big bang, keep in mind that we've found no reason to disagree with Boltzmann's suggestion that there's no objective distinction between past and future – no sense in which things "really" happen in the direction we think of as past-to-future. Without any such objective distinction, we're equally entitled to regard the big bang as the end point of a gravitational collapse. For such a collapse to produce a very smooth distribution of matter is, to put it mildly, quite extraordinary, judged by our ordinary view about how gravitating matter should behave. (Imagine throwing trillions of sticky foam pellets into a tornado, and having them settle in a perfect sheet, one pellet thick, over every square centimeter of Kansas – that's an easy trick, by comparison!)

I stress that there are two very remarkable things about this feature of the early universe. One is that it happens at all, given that it is so staggeringly unlikely, in terms of our existing theory of how gravitating matter behaves. (Penrose (1989, ch. 7) estimates the unlikeliness of such a smooth arrangement of matter at 1 in $10^{10^{123}}$.) The other is that, so far as we know, it is the *only* anomaly necessary to account for the vast range of low-entropy systems we find in the universe. In effect, the smooth distribution of matter in the early universe provides a vast reservoir of low entropy, on which everything else depends. The most important mechanism is the formation of stars and galaxies. Smoothness is necessary for galaxy and star formation, and most irreversible phenomena with which we are familiar owe their existence to the sun (for more details, see Penrose, 1989, ch. 7).

In my view, this discovery about the cosmological origins of low entropy is the most important achievement of late-twentieth-century physics. It is true that in one sense it simply moves the puzzle of temporal bias from one place to another. We now want to know *why* the early universe is so smooth. But, as I've emphasized, it's an extraordinary fact that the puzzle turns out to be capable of being focused in that place.

The puzzle of initial smoothness thus gives concrete form to the explanatory project that begins with the time asymmetry of thermodynamics. If cosmology could explain initial smoothness, the project would be substantially complete, and PTB would be substantially solved. At the moment, however, it is very unclear what form a satisfactory explanation might take. I'll say a little more below about some of the possibilities.

However, my main task in the remainder of this chapter is to defend the claim that initial smoothness needs explaining. Some philosophers argue that it is inappropriate to ask for an explanation of such an initial condition. It would be nice if this were true, for it would imply that PTB has in large part been laid to rest – that most of the work that needs to be done has been done. But I think these philosophers are mistaken, and hence that there is still a lot of work for cosmologists to do on this issue. Our role as philosophers is to help them to see the importance of the issue.

11.7 Should Cosmologists be Trying to Explain Initial Smoothness?

In the light of late-twentieth-century cosmology, then, the late-nineteenth-century puzzle of low entropy takes a new concrete form. Why is the universe smooth, soon after the big bang? Should we be looking for an answer to this question? Some philosophers say not. For example, Craig Callender suggests that "the whole enterprise of explaining global boundary conditions is suspect, for precisely the reasons Hume and Kant taught us, namely, that we can't obtain causal or probabilistic explanations of why the boundary conditions are what they are." (Callender, 1997, p. 69 – for similar concerns, see Callender's companion chapter in this volume (chapter 12); and also Callender, 1998, pp. 149–50; Sklar, 1993, pp. 311–12).

There are a number of ways we might respond to this objection. We might argue that Hume and Kant are simply wrong. We might argue that there's some different kind of explanation that it is appropriate to seek for the smooth early universe, an explanation that is neither causal nor probabilistic. Or we might argue that the objection misses its target, because the smooth early universe isn't a boundary condition in the relevant sense; but, rather, something else, something that does call for explanation. My strategy will be predominantly the third of these options, though with some elements of the first and second.

I'll proceed as follows. First, I'll appeal to your intuitions. I'll ask you to imagine a discovery that cosmology might have made about the universe, a case in which it seems intuitively clear that we would seek further explanation. I'll then argue that we have no grounds for taking this imaginary case to be different from the actual case. (On the contrary, I claim, it is the actual case, but described in a nonstandard way.) Next, I'll clarify the status of the low-entropy "boundary condition," and in particular, call attention to a sense in which its status seems necessarily to be more than that of a mere boundary condition. It is "law-like" rather than "fact-like" in nature, in a sense that I'll make more precise. (I'll argue that unless it has this status, we have no defense against the kind of skeptical challenge that proved so devastating to Boltzmann's proposed explanation of the low-entropy past.) Finally, I'll respond briefly to Callender's elucidation of the Humean objection to the project of explaining initial smoothness, in the light of this clarification of its status.

First, a note on terminology. In recent years, a number of writers have taken to calling the supposition that the early universe has low entropy the "Past Hypothesis." This phrase was introduced by David Albert, who takes it from a passage in Richard Feynman's *The Character of Physical Law*, in which Feynman says "I think it necessary to add to the physical laws the hypothesis that in the past the universe was more ordered . . . than it is today" (Feynman, 1967, p. 116). In this formulation, however, there is no special mention of cosmology. But Albert's most explicit formulation of what he means by the term refers explicitly to cosmology. According to Albert, the Past Hypothesis "is that the world first came into being in whatever particular low-entropy highly condensed big-bang sort of macrocondition it is that the normal inferential procedures of cosmology will eventually present to us" (Albert, 2000, p. 96).

This terminological point is important. Taken in Feynman's original nonspecific form, the Past Hypothesis is certainly capable of further explanation – a fit topic for cosmological investigation (as Feynman (1967, p. 116) himself notes, saying that although this hypothesis is now "considered to be astronomical history," "perhaps someday it will also be a part of physical law."). And after all, this is precisely what's happened. The abstractly characterized fact has now been explained by the smoothness of the early universe, and the issue is simply whether cosmologists should be trying to take things a stage further. Nothing in Feynman's proposal suggests that they should not.

Taken in Albert's form, however, the Past Hypothesis is by definition the final deliverance of cosmology on the matter. While it is then analytic that the Past Hypothesis itself will not be further explained, we have no way of knowing whether current cosmology is final cosmology – experience certainly suggests not! So the trivial semantic fact that the Past Hypothesis (so defined) cannot be further explained provides no reason not to try to explain the smooth early universe. Even if it that were the Past Hypothesis, we wouldn't find that out, presumably, until we'd tried to explain it further, and become convinced by persistent failure that it was the final theory. I think that similar remarks apply to the term "Past State," which Callender uses in chapter 12 – it, too, can be read in either of two ways. So to side-step these terminological confusions, I'll to avoid using these terms, and concentrate on what we actually have from cosmology, namely the smooth early universe. Is this something that we should be trying to explain?

11.8 What's Special about Initial Conditions?

Part of the usual resistance to the idea of explaining initial conditions is associated with the thought that we normally explain events in terms of *earlier* events. By definition, there is nothing earlier than the initial conditions.

In the present context, however, this preference is on shaky ground. Here's a way to make this vivid. Imagine that in recent years physics had discovered that the matter in the universe is collapsing toward a big crunch, 15 billion years or so in our future – and that as it does so, something very peculiar is happening. The motions of the individual pieces of matter in the universe are somehow conspiring to defeat gravity's overwhelming tendency to pull things together. Somehow, by some extraordinary feat of cooperation, the various forces are balancing out, so that by the time of the big crunch, matter will have spread itself out with great uniformity. A molecule out of place, and the whole house of cards would surely collapse! Why? Because as Albert (2000, 151), puts it, "the property of being an *abnormal* [i.e., entropy-reducing] microstate is extraordinarily *unstable* under small perturbations." By the lights of the Boltzmann measure, then, the tiniest disturbance to our imagined entropy-reducing universe would be expected to yield an entropy-increasing universe.

As a combination of significance and sheer improbability – the latter judged by well-grounded conceptions of how matter is expected to behave – this discovery would surely trump anything else ever discovered by physics. Should physicists sit on their hands, and not even try to explain it? (They might fail, of course, but that's always on the cards – the issue is whether it is appropriate to try.) If this discovery didn't call for explanation, what conceivable discovery ever would?

In my view, however, this state of affairs is *exactly* what physics has discovered! I've merely taken advantage, once again, of the fact that if there is no objective sense in which what we call the future is *really* the "positive" direction of time, then we can equally well describe the world by reversing the usual temporal labelling. Relabelled in this way, the familiar expansion from a smooth big bang becomes a contraction to a smooth big crunch, with the extraordinary characteristics just described. And surely if it is a proper matter for explanation described one way, it is a proper matter for explanation described the other way.

Both steps in this argument could conceivably be challenged. The first relies, as I said, on the view that there is no objective distinction between past and future, no difference between our world and a world in which exactly the same things happen, but in the opposite order. This claim is contentious. One prominent writer who rejects it is John Earman. In a classic (1974) paper on the direction of time, Earman suggests – correctly, in my view – that someone who endorses this view about time would have no grounds to reject an analogous view about spatial parity; and would thus be committed to the view that there is no objective difference between a possible world and its mirror-reversed twin. I agree, and to me, this seems the right view in that case too.

It would take us too far afield to try to settle this issue here. For the moment, the important thing is that someone who wants to say that my imagined physical discovery is different from the actual discovery made by cosmology, and that this accounts for the fact that it would call for explanation in a way that the smooth early universe does not, faces an uphill battle. First, they owe us an account of the objective difference between past and future. Secondly, they need to explain how this difference makes a difference to what needs explaining. And, thirdly, they need to explain how they know they've got things the right way round – how they know that we live in the world in which the smooth extremity does not need explaining, rather than the temporal mirror world, in which it does.

Absent such arguments, I take the lesson of this example to be as follows. Our ordinary intuitions about what needs explaining involve a strong temporal bias, a temporal bias that we should eliminate if we want our physical explanations to show reasonable invariance under trivial redescriptions of the phenomena in question. In particular, our tendency simply to take initial conditions for granted is unreliable, because the same conditions can equally well be regarded as final conditions.

It might be objected that this doesn't necessarily show that the smooth early universe calls for explanation. Perhaps the argument actually cuts the other way, showing that our intuitions about the redescribed case – the smooth "late" universe – are unreliable. Perhaps we would be wrong to try to explain a smooth big crunch. (Callender's companion chapter 12 suggests that he would take this view.) For my part, I find it hard to make sense of this possibility. As I said, if the imagined discovery did not strike us as calling for explanation (in the light of our preexisting expectations about how gravitating matter ought to behave), then it is hard to see what discovery ever would call for explanation. However, it would be nice to do better than simply trading intuitions on this point, and for this, I think, we need some additional guidelines. In particular, we need to pay closer attention to the theoretical role of the "boundary condition" in question. I'll approach this issue by considering another objection to the project of explaining the low-entropy past.

11.9 The Just Good Luck Objection

The objection in question is close to one expressed by D. H. Mellor, in a recent response to John Leslie. Mellor describes the following example from Leslie:

> Suppose you are facing a firing squad. Fifty marksmen take aim, but they all miss. If they hadn't all missed, you wouldn't have survived to ponder the matter. But you wouldn't leave it at that: you'd still be baffled, and you'd seek some further reason for your luck. (2002, p. 227)

Mellor then writes,

> Well, maybe you would; but only because you thought the ability of the firing squad, the accuracy of their weapons, and their intention to kill you made their firing together a mechanism that gave your death a very high physical probability. So now suppose there is no such mechanism. Imagine ... that our universe (including all our memories and other present traces of the past) started five minutes ago, with these fifty bullets coming past you, but with no prior mechanism to give their trajectories any physical probability, high or low. Suppose in other words that these trajectories really were among the *initial* conditions of our universe. If you thought that, should you really be baffled and seek some further reason for your luck? (2002, p. 227)

It might be argued – and Callender's companion chapter suggests that he would be sympathetic to this idea – that the smooth early universe is like this imagined case, in requiring no mechanism to bring it about. Isn't the smooth early universe just a matter of luck, like the trajectories of the bullets, in Mellor's example?

But let's think some more about Mellor's example. Let's imagine ourselves in Mellor's world, being told that another 50, or 500, or 5,000 bullets are yet to arrive. Should we expect our luck to continue? In my view, to think it's an accident that the first 50 bullets missed us just *is* to have no expectation that the pattern will continue in new cases. Perhaps we think something else about new cases, or perhaps we're simply agnostic, but either way, we don't "project" from the initial 50 cases. If the pattern does continue, say for another 500 cases, we might go on attributing it purely to luck. But we can't both expect it to continue indefinitely, and attribute that in advance merely to luck. For to take the generalization to be projectible *is* to treat it is something more than merely an accident – as something *law-like*.

Similarly, if we think that the smooth early universe is just a matter of luck, then we have no reason to expect that the luck will continue when we encounter new regions of the universe – regions previously too far away to see, for example. Again, perhaps we'll think it won't continue, or perhaps we'll be agnostic. But either way, we won't think that it will continue.

This argument is very similar to one version of the objection we encountered in section 11.5 to the Boltzmann hypothesis. There, the spatial version of the objection was that if the low-entropy past is just a statistical fluctuation, we shouldn't expect more of it than we've already discovered – we shouldn't expect to see more order, as we look further out into space. Similarly in the present case: if the smooth early universe is just a piece of luck, we shouldn't expect our luck to continue.

As actually used in contemporary cosmology, the hypothesis of the smooth early universe is not like this. It is taken to be projectible, in the sense that everyone expects it and its consequences (e.g., the existence of galaxies) to continue to hold, as we look further and further out into space. The hypothesis is thus being accorded a law-like status, rather than treated as something that "just happens."

This argument was analogous to the spatial version of the objection to the Boltzmann hypothesis. The more striking temporal version of that objection also carries over to the present case, I think. For suppose we did think of the smooth early universe as a lucky accident. The essence of the temporal objection to Boltzmann was that there are many lucky accidents compatible with what we see – almost all of them far more likely than the smooth big bang, in terms of the Boltzmann measure. So why should we think that the actual accident was a smooth early universe, rather than one of those other possibilities? The upshot is that the belief that the smooth big bang is a lucky accident seems (all but) incompatible with the belief that it actually happened!

In my view, the present state of play is this. Modern cosmology is implicitly committed to the view that the smooth big bang is not merely a lucky accident. But we don't yet understand how this can be the case. This puzzle is the twentieth century's legacy to the twenty-first – its transformation of the original nineteenth-century puzzle of temporal bias. It is not an exaggeration, in my view, to say that this is one of *the* great puzzles of contemporary physics (even if a puzzle whose importance is easily underrated, for the reasons we've already canvassed). At this point, philosophers should not be encouraging physicists to rest on their laurels (or laureates). On the contrary, we should be helping them to see the full significance of this new puzzle, and encouraging them to get to work on it!

In what directions should they be looking? I'll say a little about this issue in a moment, but before that, I want to respond briefly to another aspect of the challenge from Hume, Kant, and Callender to the project of explaining initial conditions.

11.10 The Only One Universe Objection

In his companion chapter, Callender cites Hume's famous objection to the project of explaining "the generation of a universe," as Hume puts it. As Callender says, Hume's point "is that since the cosmos happens only once, we cannot hope to gain knowledge of any regularities in how it is created."

I offer three responses to this objection. First, the required boundary condition is not necessarily unique, because the universe may contain other relevantly similar singularities. Secondly, even if it were unique, there is an important and familiar sense in which its components provide generality. And, thirdly, explanations of unique states of affairs are perfectly normal – in fact, unavoidable – at least in the case of laws.

11.10.1 Not necessarily unique

It is far from clear that the required boundary condition is unique. The expanding universe may eventually recollapse, in which case there will be a big crunch in our future, as well as a big bang in our past. It is true that the trend of recent astronomical evidence has been against this possibility, but we are here canvassing possibilities, and should certainly leave this one on the table. In any case, even if whole universe doesn't recollapse, it is thought that parts of it will, as large accumulations of matter form black holes. As writers such as Hawking (1985, p. 2491) and Penrose (1979, pp. 597–8) have pointed out in this context, this process is very much like a miniature version of collapse of the entire universe. In some respects, then, the big bang is one of a general class of events, of which the universe may contain many examples. The big bang may have special significance, but it far from clear that its properties could not be derivable from some general theory of singularities, a theory testable in principle by observation of multiple instances of the phenomena that it describes.

11.10.2 Even one case provides generality

If matter in the universe as a whole is smoothly distributed after the big bang, this implies that the following is true of the matter in every individual region of the universe. As we follow the matter in that region backward in time, toward the big bang, we find irregular accumulations of matter disappearing. Somehow, the particular chunk of the matter in the region in question manages to spread itself out – interacting with other chunks as it does so, but not presumably with all other chunks, since some of them are too far away. As aliens from another dimension, as it were, we could select chunks at random, and discover that this same behavior was characteristic of all of them. Wouldn't this count as a generalization, if anything does?

Here's an analogy. Recent observational evidence suggests that the expansion of the universe is actually accelerating. We don't think it inappropriate to seek to explain why this should be so. On the contrary, this is widely regarded as a fascinating project, likely to require new physical theories. But in one sense, this expansion too is just one unique case. There's just one universe, and it has to behave in some way, so why not in this way? Again, part of a proper response to this challenge seems to be that we find the same thing happening in many parts of the universe, and that this suggests some unifying underlying explanation. That seems to me to be precisely what we find in the case of the smooth early universe too.

11.10.3 Even unique things get explained

The laws of nature are unique, in the sense that there is only one world of which they are the laws. Yet we often think it proper to explain laws, by showing that they follow from more fundamental laws. It isn't always clear where this is appropriate or needed, but there's certainly a good deal of consensus on these things. I've argued above that the low-entropy early universe needs to be regarded as a law-like hypothesis, if we are to avoid objections analogous to those that afflict the Boltzmann hypothesis. It seems a reasonable project to seek some deeper understanding of this hypothesis – to hope to show how it follows from something more fundamental. (Again, we may fail, but the question is whether we should try.)

11.11 What Might Explanations Look Like?

I want to finish by mentioning some strategies for seeking to explain the smooth early universe. I don't think any of these strategies is unproblematic as it stands, but they do give some sense of both the options and the problems facing this important theoretical task.

11.11.1 The appeal to inflation

The first approach stems from what cosmologists call the inflationary model. This model is a kind of front end to the standard big bang model, describing what might have happened to the universe in its extremely early stages. The proposal is that when the universe is extremely small – perhaps simply the product of some quantum fluctuation – the physical forces in play are different from those with which we are familiar. In particular, gravity is repulsive, rather than attractive, and the effect is that the universe experiences a period of exponential expansion. As it grows it cools, and at a certain point undergoes a "phase transition." The forces change, gravity becomes attractive, and the universe settles into the more sedate expansion of the "ordinary" big bang (for an introduction, see Linde, 1987).

Since it was first proposed in the 1980s, one of the main attractions of the inflationary model has been that it seems to explain features of the early universe that the standard big bang model simply has to take for granted. One of these features, it is claimed, is the smoothness of the universe after the big bang. However, the argument that inflation explains smoothness is essentially statistical. The crucial idea is that during the inflationary phase the repulsive gravity in will tend to "iron out" inhomogeneities, leaving a smooth universe at the time of the transition to the classical big bang. Presenting the argument in *Nature* in 1983, Paul Davies concludes that:

> the Universe . . . began in an arbitrary, rather than remarkably specific, state. This is precisely what one would expect if the Universe is to be explained as a spontaneous random quantum fluctuation from nothing. (1983, p. 398)

But this argument illustrates the temporal double standard that commonly appears in discussions of these problems. After all, we know that we might equally well view the problem in reverse, as a gravitational collapse toward a big crunch. In statistical terms, this collapse may be expected to produce *inhomogeneities* at the time of any transition to an inflationary phase. Unless one temporal direction is already privileged, the statistical reasoning is as good in one direction as the other. Hence in the absence of a justification for the double standard – a reason to apply the statistical argument in one direction rather than the other – the appeal to inflation doesn't seem to do the work required of it.

Davies misses this point. Indeed, he also argues that:

> a recontracting Universe arriving at the big crunch would not undergo "deflation," for this would require an exceedingly improbable conspiracy of quantum coherence to reverse-tunnel through the phase transition. There is thus a distinct and fundamental asymmetry between the beginning and the end of a recontracting Universe. (1983, p. 399)

But as Page (1983) points out, this conflicts with the argument he has given us concerning the other end of the universe. Viewed in reverse, the transition from the ordinary big bang to the inflationary phase involves exactly this kind of "improbable conspiracy." If deflation is unlikely at one end, then inflation is unlikely at the other.

For these reasons, amongst others, it is far from clear that the inflationary approach works as it stands to explain the smooth early universe. Nevertheless, it illustrates a possible strategy for doing so – an approach that involves making early smoothness probable by showing that, under plausible constraints, all or most possible universes compatible with those constraints have the feature in question.

11.11.2 The anthropic strategy

Perhaps the reason that the universe looks so unusual to us is simply that we can only exist in very unusual bits of it. We depend on the entropy gradient, and could not survive in a region in thermodynamic equilibrium. Could this explain why we find ourselves in a low-entropy region?

This is the anthropic approach, already encountered in the form of the Boltzmann hypothesis. As in that case, the idea is interesting, but faces severe difficulties. For one thing, it depends on there being a genuine multiplicity of actual bits of a much larger universe, of which our bit is simply some small corner. It is no use relying on other merely *possible* worlds, since that would leave us without an explanation for why ours turned out to be the real world. (If it hadn't turned out this way, we wouldn't have been around to think about it, but this doesn't explain why it did turn out this way.) So the anthropic solution is very costly in ontological terms. It requires that there be vastly more "out there" than we otherwise expect.

All the same, this would not be a disadvantage if the cost was one that we were committed to bearing anyway. It turns out that according to some versions of the inflationary model, universes in the normal sense are just bubbles in some vast foam of universes. So there might be independent reason to believe that reality is vastly more inclusive than it seems. In this case, the anthropic view does not necessarily make things any worse.

The second difficulty is that, as Penrose (1979, p. 634) emphasizes, there may well be much cheaper ways to generate a sufficient entropy gradient to support life. The observed universe seems vastly more unlikely than intelligent life requires. Again, this is close to an objection to Boltzmann's view. We noted that Boltzmann's suggestion implies that, at any given stage, we should not expect to find more order than we have previously observed. The same seems to apply to the contemporary argument. Life as we know it doesn't seem to require an early universe that is smooth everywhere, but only one that is smooth over a sufficiently large area to allow a galaxy or two to form (and to remain relatively undisturbed while intelligent life evolves). This would be much cheaper in entropy terms than global smoothness.

However, the inflationary model might leave a loophole here too. If the inflationary theory could show that a smooth universe of the size of ours is an all-or-nothing matter, then the anthropic argument would be back on track. The quantum preconditions for inflation might be extremely rare, but this would not matter, so long as there is enough time in some background grand universe for them to be likely to occur eventually, and it is guaranteed that when they do occur a universe of our sort arises, complete with its smooth boundary.

So, the anthropic strategy cannot be excluded altogether. It depends heavily on the right sort of assistance from cosmological theory, but if that were forthcoming, this approach might explain why we find ourselves in a universe with a low-entropy history. If so, however, then there is hugely more to reality than we currently imagine, and the concerns of contemporary astronomy pale into insignificance in comparison.

11.11.3 Penrose's Weyl hypothesis

The writer who has done most to call attention to the importance and specialness of the smooth big bang is Roger Penrose. Penrose himself proposes that there must be an additional law of nature, to the effect that the initial extremities of the universe obey what amounts to a smoothness constraint. In technical terms, his hypothesis is that the so-called Weyl curvature of spacetime approaches zero in this initial region (see Penrose, 1979; and particularly Penrose, 1989, ch. 7).

We might object to the use of the term "initial" here. Isn't Penrose presupposing an objective distinction between past and future? But the difficulty is superficial. Penrose's claim need only be that it is a physical law that there is one temporal direction in which the Weyl curvature always approaches zero toward the universe's extremities. The fact that conscious observers inevitably regard that direction as the past will then follow from the sort of argument already made by Boltzmann.

Another objection might be that Penrose's proposal does little more than simply redescribe the smooth early universe. There is some justice in this comment, but the proposal might nevertheless constitute theoretical progress. By characterizing what needs to be explained in terms of the Weyl curvature, it might provide the right focus for further and deeper theoretical explanation – say, from quantum cosmology.

One important issue is whether, as Penrose thinks, his proposal needs to be time-asymmetric, or whether the Weyl curvature might approach zero toward the extremities of the universe in both directions. This alternative would do just as well at explaining the smoothness of the big bang, and have the advantage of not introducing a new time asymmetry into physics. However, it would imply that entropy would eventually decrease if the universe recontracts in the distant future. This is an unpopular view, but it turns out that most of the arguments that cosmologists give for rejecting it are rather weak. They amount to pointing out that such an outcome would be unlikely – as indeed it would be, by ordinary Boltzmann lights. But, as in the case of the past low-entropy condition, the whole point of the extra condition would be that the Boltzmann measure is not the last word, once cosmological factors are taken into account. So there seems little reason to prefer Penrose's asymmetric hypothesis to the symmetric version of the same thing. (One interesting question is whether a future low-entropy condition would have observable present effects. For more on this and related issues, see Price (1996, ch. 4).)

11.12 Conclusion

The above examples give some sense of how physics might come to regard the smooth early universe as a consequence of something more basic. It is true that the project remains rather vague, but isn't this what we should expect? Looking back into the history of modern physics, we can see that the recent discovery of the smooth early universe represents a huge advance in our understanding of a puzzle that was only coming dimly into view a century ago. The size of the advance ought to remind us how hard it is to look forward, and predict the course of future physics. True, we know that in some way or other, future physics will incorporate much of current physics (much of which is surely right, so far as it goes). But we know almost nothing about the form the incorporation will take, or the nature and extent of the novelty – the new framework, within which the incorporation of the old will take place.

Concerning the smooth early universe itself, we can be reasonably confident that it, or something like it, will remain a part of future physics – and an important part, given its centrality to explanation of something so crucial as the temporal bias of thermodynamics to the nature of the world in which we find ourselves. But whether it will remain fundamental in its own right, something not further explained elsewhere in our new theories, we simply don't know. The best we can do is to trust our intuitions, and see what we can find.

I've argued that these intuitions benefit from some philosophical therapy, to prevent us from taking too seriously some old concerns about the project of explaining initial conditions. With the benefit of such therapy, I think that most physicists' intuitions, like mine, will be that there is an important explanatory project here, that there is likely to be something interesting to find. As to whether we're right, of course, only time itself will tell – but only if we try!

Bibliography

Albert, D. 1994: The foundations of quantum mechanics and the approach to thermodynamic equilibrium. *The British Journal for the Philosophy of Science*, 45, 669–77.

— 2000: *Time and Chance*. Cambridge, MA: Harvard University Press.

Barbour, J. 1999: *The End of Time*. London: Weidenfeld and Nicholson.

Boltzmann, L. 1877: Über die Beziehung zwischen des zweiten Hauptsatze der mechanischen der Wärmetheorie. *Sitzungsberichte de kaiserlichen Akademie der Wissenschaften*, Wien, 75, 67–73. Reprinted in translation in Brush (1966, pp. 188–93).

— 1895: On certain questions of the theory of gases. *Nature*, 51, 413–15.

— 1897: Zu Hrn. Zermelo's Abhandlung Über die mechanische Erklärung irreversibler Vorgänge. *Annalen der Physik,* 60, 392–8.

Brush, S. 1966: *Kinetic Theory*, vol. 2: *Irreversible Processes*. Oxford: Pergamon Press.

Burbury, S. H. 1894: Boltzmann's minimum function. *Nature,* 51, 78.

— 1895: Boltzmann's minimum function. *Nature,* 51, 320.

Callender, C. 1997: Review of H. Price, *Time's Arrow and Archimedes' Point. Metascience,* 11, 68–71.

— 1998: The view from no-when. *The British Journal for the Philosophy of Science*, 49, 135–59.

Davies, P. 1983: Inflation and time asymmetry in the universe. *Nature*, 301, 398–400.

Earman, J. 1974: An attempt to add a little direction to the problem of the direction of time. *Philosophy of Science,* 41, 15–47.

Feynman, R. 1967: *The Character of Physical Law*. Cambridge, MA: The MIT Press.

Hall, E. H. 1899: Review of S. H. Burbury, *The Kinetic Theory of Gases* (Cambridge: Cambridge University Press, 1899). *Science, New Series*, 10, 685–8.

Hawking, S. W. 1985: Arrow of time in cosmology. *Physical Review, D,* 33, 2, 489–95.

Linde, A. 1987: Inflation and quantum cosmology. In S. W. Hawking and W. Israel (eds.), *Three Hundred Years of Gravitation*. Cambridge: Cambridge University Press, 604–30.

Mellor, D. H. 2002: Too many universes. In N. A. Manson (ed.), *God and Design: The Teleological Argument and Modern Science*. London: Routledge, 221–8.

Page, D. 1983: Inflation does not explain time asymmetry. *Nature*, 304, 39–41.

Penrose, R. 1979: Singularities and time-asymmetry. In S. W. Hawking and W. Israel (eds.), *General Relativity: An Einstein Centenary*. Cambridge: Cambridge University Press, 581–638.

— 1989: *The Emperor's New Mind*. Oxford: Oxford University Press.

Price, H. 1996: *Time's Arrow and Archimedes' Point: New Directions for the Physics of Time*. New York: Oxford University Press.

— 2002a: Boltzmann's time bomb. *The British Journal for the Philosophy of Science*, 53, 83–119.

— 2002b: Burbury's last case: the mystery of the entropic arrow. In C. Callender (ed.), *Time, Reality and Experience*. Cambridge: Cambridge University Press, 19–56.

Sklar, L. 1993: *Physics and Chance: Philosophical Issues in the Foundations of Statistical Mechanics*. Cambridge: Cambridge University Press.

Further reading

Callender, C. 1999: Reducing thermodynamics to statistical mechanics: the case of entropy. *Journal of Philosophy*, 96, 348–73.

— 2001: Thermodynamic time asymmetry. In the *Stanford Online Encyclopedia of Philosophy*, http://plato.stanford.edu

Goldstein, S. 2001: Boltzmann's approach to statistical mechanics. In J. Bricmont, D. Durr, M. C. Galavotti, G. Ghirardi, F. Pettrucione, and N. Zanghi (eds.), *Chance in Physics*. Berlin: Springer-Verlag, 39–54.

Hawking, S. W. and Penrose, R. 1996: *The Nature of Space and Time*. Princeton, NJ: Princeton University Press.

CHAPTER TWELVE

There is no Puzzle about the Low-Entropy Past

Craig Callender

12.1 A Story

Suppose that God or a demon informs you of the following future fact: despite recent cosmological evidence, the universe is indeed closed and it will have a "final" instant of time; moreover, at that final moment, all 49 of the world's Imperial Fabergé eggs will be in your bedroom bureau's sock drawer. You're absolutely certain that this information is true. All of your other dealings with supernatural powers have demonstrated that they are a trustworthy lot.

After getting this information, you immediately run up to your bedroom and check the drawer mentioned. Just as you open the drawer, a Fabergé egg flies in through the window, landing in the drawer. A burglar running from the museum up the street slipped on a banana peel, causing him to toss the egg up in the air just as you opened the drawer. After a quick check of the drawer, you close it. Reflecting on what just happened, you push your bed against the drawer.

You quit your job, research Fabergé eggs, and manage to convince each owner to place a transmitter on his egg, so that you can know the egg's whereabouts from the radar station in your bedroom. Over time you notice that, through an improbable set of coincidences, they're getting closer to your house. You decide to act, for the eggs are closing in and the news from astronomers about an approaching rapid contraction phase of the universe is gloomy. If – somehow – you can keep the eggs from getting into the drawer, perhaps you can prevent the world's demise. (Already eight eggs are in the drawer, thanks to your desire to peek and your need for socks.) Looking out your window, you can actually see eggs moving your way: none of them breaking laws of nature, but each exploiting strange coincidences time and again. Going outside, you try to stop them. You grab them and throw them away as far as you can, but always something – a bird, a strange gust of wind – brings the egg back. Breaking the eggs has proved impossible for the same kinds of reasons. You decide to steal all of the eggs, seal them in a titanium box and bury it in Antarctica. That, at least, should buy some time, you think. Gathering all the eggs from outside, you go upstairs to get the ones from the drawer. The phone rings. It's a telemarketer selling life insurance. You decide to tell the telemarketer that their call is particularly ill timed and absurd, given that the universe is about to end. Absent-mindedly, you sit down, start speaking, put the eggs down in the open bureau drawer . . . and the universe ends.

Drop God or the demon from the story. Perhaps science posits this strange future state as a way of explaining the otherwise miraculous coincidences found among the eggs. After all, the hypothesis has some power in explaining the whereabouts of these eggs; it is also very simple. Conditionalizing on it makes all sorts of improbable events probable.

If the standard Boltzmannian explanations of entropy increase and the direction of time are right, then contemporary science posits something *vastly* more surprising than the Fabergé egg hypothesis in the story. It posits what is sometimes called the "Past Hypothesis," the claim that the global entropy at the beginning of the universe is very low. Viewed backward in time, each fundamental physical system in the world is carefully orchestrating its movements to evolve to a low-entropy state roughly 15 billion years ago. The Past Hypothesis demands behavior that is more unlikely than trillions of chicken eggs finally surrounding your bedroom.

Surely this monstrously improbable state – call it the "Past State" – deserves explanation. In his companion chapter (chapter 11), Huw Price argues, in accord with our initial intuitions, that it does. Here, I will argue that when one sees what an explanation of this state involves, it is not at all clear that it can or should be explained. By positing the Past State, the puzzle of the time asymmetry of thermodynamics is solved, to all intents and purposes. (Although I am here claiming that the puzzle of time asymmetry in thermodynamics is effectively solved, there are many other related issues that are still unresolved. For an entry into the foundations of statistical mechanics and time asymmetry literature see, for example, Callender (1999, 2001), Albert (2000), Goldstein (2001), Price (2002), Sklar (1993), and references therein. For a more detailed version of this argument and related issues, see Callender (2003).)

12.2 Thermodynamics and Probability

Classical phenomenological thermodynamics is an amazing science. With simple functional relationships among a few macroscopic predicates, it is able to make successful predictions about all thermal phenomena. Within its domain, there has not been a single exception found to its principal laws. One of these laws, the so-called "second law," has attracted much attention from philosophers. There is, as Price describes in his chapter, a puzzle about how to reconcile the second law with the underlying laws of physics. Gases, for instance, spontaneously relax to equilibrium, always filling their available volumes and never spontaneously returning to nonequilibrium. Why does this happen when it's perfectly possible, according to classical or quantum mechanics, for gases *not* to fill their available volumes? Why, for that matter, does heat always flow from hot to cold and not vice versa, despite cold to hot heat transitions being possible according to classical and quantum mechanics? From the mechanical perspective, these regularities just appear to be coincidences.

The answers to these questions are still controversial. However, there is a majority consensus that Boltzmann's answer is basically right – and right about similar questions regarding any other process governed by the second law of thermodynamics. Roughly, the answer goes as follows. First, separate the macroscopic scale from the microscopic scale. The macroscopic scale will be defined via properties such as pressure, temperature, entropy, volume, and so on. The microscopic scale will be defined via the positions and momenta of all the particles (in classical mechanics) or the quantum wavefunction (in quantum mechanics). Consider the fact that many different microstates can realize the very same macrostate. Slightly changing the position of one proton in your shoe, for instance, will not alter the temperature of the shoe. Boltzmann's insight was to see that those macrostates we call "equilibrium" macrostates are such that they are realized by many more microstates than those we call "nonequilibrium" macrostates. And, in general, higher-entropy macrostates can be realized by more microstates (classical or quantum) than those with lower entropy.

The idea is familiar from playing cards. In a game of poker, with 52 cards and five-card hands, there are over 2.5 million possible hands. Consider the "macrostates" called *royal flush* and *no points*. There are four hands, or "microstates," that can give one a *royal flush*, yet there are over one-and-a-half million hands that correspond to *no points*. Not surprisingly, if the deck is fair, *royal flush* is rare and *no points* is the norm. The explanation of the approach to equilibrium relies on a similar asymmetry: nonequilibrium macrostates are rare, so if a microstate is in a state corresponding to a low-entropy state, it will most likely evolve to a state closer to equilibrium. In this way, one says that it's more likely for temperatures to be uniform throughout the joint system than not, and thus more likely for heat to flow from hot to cold than from cold to hot.

Notoriously, this Boltzmannian explanation works in both temporal directions. As Price nicely explains, neither the combinatorial arguments nor the laws of physics introduce a temporal asymmetry, so on this theory, entropy, which is maximized at equilibrium, would increase toward the future *and* past, contrary to the observed facts. We need to break the symmetry. How? One way is to stipulate a solution to this problem by positing a cosmological hypothesis that states that in the distant past the global macrostate is one of very low entropy. How low? *Really* low: low enough to make thermodynamic generalizations applicable for the roughly 15+ billion years we think these generalizations held or will hold. This hypothesis, suggested by Boltzmann and adopted by Schrödinger, Feynman, and others, is the "Past Hypothesis" mentioned above (coined as such by Albert, 2000) and we'll call the state it posits the "Past State." If the Past Hypothesis is true, then the most probable history of the universe is one wherein entropy rises because it started off so low.

Huw Price (1996) argues that when one appreciates the above situation, the appropriate question to ask in foundations of statistical mechanics is no longer "Why does entropy rise?" but, rather, "Why was it ever low to begin with?" I agree with him that we don't really need to ask why it rises, for rising is what is "natural" for entropy to do if a system is out of equilibrium. Should we explain why it was low to begin with? To me, that sounds like asking for an explanation of the initial condition of the uni-

What does "more likely" mean here? Read one way, this question is a tricky issue in philosophy of probability (see, e.g., Sklar, 1993). But the question also arises in a straightforward mathematical sense too. Our coarse measurements of temperature, pressure, and so on are only good up to finite errors. Yet classical or quantum systems take their values – say, for position – from the continuous infinity of values associated with the real number line. Really there are infinitely many microstates that can realize any given actual macrostate. Assuming they are all equally probable, this seems to imply that each state has zero probability. Technically, this is a nonstarter. Yet giving different probabilities as a function of the number of microstates that "look like" a certain macrostate is crucial to statistical mechanics. Enter measure theory, developed originally by the French mathematician H. Lebesque. Measure theory provides a rigorous way of understanding the length or size of a set, even of sets that intuitively have no length (it also is important for the idea of an integral and many concepts in analysis, topology, and more). Importantly, it solves our problem with infinities of microstates. To see how, consider a very simple case, the interval [0, 1]. We can determine the size of the proper subset [0, 0.5] if we define the size as simply 0.5 − 0. We can therefore say the size of [0, 0.5] is greater than the size of the interval between [0.4, 0.6] despite the fact that each contains a continuous infinity of points. We are here just using the simple rule that if the interval is [a,b], then the size is given by [b − a]. Matters become much more complicated with different kinds of sets and a more rigorous and general definition of measure is then needed. But from this example one can see how the concept of measure is important in defining nontrivial probability distributions on continuously infinite sets. Measures (sizes) are defined of sets of microstates and probabilities are a function of these sizes. Note that the rule we used ([b − a]) is a very "natural" one, but in fact, formally speaking, there is an indefinite number of alternative rules. If we had used the bizarre rule [b − a³], then it would say [0, 0.5] is smaller than [0.4, 0.6].

verse, and we need not explain those for reasons that I'll describe (see also Sklar, 1993, pp. 311–18; Callender, 1998, 2003).

12.3 What We Really Want Explained

Suppose, for simplicity, that the universe is closed and has what we might call a beginning and an end. In fact, recent evidence suggests that it may keep expanding forever, but there are no laws of nature saying that it must. Suppose also that the spacetime is well behaved enough causally that we can foliate it via a succession of distinct spatial hypersurfaces at different times. Then we can say that the universe has initial and final slices of time.

Is there reason to think that the labels "initial" and "final" carry any significance when it comes to explanations? In particular, is there reason to think that final conditions need explanation but initial ones do not?

One can imagine reasons for thinking so. For instance, if the laws of nature were not time-reversal invariant (i.e., if they cared about the direction of time), then only one temporal ordering of slices of time of the universe would be allowed by the laws – the opposite ordering would break the laws. This law-like asymmetry may sanction an explanatory asymmetry. Consider the state of some system S at time t_1 and another state S^* at t_2. If the laws of nature tell us the probability of evolving from S at t_1 to S^* at t_2 *but not vice versa*, then we'll have explanations from t_1 to t_2 but not vice versa. Given t_2, the laws entail *nothing* about the state at t_1. It would therefore be difficult to know what considerations to use in explaining t_1 in terms of what happens at t_2. At any rate, except where noted, the present debate operates under the assumption, for better or worse, that the fundamental laws are time-reversal invariant (on whether they are, see Callender, 2000).

Another reason for treating final and initial conditions differently – one that I very much want to resist – is the thought that it follows from what we mean by "explanation," or more specifically, "causation." Many philosophers are attracted to the idea that good scientific explanations are causal explanations: an event E is explained by citing one or more of its causes C. But causation is typically temporally asymmetric. Usually, if C causes E, then C precedes E. On this view, final conditions can be explained because there are possible causes that precede final conditions, but initial conditions cannot be explained because there are no times that precede them. I do *not* want to make this argument. To do so would be to commit what Price calls a "temporal double standard." Recall the Fabergé egg story. Doesn't knowing the final state of the universe explain why the eggs were moving ever closer to the sock drawer? I am loath not to count this as an explanation merely due to its unusual time direction. Doing so without saying more would clearly beg the question in this context, as Price's chapter shows.

I want to treat initial and final conditions the same way. Hence, in claiming that we shouldn't explain initial conditions of the universe, I should also state that we shouldn't explain final conditions of the universe. But isn't this an odd thing to say? Why shouldn't we try to explain why that state is the way it is?

The key to seeing the answer is to see that the issue is not really about initial conditions versus final conditions, nor even about these versus what lies between. In one sense of "explain," we certainly can explain the initial or final state. Just grab some state before the final state, evolve it in accord with the laws of nature and show what it yields for the final state. In figure 12.1, we might show how the final state arises from S(B) and the laws of nature. Alternatively, we can also explain the initial state of the universe in terms of a state after it, assuming that the laws are time-reversal invariant. For instance, we might show how the initial state arises from S(A) and the laws. The only objection that one might have to this explanation would be one committing Price's temporal double standard.

I trust that Price and others are after something much more than this kind of explanation. When people ask for an explanation of the Past Hypothesis, they are not merely asking whether the Past State is consistent with the laws and conditions on some other time slice. All hands agree that this is at least possible. Rather, the feature that cries out for explanation is that the Past State is a state that is incredibly *improbable* according to the standard measure used in statistical mechanics. Penrose (1989, p. 344) estimates that the probability of this particular type of past state occurring is 1 out of $10^{10^{123}}$ and Kiessling (2001) estimates that it is infinitely improbable!

I have worries about both calculations. But however it is calculated, one must admit that the Past State is going to be monstrously unlikely if the calculation is done with

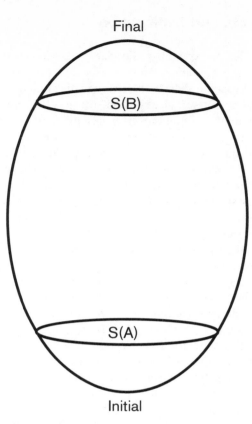

Figure 12.1 A closed time-symmetric universe.

the standard measure. If we want to understand 15 billion years of dynamical evolution as evolution always to more probable states, that first state will need to be very improbable.

Can anything explain this unlikely state? Price sometimes says he wants "some sort of lawlike narrowing of the space of possibilities, so that such a universe [one with a Past Hypothesis] no longer counts as abnormal" (2002, p. 116). I'm skeptical that one might do this while retaining Boltzmann's story and, more importantly, I'm skeptical about the initial motivation to explain the Past State. Here it is interesting to note that scientists also appear to disagree about whether it should be explained. Boltzmann writes "That in nature the transition from a probable to an improbable state does not take place as often as the converse, can be explained by assuming a very improbable initial state of the entire universe surrounding us. This is a reasonable assumption to make, since it enables us to explain the facts of experience, and *one should not expect to be able to deduce it from anything more fundamental*" (1897; my emphasis). By contrast, Kiessling and others think that it points to a "need for a deeper postulate" (2001, p. 86). As I'll show, this tension within science and philosophy about explanation has echoes in many other areas as well.

12.4 Brute Facts and Explanation

My objection to the idea of explaining boundary conditions originates in arguments by David Hume. Consider St. Thomas Aquinas's classic cosmological argument for the existence of God. We assume that every effect in the universe must have a cause. Otherwise there would be no "sufficient reason" for the effect. But if every effect must have a cause, said Aquinas, we find ourselves in a dilemma: either there was an infinite chain of causes and effects or there was a first cause, the Uncaused Cause (God). Not believing that an infinite chain of causation would be explanatory (for reasons that are not entirely compelling now), Aquinas concluded that there was an Uncaused Cause. Similar arguments from motion yielded an Unmoved Mover. There are several objections to these classic arguments. One reaction popular among students is to ask, as Hume did, what caused or moved God? This question raises many more. Should we posit an infinite regress of gods, in keeping with the original explanatory demand? Or should we "bend" the explanatory demand so that in the case of God, He doesn't have to be caused by something distinct from Himself? But then, one thinks, if it's acceptable for something to cause itself or to selectively apply the explanatory demand, we have gone a step too far in positing God as the causer or mover of the universe. Just let the universe itself or the big bang be the "first" mover or cause and be done with it.

Although the situation with the Past Hypothesis is more complicated, at root the above is my criticism of Price. What would explain a low-entropy past state? The most natural thing to say is that an even lower entropy state just before the Past State would explain it. The natural "tendency" of systems is to go to equilibrium, after all. The original low-entropy past state would naturally and probably evolve from an earlier and lower-entropy state. But now that lower-entropy state is even more unlikely than the original. Either we just keep going, explaining low-entropy states in terms of lower ones *ad infinitum*, or we stop. And when we stop, should we posit a first Unlow Low-Entropy State (or in Price's terminology, a Normal Abnormal State)? No. We should just posit the original low-entropy state and be done with it.

Are there different theoretical explanations of the Past State, ones not appealing to earlier low-entropy states? Maybe, but here I am skeptical for reasons again enunciated by Hume. In his *Dialogues Concerning Natural Religion*, Hume has Philo argue:

> ... the subject in which you [Cleanthes] are engaged exceeds all human reason and inquiry. Can you pretend to show any such similarity between the fabric of a house and the generation of a universe? Have you ever seen Nature in any situation as resembles the first arrangement of the elements? Have worlds ever been formed under your eye ...? If [so] ... then cite your experience and deliver your theory. (Hume, 1980, p. 22)

His point is that since the cosmos happens only once, we cannot hope to gain knowledge of any regularities in how it is created. This, I take it, implies that we will not be able to defend any grand principle of how contingent matter–energy sources are distributed at the boundaries of the universe, for what justification would we ever have for such a principle?

There are at least two worries buried in this discussion. One is an empiricist worry about the justification that one would have for any grand principle that would explain why the initial conditions are what they are. The second is a more general question about judging when certain basic facts need explanation and when they don't. The design argument assumes that some purported basic facts, such as the big bang, are facts in need of explanation, whereas other purported basic facts, such as God, are not. But what is the difference? Are some basic facts acceptable and others not? Is there a criterion that separates the facts that need explanation from those that do not? What makes the "new" basic fact better than the old?

The two worries are often linked. Consider an old chestnut in the history and philosophy of science; namely, the example of scientists rejecting Newton's gravitational theory because it posited an action-at-a-distance force. Such a force could not be basic because it was judged to be not explanatory. But *a priori*, why are nonlocal forces not explanatory and yet contact forces explanatory? This is the second objection above. Furthermore, note that believing Newton's action-at-a-distance to be problematic stimulated scientists to posit all manner of mechanisms that would restore contact forces. Not only were these efforts ultimately in vain, but many of the posits came at the price of these mechanisms not being independently testable. Thus enters the first objection.

I see the same problem in Price's claim that the Past State needs to be explained. What is it about the Past State that makes it needy of further explanation? Why can't it simply be a brute fact or the Past Hypothesis be a fundamental law? One answer might be to accept that the Past State plus laws are empirically adequate yet find fault with them for lacking some theoretical virtue or other. Empiricists – those who see empirical adequacy as the only criterion that really matters – will not like this, but others will. Which theoretical virtue is the Past State lacking? It is simple, potentially unifying with cosmology, and it has mountains of indirect evidence via our evidence for thermodynamics and whatever mechanics we're considering. But still, it is highly improbable. Although we can reasonably worry about what exactly it means to say that a state of the entire universe is improbable, we can postpone such worries here, since that is not the source of Price's problem. The standard probability distribution pulls its weight in science and seems to be a successful theoretical posit in science. Can the improbability of the state mean that it can't be true or that it is needy of explanation? Well, the Past State can certainly be true; virtually everything that happens is unlikely. What about explanation? I don't think that explanation and probability have such a tidy relationship. Lots of low-probability events occur and not all of them demand explanation. Arguably, low-probability events can even function as the explananda, not merely the explanans. For example, an asteroid strike in the Yucatan region might explain the death of the dinosaurs, even though (arguably) the prior probability of the asteroid strike is lower than that of the dinosaurs' extinction (see Lange, 2002, p. 108). It is far from automatic that low- probability events all deserve explanation. Furthermore, the sorts of explanations of the Past Hypothesis that Price (1996) envisions seem to me to be examples that are wildly speculative, potentially untestable, and not obviously more probable.

My own view is that there is not some feature of facts that makes them potentially acceptably brute or self-explanatory, that makes some facts acceptable as brute and others not. Instead, we look at the theoretical system as a whole and see how it fares empirically, and if there are ties between systems then we look to various theoretical virtues to decide (if one is realist). What we don't want to do is posit substantive truths about the world *a priori* to meet some unmotivated explanatory demand – as Hegel did when he notoriously said there *must* be six planets in the solar system. In the words of John Worrall (1996),

> the worst of all possible worlds is one in which, by insisting that some feature of the universe cannot just be accepted as "brute fact", we cover up our inability to achieve any deeper, testable description in some sort of pseudo-explanation – appealing without any independent warrant to alleged a priori considerations or to designers, creators and the rest. That way lies Hegel and therefore perdition. (p. 13)

Price and the scientists he mentions are of course free to devise alternative theoretical systems without the Past Hypothesis. So far, there is not much on the table. And what is on the table doesn't look very explanatory, at least according to my (perhaps idiosyncratic) intuitions about explanation. (I try to back up this claim in section 12.6.)

There is an echo of our debate in quantum field theory. Famously, many physicists complain that the standard model in particle physics contains too many fundamental parameters. Here is the physicist Sheldon Glashow:

> Although (or perhaps, because) the standard model works so well, today's particle physicists suffer a peculiar malaise. Imagine a television set with lots of knobs; for focus, brightness, tint, contrast, bass, treble, and so on. The show seems much the same whatever the adjustments, within a large range. The standard model is not like that. . . . The standard model has about 19 knobs. They are not really adjustable: they have been adjusted at the factory. Why they have their values are 19 of the most baffling metaquestions associated with particle physics. (Glashow, 1999, p. 80)

Feeling that these 19 "knobs" are too *ad hoc*, physicists strive in many directions. Some seek to prove the standard model false in an experiment, others search for some "meaningful pattern" among these parameters, and yet others devise theories such as superstring theory to deal with the problem. But why can't these 19 knobs be brute? Why can't, to take an example, the muon just be 200 times as heavy as the electron and that be that? Glashow even asks why is there a muon at all. But is *everything* to be explained? Will all models of the universe be deficient until physics answers why there is something rather than nothing? Surely that is too strong an explanatory demand to impose on physics.

Again, it seems to me that it's perfectly within the rights of the physicist to find something ugly about the standard model and want to devise an alternative without so many knobs. When that alternative exists and is shown to be empirically adequate, we can then compare the two. It may well be that it is superior in various empirical (one would hope) *and* theoretical ways. But to know *beforehand*, as it were, that the existence of muons can't be brute seems to me too strong. Similarly, knowing beforehand that the Past Hypothesis needs explanation seems too strong.

I now want to turn to two tasks. First, in agreement with Price, I want to show that the Past Hypothesis operates as a fundamental law. If one agrees that it is a law, then it is particularly puzzling to me to insist that it demands explanation, as if laws wear on their sleeve their appropriateness as fundamental laws. Secondly, I then want to sketch why none of the ways I can imagine of explaining the Past State really count as explaining the Past State.

12.5 The Past Hypothesis as Law

There are many different conceptions of what a law of nature is. See, for example, chapters 7 and 8 of this volume, by John Roberts and Harold Kincaid. According to some, the Past Hypothesis counts as a law and according to others it probably doesn't. Here, I simply want to point out that according to one view of laws of nature, one attractive to some contemporary empiricists, the Past Hypothesis would count as a law. This conception is known as the "Ramsey–Lewis" account of laws, after the philosophers Frank Ramsey and David Lewis. The Ramsey–Lewis theory would seem to abruptly end our debate, for it will declare that the Past Hypothesis doesn't call for explanation. Why was entropy low in the past? "Because it's physically impossible for it not to be," a real conversation stopper, answers this question.

Lewis describes the theory as follows:

> Take all deductive systems whose theorems are true. Some are simpler and better systematized than others. Some are stronger, more informative than others. These virtues compete: An uninformative system can be very simple, an unsystematized compendium of miscellaneous information can be very informative. The best system is the one that strikes as good a balance as truth will allow between simplicity and strength. How good a balance that is will depend on how kind nature is. A regularity is a law iff it is a theorem of the best system." (Lewis, 1994, p. 478)

Roughly, the laws of nature are the axioms of those true deductive systems with the greatest balance of simplicity and strength. Imagine that you are God the Programmer and you want to write a program for the universe. Merely listing every fact would make for a very long program. And simply writing "anything goes," while short and sweet, doesn't allow one to infer much in particular. The laws of nature are those lines you would write in the program. Loewer (2001) develops the idea of the "best system" further, showing how it can reconcile chance with an underlying determinism. Great gains in simplicity and strength can be achieved by allowing probabilities of world histories given the system of axioms.

Look at our world with this idea of law in mind. We try to find the simplest most powerful generalizations we can. Dynamical laws such as Newton's second law and Schrödinger's equation are great, because they're remarkably strong and simple. But many more patterns are detectable in the world that do not follow from such laws, such as all thermal regularities. So, we might also introduce some special science laws and probabilistic laws to capture these regularities.

Suppose, as seems to be the case, that capturing these thermal regularities implies positing a special state of the universe very early on. Since this special state is so fantastically powerful, allowing us to make predictions in thermodynamics and else-where, and yet so simply state-able, it seems very likely that the Past Hypothesis would be among the best system's axioms. Lewis himself seems to be reluctant to call axioms "laws" if they happen only once. He writes, "the ideal system need not consist entirely of regularities; particular facts may gain entry if they contribute enough to collective simplicity and strength. (For instance, certain particular facts about the big bang might be strong candidates.) But only the regularities of the system are to count as laws" (1983, p. 367). Apart from the possible ordinary-language uneasiness of calling an initial state a "law," I can't see any motivation for claiming that some of the axioms of the Best System are laws but not others. In any case, it would have the same kind of status – axiom status – as the laws.

Price is right to point out that we use the Past Hypothesis in a law-like manner. We don't continually rewrite the Past Hypothesis as we notice that entropy keeps increasing. We posit entropy to be low enough for the observed past *and* the inferred future. We don't expect entropy to start decreasing in the next moment. But this just means that the Past Hypothesis is functioning as a brute explainer in the best sys-tematization that we have of the world.

12.6 Explaining the Past State

What would explain the Past State? The answer to this question hangs in large part on what we mean by "explanation" (and even "we" and "can"), so there is plenty of room here for people to talk past one another. Not only are there many different the-ories of what scientific explanation is, but there are many different contexts in which explanation is needed. I can't hope to – nor do I aspire to – show that the Past State can't be explained according to any conception of explanation. What I can do, however, is show that none of the ways you might have thought of to explain it are so promising or so explanatory. In light of the previous discussion, I hope this makes the reader see that being stuck with a Past State is not so bad.

Let's suppose the Past State is the very first state of the universe. I've argued else-where (Callender, 2003) that it doesn't make a difference if it isn't. Then I believe that there are essentially three different ways to explain the Past State. One could (a) rewrite the dynamics so that the Past State would be generic in the solution space of the new dynamics, (b) add a new nondynamical law of nature, or (c) eliminate the measure that makes such states abnormal. In other words, we could make the Past State likely (as a result of new dynamics or not) or make it "a-likely." Let's take these options in turn.

12.6.1 Dynamical explanations

Consider the so-called "flatness problem" in standard big bang cosmology. It is a problem structurally very similar to the one under consideration. What is the problem? Our universe right now appears to be very nearly flat; that is, spacetime is curved, but only very slightly. Cosmologists are interested in a parameter known as Ω, the so-called critical density. Near-flatness in our universe means that Ω must now be very close to 1. The trouble is that the laws of nature governing the dynamics of spacetime – Einstein's field equations – are such that (using a model known as the Friedman model) departures from "flat" curvature should grow larger with time. If the universe began with even minute curvature irregularities in it, these should be greatly enlarged by now. But the universe presently appears to be roughly flat, so the Friedman equations demand that it must have been even closer to flat much earlier. In terms of Ω, for Ω to now fall anywhere near 1, say even in the range $0.1 \leq \Omega \leq 10$, near the beginning of the universe Ω must have been equal to 1 within 59 decimal places. So-called "inflationary scenarios" are the main response to the flatness problem and other similar problems. Inflation posits a period of very fast universal expansion that would, if correct, reduce the "specialness" of the initial conditions needed for long-lasting near-flatness of the universe. (Note that, unlike in the statistical mechanical case, there is considerable controversy regarding which measure is appropriate – and even whether there exist well-defined measures on the relevant spaces of cosmological histories. For more about this, see Ellis (1999, p. A61), Evrard and Coles (1995), and Callender (2003).)

Note the similarity to our problem. A vast regularity (entropy increase, flatness) is compatible with the laws of nature only if the universe began in an immensely unlikely state. Physicists do seem to think that this is a problem. The counterpart of the Past Hypothesis, namely the "past is *really* flat hypothesis," demands explanation, they think, and so they posit inflation. I'm not so sure that it does (for some of the reasons, see Earman, 1995). Even if I agreed, however, consider an important difference between what the physicists want and what Price wants. With inflation, physicists would *explain away* the "past is really flat hypothesis." Price, by contrast, wants to keep the Past Hypothesis and explain why that Past State obtains. Yet the counterpart of inflation in the present case is modifying or replacing classical or quantum dynamics so that the Past State itself emerges as the natural product of the dynamics. That is not what Price is after. It seems that he has lost sight of the explanandum. Originally, we wanted to explain *thermal phenomena*. But a dynamics that makes thermal phenomena inevitable or generic would *remove the need* for a Past Hypothesis.

If we stick with Price's new explanandum, explaining the Past State itself, and it is the first state, then any kind of novel causal process that brings this state about would seem to require referring to a mechanism outside of spacetime – or adding more spacetime to current models than is warranted by empirical evidence. But that seems to me suspiciously akin to adding some untestable mechanism merely to satisfy one's *a priori* judgment of what facts can be brute. Worse, why should I prefer this brute element over the previous one?

Price would say that we know that the Past State, as described, occurred independently of the inference we make from present thermal phenomena, unlike the special state required by noninflationary cosmology to explain flatness and temperature.

I have two replies to this response. First, do we really have such strong independent evidence that in the distant past entropy was low? Price is thinking that we already independently know that the Past State was of low entropy, for we have lots of evidence for a relativistic hot big bang model of the universe in the usual interpretations of cosmic red shifts and the cosmic background radiation. There is no doubt that the model predicts a highly concentrated past state. Going backward in time, we see the density decrease until the big bang. Price's claim hangs on this initial concentrated state being rightly described as a state of low thermodynamic entropy. It is, of course, but I'm not sure that we know this independently solely from Einstein's equations. As I understand matters, the expansion of the universe is what is called an "adiabatic" process, one that is "isentropic" – that is, its entropy doesn't change – and reversible. Sometimes people talk of "gravitational entropy," but this is very speculative and not obviously related to the thermodynamic entropy. Secondly, and putting details aside, even if we know the big bang state is one of low entropy, there is simply no reason to explain it rather than anything else unless it is low according to a measure that is actually being used in science. That is, only the Boltzmannian explanation in statistical mechanics gives us reason to think that it's improbable in any objective sense. But if the new dynamics makes the Past State likely, then we're no longer using the Boltzmann explanation.

12.6.2 New nondynamical law?

In the sense described in section 12.5, I believe that this is the right option. A good theory of laws of nature tells us that we ought to count the Past Hypothesis itself as a law. Does this count as an explanation of the Past Hypothesis? Clearly not. This move doesn't so much explain the Past Hypothesis as state that it doesn't need explanation because it is nomic.

Aren't laws sometimes explained in terms of other laws? Yes; indeed, we're currently operating under the assumption that the "laws" of thermodynamics follow from the laws of mechanics, the law of low past entropy, and a statistical postulate. Laws can help explain other laws. So if we imagined some new more fundamental laws it might be the case that these explain the Past State in some sense. It is entirely possible that physics be completely "re-packaged" in the distant future.

Before getting carried away with this idea, however, think carefully about what would actually count as explaining the Past State. If this re-packaging of physics gets rid of the Boltzmann story, then there may be no need to think of the Past State as unlikely; hence there would be no reason to deem it needy of explanation (if unlikeliness alone even counts as a reason). We will then have "explained" the state only by removing the idea that it is an unlikely state.

Concentrating on new nondynamical laws, we're imagining a possibility that the new law is nondynamical and yet not simply the statement that the Past Hypothesis is law-like. Price often seems to have such a picture in mind when he considers Penrose's (1989) Weyl Curvature Hypothesis. This hypothesis says, loosely, that a particular measure of spacetime curvature known as the Weyl curvature vanishes near the initial singularity and that this vanishing implies low entropy. I have three points to make about Price's treatment of this proposal. First, Penrose understands the Weyl Curvature Hypothesis as a time-asymmetric law, a claim about *past* singularities having low entropy. To me, it seems a bit of a stretch to say that such a principle explains the Past Hypothesis. *If* the Weyl curvature is related to the thermodynamic entropy, then the Weyl Curvature Hypothesis just is the Past Hypothesis, but dressed in fancy clothes. The Past Hypothesis would follow *logically*, not causally, from the Weyl Curvature Hypothesis, so I wouldn't say that the second explains the first in any interesting sense (no disrespect to logical explanations intended – they're great, especially in logic). Secondly, Price (1996) dislikes Penrose's time-asymmetric law and proposes instead a time-symmetric version of the Hypothesis, a law to the effect that singularities, wherever and whenever they are found, have low entropy. (He seems to think, *a priori*, that the laws of nature must be time-symmetric – something for which I've criticized him before: see Callender (1998).) This more risky hypothesis does make different predictions than Penrose's; for example, it predicts that any singularity that we meet in the future will be a source of low entropy. This hypothesis, I grant, is much stronger than the Past Hypothesis, but note that it goes well beyond the empirical evidence. Based on our experience of one singularity, the big bang, we simply have no idea whether or not other singularities will be low-entropy sources. If it turns out that they are, I'm happy to change the Past Hypothesis to what would essentially be the Singularity Hypothesis. We could then call this an explanation of the Past Hypothesis and I wouldn't quibble with that. I just want to insist that without any mechanism linking singularities with low entropy, this new hypothesis would be just as mysterious as the old one. Finally, note that Penrose himself does not want the Weyl Curvature Hypothesis to have the kind of stipulative status that it appears to get from Price. Rather, this hypothesis is a conjecture that it will follow, presumably dynamically, from some new quantum theory of gravity for Penrose.

12.6.3 Eliminating the initial probability distribution

There are a variety of programs that would eliminate the need for an initial probability distribution. If there is a problem with the Past State, I take it that the problem is that it is so unlikely. So if new physics could thrive without calling the Past State unlikely, that might explain the Past State by removing its improbability. There have been, and continue to be, programs that have this effect. Had Boltzmann's original H-theorem worked, every nomically possible initial condition would subsequently have its entropy increase. And, recently, Albert (2000) has claimed that this option follows from the Ghirardi–Rimini–Weber (GRW) interpretation of quantum mechanics. In GRW, the statistical-mechanical probabilities seem to be corollaries of the quantum-mechanical probabilities. Although a probability measure is used in the stochastic temporal development of quantum states, GRW doesn't require a measure over initial conditions to account for statistical mechanics. (One can also imagine adding a stochastic "kick" term to classical mechanics that would have the same results as GRW classically.)

From our point of view GRW is an interesting case, for it still requires the same past cosmological state that the Boltzmannian does – just not the probability measure over initial conditions. That is, since we know that entropy has been increasing throughout history (history is vastly more interesting than merely equilibrium-to-equilibrium transitions) it must have been much lower much earlier. So we still have a Past State, only now we don't need to say it's unlikely because the stochastic development of states will make entropy increase likely for all initial conditions. We know, however, that there is a natural well defined measure over initial conditions, namely the standard one; moreover, we know that according to this measure this past cosmological state is of extremely small size. Yet since this probability measure now is no longer needed to explain thermal phenomena (GRW does that instead) we should *not* think of this small size as an epistemic fault with the theory. The Past State, in this context, becomes a problem on the order of receiving surprising hands in card games.

12.7 Conclusion

The only types of explanation that I believe I've left out are so-called "anthropic" explanations. These explanations posit an ensemble of real worlds, each one corresponding to a possible initial condition of the universe. The Past State is then explained as follows. Sure, we are told, the Past State is unlikely, but if we factor in the (alleged) fact that intelligent life can only exist in worlds with Past States, we should not be surprised to find ourselves in a world with a Past State. We should be no more surprised than a fish caught by a net with a one foot hole in it should be surprised at finding himself in a bucket of one foot or longer fish.

This type of explanation, which has been criticized elsewhere (see Worrall, 1996), is making precisely the mistake that I warned against earlier. It is positing a substantive – and enormously extravagant – claim about the world in order to satisfy an explanatory itch that does not need to be scratched. Those seeking to explain the Past State who are not motivated by empirical considerations or inconsistencies need a reconsideration of scientific methodology more than they do the ensembles of worlds, imaginary times, and recurring universes to which they appeal.

Bibliography

Albert, D. 2000: *Time and Chance.* Cambridge, MA: Harvard University Press.

Boltzmann, L. 1897: Zu Hrn. Zermelos Abhandlung über die mechanische Erklärung irreversiler Vorgänge. *Annalen der Physik*, 60, 392–8. Reprinted and translated in Brush, S. G. 1966: *Kinetic Theory.* Oxford: Pergamon Press.

Bricmont, J., Durr, D., Galavotti, M. C., Ghirardi, G., Pettrucione, F., and Zanghi, N. (eds.) 2001: *Chance in Physics.* Berlin: Springer-Verlag.

Callender, C. 1998: The view from no-when: Price on time's arrow. *The British Journal for the Philosophy of Science*, 49(1), 135–59.

—— 1999: Reducing thermodynamics to statistical mechanics: the case of entropy. *Journal of Philosophy*, XCVI, 348–73.

—— 2000: Is time "handed" in a quantum world? *Proceedings of the Aristotelian Society*, June, 247–69.

—— 2001: Thermodynamic time asymmetry. In the *Stanford Online Encyclopedia of Philosophy*, http://plato.stanford.edu

—— 2003: Measures, explanations and the past: should "special" initial conditions be explained? *The British Journal for Philosophy of Science*, in press.

Earman, J. 1995: *Bangs, Crunches, Whimpers and Shrieks: Singularities and Acausalities in Relativistic Spacetimes.* New York: Oxford University Press.

Ellis, G. F. R. 1999: 83 years of general relativity and cosmology. *Classical and Quantum Gravity*, 16, A37–76.

Evrard, G. and Coles, P. 1995: Getting the measure of the flatness problem. *Classical and Quantum Gravity*, 12, L93–7.

Feynman, R. 1967: *The Character of Physical Law*. Cambridge, MA: The MIT Press.

Glashow, S. L. 1999: Does quantum field theory need a foundation? In T. Cao (ed.), *Conceptual Foundations of Quantum Field Theory*. Cambridge: Cambridge University Press, 74–88.

Goldstein, S. 2001: Boltzmann's approach to statistical mechanics. In Bricmont et al., op. cit., pp. 39–54.

Hume, D. 1980: *Dialogues Concerning Natural Religion*, ed. R. H. Popkin. Indianapolis: Hackett.

Kiessling, M. 2001: How to implement Boltzmann's probabilistic ideas in a relativistic world? In Bricmont et al., op. cit., pp. 83–102.

Lange, M. 2002: *An Introduction to the Philosophy of Physics: Locality, Fields, Energy and Mass*. Oxford: Blackwell.

Lewis, D. 1983: New work for a theory of universals. *Australasian Journal of Philosophy*, 61, 343–77.

Lewis, D. K. 1994: Humean supervenience debugged. *Mind*, 103, 473–90.

Loewer, B. 2001: Determinism and chance. *Studies in History and Philosophy of Modern Physics*, 32(4), 609–20.

Penrose, R. 1989: *The Emperor's New Mind*. New York: Oxford University Press.

Price, H. 1996: *Time's Arrow and Archimedes' Point: New Directions for the Physics of Time*. New York: Oxford University Press.

—— 2002: Boltzmann's time bomb. *The British Journal for the Philosophy of Science* 53(1), 83–119.

Sklar, L. 1993: *Physics and Chance*. Cambridge: Cambridge University Press.

Worrall, J. 1996: Is the idea of scientific explanation unduly anthropocentric? The lessons of the anthropic principle. Discussion Paper Series, 25/96. LSE Centre for the Philosophy of Natural and Social Sciences.

Further reading

Bricmont, J. 1995: Science of chaos or chaos in science? *Physicalia Magazine*, 17(3–4), 159–208. (Available online at *http://dogma.free.fr/txt/JB-Chaos.htm*.)

Horwich, P. 1987: *Asymmetries in Time*. Cambridge, MA: The MIT Press.

Reichenbach, H. 1956: *The Direction of Time*. Berkeley, CA: UCLA Press.

Savitt, S. (ed.) 1995: *Time's Arrow Today*. Cambridge: Cambridge University Press.

DO GENES ENCODE INFORMATION ABOUT PHENOTYPIC TRAITS?

It is has become common, not merely among scientists but also among science reporters and interested bystanders, to speak of a "genetic code," or of "genetic information." What is conveyed by these metaphors? Genes manufacture protein molecules, and in so doing affect developmental processes. Our genes play a causal role in determining how tall we will be, how intelligent we will be, whether we will develop a specific form of cancer, and so on. But our environment also has a causal hand in these matters, and we do not speak of "environmental coding." So the metaphor must convey something about the particular way in which genes affect our phenotypic traits. Morse code has a basic alphabet of "dot," "dash," and "pause." A short sequence of dots and dashes, bracketed by two pauses, "codes for" a letter of the alphabet. In this way, a sequence of dots, dashes, and pauses can be used to transmit information. The metaphors presumably suggest that genes work in an analogous fashion. DNA forms a natural alphabet: there are four base nucleotides, normally represented by the letters *A, C, G,* and *T.* But the metaphor is not vindicated until we understand what it is that genes *code for,* what *information* they transmit. Sahotra Sarkar and Peter Godfrey-Smith both agree that the best-known mathematical account of information – that elucidated by Shannon and Weaver – is not particularly helpful for elucidating the relationship between genes and phenotypic traits. Sarkar provides us with an introduction to Shannon–Weaver information theory, and then develops an alternative account of information. He argues that genes code for proteins directly, and hence indirectly encode information for phenotypic traits. Peter Godfrey Smith attempts to understand the metaphors by contrasting the actual mechanism by which genes affect phenotypic traits with alternative possible mechanisms. This leads him to conclude that it is appropriate to talk of genes encoding information for protein molecules, but not for anything "higher up" than that. This discussion provides a window on the way in which metaphors and other linguistic shortcuts may be used both to convey and to distort complicated scientific truths.

Genes Encode Information for Phenotypic Traits

Sahotra Sarkar

13.1 Introduction

According to the *Oxford English Dictionary*, the term "information" (though spelt "informacion") was first introduced by Chaucer in 1386 to describe an item of training or instruction. In 1387, it was used to describe the act of "informing." As a description of knowledge communicated by some item, it goes back to 1450. However, attempts to quantify the amount of information contained in some item only date back to R. A. Fisher in 1925. In a seminal paper on the theory of statistical estimation, Fisher argued that "the intrinsic accuracy of an error curve may . . . be conceived as the amount of information in a single observation belonging to such a distribution" (1925, p. 709). The role of information was to allow discrimination between consistent estimators of some parameter; the amount of "information" gained from a single observation is a measure of the efficiency of an estimator. Suppose that the parameter to be estimated is the mean height of a human population; potential estimators can be other statistical "averages" such as the median and the mode. Fisher's theory of information became part of the standard theory of statistical estimation, but is otherwise disconnected from scientific uses of "information." The fact that the first successful quantitative theory of "information" is irrelevant in scientific contexts outside its own narrow domain underscores an important feature of the story that will be told here: "information" is used in a bewildering variety of ways in the sciences, some of which are at odds with each other. Consequently, any account of the role of informational thinking in a science must pay careful attention to exactly what sense of "information" is intended in that context.

Shortly after Fisher, and independently of him, in 1928 R. V. L. Hartley provided a quantitative analysis of the amount of information that can be transmitted over a system such as a telegraph. During a decade in which telecommunication came to be at the forefront of technological innovation, the theoretical framework it used proved to be influential. (Garson (2002) shows how this technological context led to the introduction of informational concepts in neurobiology during roughly the same period.) Hartley recognized that, "as commonly used, information is a very elastic term," and proceeded "to set up for it a more specific meaning" (1928, p. 356). Relying essentially on a linear symbolic system of information transmission (for instance, by a natural language), Hartley argued that, for a given message, "inasmuch as the precision of the information depends upon what other symbol sequences might have been chosen it would seem reasonable to hope to find in the number of these sequences the desired quantitative measure of information" (p. 536). Suppose that a telegraphic message is n symbols long with the symbols drawn from an alphabet of size s. Through an ingenious argument, similar to the one used by Shannon (see below), Hartley showed that the appropriate measure for the number of these sequences is $n \log s$. He identified this measure with the amount of information contained in the message. (Even earlier, H. Nyquist (1924) had recognized that the logarithm function is the appropriate mathematical function to be used in this context. Nyquist's "intelligence" corresponds very closely with the modern use of "information.")

Using the same framework as Hartley, in 1948, C. E. Shannon developed an elaborate and elegant mathematical theory of communication that came to be called "information theory" and constitutes one of the more important developments of applied mathematics in the twentieth century. The theory of communication will be briefly analyzed in section 13.2 below, with an emphasis on its relevance to genetics. The assessment of relevance will be negative. When, for instance, it is said that the hemoglobin-S gene contains information for the sickle cell trait, communication-theoretic information cannot capture such usage. (Throughout this chapter, "gene" will be used to refer to a segment of DNA with some known function.) To take another example, the fact that the information contained in a certain gene may result in polydactyly (having an extra finger) in humans also cannot be accommodated by communication-theoretic information. The main problem is that, at best, communication-theoretic information provides a measure of the amount of information in a message but does not provide an account of the content of a message, its specificity, what makes it *that* message. The theory of communication never had any such pretension. As Shannon bluntly put it at the beginning of his paper: "These semantic aspects of communication are irrelevant to the engineering problem" (1948, p. 379).

Capturing *specificity* is critical to genetic information. Specificity was one of the major themes of twentieth-century biology. During the first three decades of that century, it became clear that the molecular interactions that occurred within living organisms were highly "specific" in the sense that particular molecules interacted with exactly one, or at most a very few, reagents. Enzymes acted specifically on their substrates. Mammals produced antibodies that were highly specific to antigens. In genetics, the ultimate exemplar of specificity was the "one gene–one enzyme" hypothesis of the 1940s, which served as one of the most important theoretical principles of early molecular biology. By the end of the 1930s, a highly successful theory of specificity, one that remains central to molecular biology today, had emerged. Due primarily to L. Pauling (see, e.g., Pauling, 1940), though with many antecedents, this theory claimed: (i) that the behavior of biological macromolecules was determined by their shape or "conformation"; and (ii) what mediated biological interactions was a precise "lock-and-key" fit between shapes of molecules. Thus the substrate of an enzyme had to fit into its active site. Antibodies recognized the shape of their antigens. In the 1940s, when the three-dimensional structure of not even a single protein had been experimentally determined, the conformational theory of specificity was still speculative. The demonstration of its approximate truth in the late 1950s and 1960s was one of early molecular biology's most significant triumphs.

Starting in the mid-1950s, assumptions about information provided an alternative to the conformational theory of specificity, at least in the relation between DNA and proteins (Lederberg, 1956). This relation is the most important one, because proteins are the principal biological interactors at the molecular level: enzymes, antibodies, molecules such as hemoglobin, molecular channel components, cell membrane receptors, and many (though not most) of the structural molecules of organisms are proteins. Information, as F. H. C. Crick defined it in 1958, was the "*precise* determination of sequence, either of bases in the nucleic acid or of amino acid residues in the protein" (1958, p. 153; emphasis in the original). Genetic information lay in the DNA sequence. The relationship between that sequence and the sequence of amino acid residues in a protein was mediated by the genetic "code," an idea that, though originally broached by E. Schrödinger in 1943, also dates from the 1950s. The code explained the specificity one gene–one enzyme relationship elegantly: different DNA sequences encoded different proteins, as can be determined by looking up the genetic code table (Godfrey-Smith explains the relevant biology in chapter 14 of this volume). Whatever the appropriate explication of information for genetics is, it has to come to terms with specificity and the existence of this coding relationship. Communication-theoretic information neither can, nor was intended to, serve that purpose.

Surprisingly, a comprehensive account of a theory of information appropriate for genetics does not exist. In the 1950s there were occasional attempts by philosophers – for instance, by R. Carnap and Y. Bar-Hillel (1952) – to explicate a concept of "semantic" information distinct from communication-theoretic information. However, these attempts were almost always designed to capture the semantic content of linguistic structures and are of no help in the analysis of genetic information. Starting in the mid-1990s, there has been considerable skepticism, at least among philosophers, about the role of "information" in genetics. For some, genetic information is no more than a metaphor masquerading as a theoretical concept (Sarkar, 1996a,b; Griffiths, 2001). According to these criticisms, even the most charitable attitude toward the use of "information" in genetics can only provide a defense of its use in the 1960s, in the context of prokaryotic genetics (i.e., the genetics of organisms without compartmentalized nuclei in their cells). Once the "unexpected complexity of eukaryotic genetics" (Watson, Tooze, and Kurtz, 1983, ch. 7) – that is, the genetics of organisms with compartmentalized nuclei in their cells – has to be accommodated, the loose use of "information" inherited from prokaryotic genetics is at least misleading (Sarkar, 1996a). Either informational talk should be abandoned altogether or an attempt must be made to provide a formal explication of "information" that shows that it can be used consistently in this context and, moreover, is useful.

Section 13.3 gives a sketch of one such attempted explication. A category of "semiotic" information is introduced to explicate such notions as coding. Semiotic information incorporates *specificity* and depends on the possibility of *arbitrary* choices in the assignment of symbols to what they symbolize as, for instance, exemplified in the genetic code. Semiotic information is not a semantic concept. There is no reason to suppose that any concept of *biological* information must be "semantic" in the sense that philosophers use that term. Biological interactions, at this level, are about the rate and accuracy of macromolecular interactions. They are not about meaning, intentionality, and the like; any demand that such notions be explicated in an account of biological information is no more than a signifier for a philosophical agenda inherited from manifestly nonbiological contexts, in particular from the philosophy of language and mind. It only raises spurious problems for the philosophy of biology.

Section 13.3 also applies this framework to genetics at the macromolecular level of DNA and proteins. It concludes that there is a sense in which it is appropriate and instructive to use an informational framework for genetics at this level. However, proteins are often far removed from the traits that are usually studied in organismic biology; for instance, the shape, size, and behavior of organisms. Section 13.4 explores the extent to which informational accounts carry over to the level of such traits. Much depends on how "trait" is construed, and there is considerable leeway about its definition. Given a relatively inclusive construal of "trait," section 13.4 concludes that, to the extent that a molecular etiology can at all be attributed to a trait, a carefully circumscribed informational account remains appropriate.

Finally, section 13.5 cautions against any overly ambitious interpretation of the claims defended earlier in this chapter. They do not support even a mild form of genetic reductionism (that genes alone provide the etiology of traits), let alone determinism. They do not support the view that DNA alone must be the repository of biological information. Perhaps, most importantly, they do not support the view that the etiology of traits can be fully understood in informational terms from a predominantly genetic basis.

13.2 The Theory of Communication

Shannon conceived of a communication system as consisting of six components:

1 An information source that produces a raw "message" to be transmitted.
2 A transmitter that transforms or "encodes" this message into a form appropriate for transmission through the channel.
3 The channel through which the encoded message or "signal" is transmitted to the receiver.
4 The receiver that translates or "decodes" the received signal back into what is supposed to be the original message.
5 The destination or intended recipient of the message.
6 Sources of noise that act on the channel and potentially distort the signal or encoded message (obviously this is an optional and undesirable component, but one that is unavoidable in practice).

The most important aspect of this characterization is that it is abstracted away from any particular protocol for coding as well as any specific medium of transmission.

From the point of view of the theory of communication, information is conceived of as the choice of one message from a set of possible messages with a definite probability associated with the choice: the lower this probability, the higher is the information associated with the choice, because a greater uncertainty is removed by that choice. The central problems that the theory of communication sets out to solve include the efficiency (or relative accuracy) with which information can be transmitted through a channel in the presence of noise, and how the rate and efficiency of transmission are related to the rate at which messages can be encoded at the transmitter and to the capacity of the channel.

To solve these problems requires a quantitative measure of information. Suppose that a message consists of a sequence of symbols chosen from a basic symbol set (often called an "alphabet"). For instance, it can be a DNA sequence, with the basic symbol set being $\{A, C, G, T\}$. In a sequence of length n, let p_i be the probability of occurrence of the ith symbol in that sequence. Then the information content of the message is $H = -\Sigma_{i=1}^{n} p_i \log p_i$. (In what follows, this formula will be called the "Shannon measure of information.") Consider the sequence "$ACCTCGATTC$." Then, at the first position, $H_1 = p_1 \log p_1$, where $p_1 = \frac{2}{10} = 0.2$ because "A" occurs twice in the sequence of ten symbols. $H = \Sigma_{i=1}^{10} H_i$, computed in this way, is the information content of the entire sequence.

Shannon justified this choice for the measure of information by showing that $-K\Sigma_{i=1}^{n} p_i \log p_i$, where K is a constant, is the only function that satisfies the following three reasonable conditions: (i) the information function is continuous in all the p_i; (ii) if all the p_i are identical and equal to $1/n$, then the function is a monotonically increasing function of n; and (iii) if the choice involved in producing the message can be decomposed into several successive choices, then the information associated with the full choice is a weighted sum of each of the successive choices, with the weights equal to the probability of each of them. K is fixed by a choice of units. If the logarithm used is of base 2, then K is equal to 1.

Formally, the Shannon measure of information is identical to the formula for entropy in statistical physics. This is not surprising since entropy in statistical physics is a measure of disorder in a system and Shannon's information measures the amount of uncertainty that is removed. Over the years, some have held that this identity reveals some deep feature of the universe; among them are those who hold that physical entropy provides a handle on biological evolution (e.g., Brooks and Wiley, 1988). Most practitioners of information theory have wisely eschewed such grandiose ambitions (e.g., Pierce, 1962), noting that the identity may be no greater significance than, for instance, the fact that the same bell curve (the normal distribution) captures the frequency of measurements in a wide variety of circumstances, from the distribution of molecular velocities in a gas to the distribution of heights in a human population. However, the entropic features of the Shannon measure have been effectively used by T. D. Schneider (see, e.g., Schneider, 1999) and others to identify functional regions of DNA sequences: basically, because of natural selection, these regions vary less than others and relatively invariant sequences with high probabilities and, therefore, low information content is expected to be found in these regions.

Shannon's work came, as did the advent of molecular biology, at the dawn of the computer era when few concepts were as fashionable as that of information.

The 1950s saw many attempts to apply information theory to proteins and DNA; these were an unmitigated failure (Sarkar, 1996a). Within evolutionary genetics, M. Kimura (1961) produced one intriguing result in 1961: the Shannon measure of information can be used to calculate the amount of "information" that is accumulated in the genomes of organisms through natural selection. A systematic analysis of Kimura's calculation was given by G. C. Williams in 1966. Williams first defined the gene as "that which segregates and recombines with appreciable frequency" (1966, p. 24) and then assumed, offering no argument, that this definition is equivalent to the gene being "any hereditary information for which there is a favorable or unfavorable selection bias equal to several or many times its rate of endogenous change [mutation]" (p. 25). From this, Williams concluded that the gene is a "cybernetic abstraction" (p. 33). Williams' book lent credence to the view that the gene, already viewed informationally by molecular biologists, can also be so viewed in evolutionary contexts.

Nevertheless, interpreting genetics using communication-theoretic information presents insurmountable difficulties even though the most popular objection raised against it is faulty:

1 The popular objection just mentioned distinguishes between "semantic" and "causal" information (Sterelny and Griffiths, 1999; Maynard Smith, 2000; Sterelny, 2000; Griffiths, 2001; see also Godfrey-Smith's chapter 14 in this volume). The former is supposed to require recourse to concepts such as intentionality. As mentioned in the last section, and as will be further demonstrated in the deflationary account of information given in the next section, such resources are not necessary for biological information. (It is also not clear why a notion of information based on intentionality should be regarded as "semantic" if that term is supposed to mean what logicians usually take it to mean.) Flow of the latter kind of information is supposed to lead to covariance (in the sense of statistical correlation) between a transmitter and receiver. Communication-theoretic information is supposed to be one type of such "causal" information. (It is also misleading to call this type of interaction "causal," since it presumes nothing more than mere correlation.) With this distinction in mind, the objection to the use of "causal" information in genetics goes as follows: in genetics, the transmitter is supposed to consist of the genes (or the cellular machinery directly interpreting the DNA – this choice does not make a difference), the receiver is the trait, and the cytoplasmic and other environmental factors that mediate the interactions between the genes and the trait constitute the channel. Now, if environmental conditions are held constant, the facts of genetics are such that there will indeed be a correlation between genes and traits. However, one can just as well hold the genetic factors constant, treat the environment as the transmitter, and find a correlation between environmental factors and traits. The kernel of truth in this argument is that, depending on what is held constant, there will be a correlation between genetic or environmental factors and traits. In fact, in most circumstances, even if neither is held entirely constant, there will be correlations between both genetic and environmental factors and traits. All that follows is that, if correlation suffices as a condition for information transfer, both genes and environments carry information for traits. Thus, if genes are to be informationally privileged, more than correlation

is required. However, mathematical communication or information theory is irrelevant to this argument. The trappings of Shannon's model of a communication system are extraneously added to a relatively straightforward point about genetic and environmental correlations, and do no cognitive work. This argument does not establish any argument against the use of communication-theoretic information to explicate genetic information.

2 The main reason why communication-theoretic information does not help in explicating genetic information is that it does not address the critical point that genetic information explains biological specificity. This is most clearly seen by looking carefully at what the various components Shannon's model of a communication system do. Three of these are relevant here: the transmitter, the channel, and the receiver. There is a symmetry between the transmitter and the receiver; the former encodes a message to be sent, the latter decodes a received message. Specific relations are only at play in Shannon's model during encoding and decoding. In most cases, though not all, encoding and decoding consists of something like translation using a dictionary. The transmitter obtains from the source a message in some syntactic form. The encoding process uses this as an input and produces as output some sequence of entities (another syntactic form) that is physically amenable for propagation through the channel. This process may consist of using the value of some syntactic object in the input (for instance, the intensity of light of a particular frequency) to select an appropriate output; however, most often, it consists of using a symbolic look-up table. In either case, there is a specific relationship between input and output. (In information theory, this is the well-studied "coding" problem.) A similar story holds for decoding. The critical point is that the Shannon measure of information plays no role in this process. That measure only applies to what happens along the channel. If a highly improbable syntactic object (the signal) is the input and, because of noise, a less improbable entity is the output, information has been lost. Specificity has nothing to do with this.

3 Even leaving aside specificity, note that high communication-theoretic information content is the mark of a message of low probability. Yet, due to natural selection, DNA sequences that most significantly affect the functioning of an organism, and are thus most functionally informative, become increasingly frequent in a population. Thus, communication-theoretic information is negatively correlated with functional information. This conclusion raises a peculiar possibility. Communication-theoretic information may be a guide to functional information if, as a measure of functional information, its inverse is used. Now if functional information is taken to be a measure of genetic information (which is reasonable since, during evolution, functional information is primarily transmitted through DNA), communication-theoretic information provides access to genetic information through this inversion. Nevertheless, this attempt to rescue communication-theoretic information for genetics fails because those genes that are not selected for are no longer carriers of information. These include not only genes that may be selected against but nevertheless persist in a population (for instance, because of heterozygote advantage or tight linkage – that is, being very close on a chromosome to a gene that is selected for) but also the many genes that are simply selectively neutral. Understanding the role of information in genetics will require a different approach altogether.

13.3 Semiotic Information

Informational talk pervades contemporary genetics with the gene as the locus of information. DNA is supposed to carry information for proteins, possibly for traits. The so-called Central Dogma of Molecular Biology states: "once 'information' has passed into protein *it cannot get out again*. In more detail, the transfer of information from nucleic acid to nucleic acid, or from nucleic acid to protein may be possible, but transfer from protein to protein, or from protein to nucleic acid is impossible" (Crick, 1958; emphasis in the original). J. Maynard Smith (2000) has claimed that DNA is the sole purveyor of biological information in organisms. Routinely, talk of information is intertwined with linguistic metaphors, from both natural and artificial languages: there is a genetic *code*, because a triplet of DNA (or RNA) nucleotides codes for each amino acid residue in proteins (polypeptide chains); there are alternative *reading frames* – DNA is *transcribed* into RNA, RNA is *translated* into protein, RNA is *edited*, and so on. The use of such talk is so pervasive that it almost seems impossible that, short of pathological convolution, the experimental results of genetics can even be communicated without these resources.

What is largely forgotten is that there is good reason to believe that such talk of information in genetics may not be necessary: "information" was introduced in genetics only in 1953, just before the DNA double-helix model (Ephrussi et al., 1953). "Information" was supposed to introduce some frugality in a field that had become unusually profligate in terminological innovation (for instance, "transformation" and "transduction") to describe genetic interactions in bacteria. The main reason why the informational framework became central to the new molecular biology of the 1950s and 1960s was the characterization of the relationship between DNA and proteins as a universal genetic code, "universal" in the sense that it is the same for all species. A triplet of nucleotide bases specifies a single amino acid residue. Since there are four nucleotide bases (*A*, *C*, *G*, and *T*) in DNA, there are 64 possible triplets. Since there are only 20 standard amino-acid residues in proteins, the code must be degenerate (or partially redundant): more than one triplet must code for the same amino-acid residue. Three factors make the informational interpretation of this relationship illuminating: (a) the relationship can be viewed as a symbolic one, with each DNA triplet being a symbol for an amino acid residue; (b) the relationship is combinatorial, with different combinations of nucleotides potentially specifying different residues; and, most importantly, (c) the relationship is arbitrary in an important sense. Functional considerations may explain some features of the code – why, for instance, an arbitrary mutation tends to take a hydrophilic amino-acid residue to another such residue – but it does not explain why the code is specifically what it is. The physical mechanisms of translation do not help either. The genetic code is *arbitrary*. Along with specificity, this arbitrariness is what makes an informational account of genetics useful.

To explain how this account works will now require an explication of semiotic information (Sarkar 2000, forthcoming). This explication will proceed in two stages. After the structure of an information system is defined, conditions will first be laid down to establish specificity; secondly, more conditions will be imposed to capture the type of arbitrariness also required of semiotic information.

A formal information system consists of a relation, ι (the information relation), between two sets, A (for instance, a set of DNA sequences) and B (for instance, a set of polypeptide sequences) and will be symbolized as $\langle A, B, \iota \rangle$. The relation ι holds between A and B because it holds between their elements. The following simplifying assumptions will be made: (i) ι partitions A into a set of equivalence classes, with all member of an equivalence class being informationally equivalent to each other. (In the case of DNA and protein, an equivalence class consists of all the triplets that specify the same amino acid residue); and (ii) A and ι exhaust B – that is, for every element of B, there is some element in A that is related to it by ι. One consequence of this assumption is that ι partitions B into an equivalence class, with each equivalence class of B corresponding to one of A.

With this background, for an information system to allow for specificity, the most important condition on $\langle A, B, \iota \rangle$ is:

(I1) *Differential specificity*: suppose that a and a' belong to different equivalence classes of A. Then, if $\iota(a, b)$ and $\iota(a', b')$ hold, then b and b' must be different elements of B.

Condition (I1) suffices to capture the most basic concept of a specific informational relation holding between A and B.[1] An additional condition will sometimes be imposed to characterize a stronger concept of specificity:

(I2) *Reverse differential specificity*: if $i(a, b)$ and $i(a', b')$ hold, and b and b' are different elements of B, then a and a' belong to different equivalence classes in A.

If every equivalence class in A specifies exactly one element of B through ι, condition (I2) is automatically satisfied provided that (I1) is satisfied. That A and B co-vary (i.e., there are statistical correlations between the occurrences of elements of A and B) is a trivial consequence of either condition. The justification for these conditions is that they capture what is customarily meant by information in any context: for instance, they capture the sense in which the present positions and momenta of the planets carry information about their future positions.

By definition, if only condition (I1) holds, then A will carry *specific information* for B; if both conditions (I1) and (I2) hold, A *alone* carries specific information for B. In the case of prokaryotic genetics, both conditions hold: DNA alone carries information for proteins. Some care has to be taken at this stage and the discussion must move beyond formal specification to the empirical interpretation of ι in this (genetic) context. The claim that DNA alone carries information for proteins cannot be interpreted as saying that the presence of a particular DNA sequence will result in the production of the corresponding protein no matter what the cellular environment does. Such a claim is manifestly false. In certain cellular contexts, the presence of that DNA sequence will not result in the production of any protein at all. Rather, the claim must be construed counterfactually, if the presence of this DNA sequence were to lead to

---

1 Because ι may hold between a single a and several b, this relation is not transitive. A may carry information for B, B for C, but A may not carry information for C. As will be seen later, this failure of transitivity results in a distinction between encoding traits and encoding information for traits.

the production of a protein, then ι describes which protein it leads to. In general, the production of the relevant protein requires an enabling history of environmental conditions. This history is not unique. Rather, there is a set M of "standard" histories that result in the formation of protein from DNA. A complete formal account of biological information must specify the structure of this set. This is beyond what is possible given the current state of empirical knowledge: M cannot be fully specified even for the most studied bacterium, *Escherichia coli*. However, because the relevant informational relation is being construed counterfactually in the way indicated, the inability to specify M does not prevent the use of that relation. (It does mean, though, that the information content of the DNA, by itself, does not suffice as an etiology for that protein.) The structure of the relation between M and the protein set will be a lot more complicated than $\langle A, B, \iota \rangle$, where A is the DNA set and B the protein set; this is already one sense in which genes are privileged.

So far, only specificity has been analyzed; it remains to provide an account of arbitrariness. This requires the satisfaction of two conditions that are conceptually identical to those imposed by Shannon on communication systems in order to achieve a suitable level of abstraction. It shows an important commonality between communication-theoretic and semiotic information:

(A1) *Medium independence*: $\langle A, B, \iota \rangle$ can be syntactically represented in any way, with no preferred representation, so long as there is an isomorphism between the sets corresponding to A, B, and ι in the different representations.[2]

An equation representing the gravitational interaction between the earth and the sun is a syntactic representation of that relation. So is the actual earth and sun, along with the gravitational interaction. The latter representation is preferred because the former is a representation of the latter in a way in which the latter is not a representation of the former. Medium independence for information denies the existence of any such asymmetry. From the informational point of view, a physical string of DNA is no more preferred as a representation of the informational content of a gene than a string of *A*s, *C*s, *G*s, and *T*s on this sheet of paper. (The most useful analogy to bear in mind is that of digital computation, where the various representations, whether it be in one code or other or as electrical signals in a circuit, are all epistemologically on a par with each other.) This is also the sense of medium independence required in Shannon's account of communication. The second criterion is:

(A2) *Template assignment freedom*: let A_i, $i = 1, \ldots, n$, partition A into n equivalence classes, and let $\langle A_1, A_2, \ldots, A_n \rangle$ be a sequence of these classes that are related by ι to the sequence $\langle B_1, B_2, \ldots, B_n \rangle$ of classes of B. Then the underlying mechanisms allow for any permutation $\langle A_{\sigma(1)}, A_{\sigma(2)}, \ldots, A_{\sigma(n)} \rangle$ (where $\sigma(1)\ \sigma(2) \ldots \sigma(n)$ is a permutation of $12 \ldots n$) to be mapped by ι to $\langle B_1, B_2, \ldots, B_n \rangle$.

This condition looks more complicated than it is; as in the case of Shannon's communication systems, it shows that any particular protocol of coding is arbitrary.

---

2   Here, ι is being interpreted extensionally for expository simplicity.

(However, as emphasized in the last section, coding and decoding are not part of information transfer in the theory of communication.) In the case of DNA (the template), all that this condition means is that any set of triplets coding for a particular amino acid residue can be reassigned to some other residue. However, there is a subtle and important problem here: such a reassignment seems to violate a type of physical determinism that no one questions in biological organisms, namely that the chemical system leading from a piece of DNA to a particular protein is deterministic. (Indeed, conditions [I1] and [I2] of specificity require determinism of this sort.) Condition (A2) must be interpreted in the context of genetics as saying that all these different template assignments were evolutionarily possible.[3] The customary view, that the genetic code is an evolutionarily frozen accident, supports such an interpretation. By definition, if conditions (I1), (A1), and (A2) hold, then ι is a coding relation, and A *encodes* B. Thus, DNA encodes proteins; for prokaryotes, DNA alone encodes proteins.

The formal characterization given above lays down adequacy conditions for any ι that embodies semiotic information. It does not specify what ι is; that is, which a's are related to which b's. The latter is an empirical question: coding and similar relations in molecular biology are empirical, not conceptual, relations. This is a philosophically important point. That DNA encodes proteins is an empirical claim. Whether conditions (I1), (I2), (A1), or (A2) hold is an empirical question. Under this interpretation, these conditions must be interpreted as empirical generalizations with the usual *ceteris paribus* clauses that exclude environmental histories not belonging to M. The relevant evidence is of the kind that is typically considered for chemical reactions: for the coding relation, subject to statistical uncertainties, the specified chemical relationships must be shown to hold unconditionally. Now suppose that, according to ι, a string of DNA, s, codes for a polypeptide, σ. Now suppose that, as an experimental result, it is found that some σ' different from σ is produced in the presence of s. There must have been an *error*, for instance, in transcription or translation: a *mistake* has been made. All this means is that an anomalous result, beyond what is permitted due to standard statistical uncertainties, is obtained. It suggests the operation of intervening factors that violate the *ceteris paribus* clauses; that is, the reactions took place in an environmental history that does not belong to M. Thus, the question of what constitutes a mistake is settled by recourse to experiment. There is nothing mysterious about genetic information, nothing that requires recourse to conceptual resources from beyond the ordinary biology of macromolecules.

As was noted, the informational account just given allows DNA alone to encode proteins' prokaryotic genetics. Genetically, prokaryotes are exceedingly simple. Every bit of DNA in a prokaryotic genome either codes for a protein or participates in regulating the transcription of DNA. For the coding regions, it is straightforward to translate the DNA sequence into the corresponding protein. Consequently, the informational interpretation seems particularly perspicuous in this context. This is the

---

3   However, condition (A2) is stronger than what is strictly required for arbitrariness, and may be stronger than what is biologically warranted. For arbitrariness, all that is required is that there is a large number of possible alternative assignments. Biologically, it may be the case that some assignments are impossible because of developmental and historical constraints.

picture that breaks down in the eukaryotic context. Eukaryotic genetics presents formidable complexities including, but not limited to, the following (for details, see Sarkar, 1996a):

(a) the nonuniversality of the standard genetic code – some organisms use a slightly different code, and the mitochondrial code of eukaryotic cells is also slightly different;

(b) frameshift mutations, which are sometimes used to produce a variant amino-acid chain from a DNA sequence;

(c) large segments of DNA between functional genes that apparently have no function at all and are sometimes called "junk DNA";

(d) similarly, segments of DNA within genes that are not translated into protein, these being called introns, while the coding regions are called exons – after transcription, the portions of RNA that correspond to the introns are "spliced" out and not translated into protein at the ribosomes; but

(e) there is alternative splicing by which the same RNA transcript produced at the DNA (often called "pre-mRNA") gets spliced in a variety of ways to produce several proteins – in humans, it is believed that as many as a third of the genes lead to alternative splicing; and

(f) there are yet other kinds of RNA "editing" by which bases are added, removed, or replaced in mRNA, sometimes to such an extent that it becomes hard to say that the corresponding gene encodes the protein that is produced.

In the present context, points (a), (b), (d), (e), and (f) are the most important. They all show that a single sequence of DNA may give rise to a variety of amino-acid sequences even within a standard history from M. Thus, in the relation between eukaryotic DNA and proteins, condition (I2) fails. Nevertheless, because condition (I1) remains satisfied, eukaryotic DNA still carries specific information for proteins. However, this formal success should not be taken as an endorsement of the utility of the informational interpretation of genetics. As stated, condition (I1) does not put any constraint on the internal structure of the sets A (in this case, the set of DNA sequences) or B (in this case, the set of protein sequences). If both sets are highly heterogeneous, there may be little that the informational interpretation contributes. In the case of the DNA and protein sets, heterogeneity in the former only arises because of the degeneracy of the genetic code. This heterogeneity has not marred the utility of the code. For the protein set, heterogeneity arises because a given DNA sequence can lead to different proteins with varied functional roles. However, leaving aside the case of extensive RNA editing, it seems to be the case that these proteins are related enough for the heterogeneity not to destroy the utility of the informational interpretation.

The main upshot is this: even in the context of a standard history from M, while genes encode proteins and carry specific information for them, this information is not even sufficient to specify a particular protein. Knowledge of other factors, in particular so-called environmental factors, is necessary for such a specification. Whether these other factors should be interpreted as carrying semiotic information depends on whether they satisfy at least conditions (I1), (A1), and (A2). So far, there is no evidence that any of them satisfy (A1) and (A2). This is yet another sense in which genes are privileged.

13.4 Traits and Molecular Biology

Proteins are closely linked to genes: compared to organismic traits such as shape, size, and weight, the chain of reactions leading to proteins from genes is relatively short and simple. It is one thing to say that genes encode proteins, and another to say that they encode information for traits. It depends on how traits are characterized. The first point to note is that "trait" is not a technical concept within biology, with clear criteria distinguishing those organismic features that are traits from those that are not (Sarkar, 1998). In biological practice, any organismic feature that succumbs to systematic study potentially qualifies as a trait. Restricting attention initially to structural traits, many of them can be characterized in molecular terms (using the molecules out of which the structures are constructed). These occur at varied levels of biological organization from the pigmentation of animal skins to receptors on cell membranes. Proteins are intimately involved in all these structures. For many behavioral traits, a single molecule, again usually a protein, suffices as a distinguishing mark for that trait. Examples range from sickle-cell hemoglobin for the sickle-cell trait to huntingtin for Huntington's disease. Thus, for these traits, by encoding proteins, genes trivially encode information for them at the molecular level.

However, at the present state of biological knowledge it is simply not true that, except for a few model organisms, most traits can be characterized with sufficient precision at the molecular level for the last argument to go through. This is where the situation becomes more interesting. The critical question is whether the etiology of these traits will permit explanation at the molecular level, in particular, explanations in which individual proteins are explanatorily relevant. (Note that these explanations may also refer to other types of molecules; all that is required is that proteins have some etiological role. The involvement of these other molecules allows the possibility that the same protein (and gene) may be involved in the genesis of different traits in different environmental histories – this is called *phenotypic plasticity*.) If proteins are so relevant, then there will be covariance between proteins and traits although, in general, there is no reason to suppose that any of the conditions for semiotic information is fulfilled. Proteins will be part of the explanation of the etiology of traits without carrying information for traits. Allowing genes to carry information does not endorse a view that organisms should be regarded as information-processing machines. Informational analysis largely disappears at levels higher than that of proteins.

Nevertheless, because of the covariance mentioned in the last paragraph, by encoding proteins, genes encode information for traits (while not encoding the traits themselves). The critical question is the frequency with which proteins are explanatorily relevant for traits in this way. The answer is "Probably almost always." One of the peculiarities of molecular biology is that the properties of individual molecules, usually protein molecules such as enzymes and antibodies, have tremendous explanatory significance, for instance, in the structural explanations of specificity mentioned in section 13.1. Explanations involving systemic features, such as the topology of reaction networks, are yet to be forthcoming. Unless such explanations, which do not refer to the individual properties of specific molecules, become the norm of organismic biology, its future largely lies at the level of proteins.

This is a reductionist vision of organismic biology, not reduction to DNA or genes alone, but to the entire molecular constitution and processes of living organisms (Sarkar, 1998). Such a reductionism makes many uncomfortable, including some developmental biologists imbued in the long tradition of holism in that field, but there is as yet no good candidate phenomenon that contradicts the basic assumptions of the reductionist program. It is possible that biology at a different level – for instance, in the context of psychological or ecological phenomena – will present complexities that such a reductionism cannot handle. However, at the level of individual organisms and their traits, there is as yet no plausible challenge to reductionism.

13.5 Conclusion

Many of the recent philosophical attacks on the legitimacy of the use of an informational framework for genetics have been motivated by a justified disquiet about facile attributions of genetic etiologies for a wide variety of complex human traits, including behavioral and psychological traits, in both the scientific literature and in the popular media. The account of semiotic information given here does not in any way support such ambitions. While, to the extent that is known today, genes are privileged over other factors as carriers of semiotic information involved in the etiology of traits, genes may not even be the sole purveyors of such information. Moreover, genetic information is but one factor in these etiologies. Traits arise because of the details of the developmental history of organisms, in which genetic information is one resource among others. Thus, even when genes encode sufficient information to specify a unique protein (in prokaryotes), the information in the gene does not provide a sufficient etiology for a trait; at the very least, some history from M must be invoked.

Thus, an informational interpretation of genetics does not support any attribution of excessive etiological roles for genes. The two questions are entirely independent: whether genes should be viewed as information-carrying entities, and the relative influence of genes versus nongenetic factors in the etiology of traits. This is why the latter question could be investigated, often successfully, during the first half of the twentieth century, when "information" was yet to find its way into genetics. To fear genetic information because of a fear of genetic determinism or reductionism is irrational.

Acknowledgments

Thanks are due to Justin Garson and Jessica Pfeifer for helpful discussions.

Bibliography

Brooks, D. R. and Wiley, E. O. 1988: *Evolution as Entropy: Towards a Unified Theory of Biology*, 2nd edn. Chicago: The University of Chicago Press.

Carnap, R. and Bar-Hillel, Y. 1952: An outline of a theory of semantic information. Technical Report No. 247. Research Laboratory of Electronics, Massachusetts Institute of Technology.

Crick, F. H. C. 1958: On protein synthesis. *Symposia of the Society for Experimental Biology*, 12, 138–63.

Ephrussi, B., Leopold, U., Watson, J. D., and Weigle, J. J. 1953: Terminology in bacterial genetics. *Nature*, 171, 701.

Fisher, R. A. 1925: Theory of statistical estimation. *Proceedings of the Cambridge Philosophical Society*, 22, 700–25.

Garson, J. 2002: The introduction of information in neurobiology. M. A. thesis, Department of Philosophy, University of Texas at Austin.

Griffiths, P. 2001: Genetic information: a metaphor in search of a theory. *Philosophy of Science*, 67, 26–44.

Hartley, R. V. L. 1928: Transmission of information. *Bell Systems Technical Journal*, 7, 535–63.

Kimura, M. 1961: Natural selection as a process of accumulating genetic information in adaptive evolution. *Genetical Research*, 2, 127–40.

Lederberg, J. 1956: Comments on the gene–enzyme relationship. In Gaebler, O. H. (ed.), *Enzymes: Units of Biological Structure and Function*. New York: Academic Press, 161–9.

Maynard Smith, J. 2000: The concept of information in biology. *Philosophy of Science*, 67, 177–94.

Nyquist, H. 1924: Certain factors affecting telegraph speed. *Bell Systems Technical Journal*, 3, 324–46.

Pauling, L. 1940: A theory of the structure and process of formation of antibodies. *Journal of the American Chemical Society*, 62, 2643–57.

Pierce, J. R. 1962: *Symbols, Signals and Noise*. New York: Harper.

Sarkar, S. 1996a: Biological information: a skeptical look at some central dogmas of molecular biology. In Sarkar, S. (ed.), *The Philosophy and History of Molecular Biology: New Perspectives*. Dordrecht: Kluwer, 187–231.

— 1996b: Decoding "coding" – information and DNA. *BioScience*, 46, 857–64.

— 1998: *Genetics and Reductionism*. New York: Cambridge University Press.

— 2000: Information in genetics and developmental biology: comments on Maynard-Smith. *Philosophy of Science*, 67, 208–13.

— forthcoming: Biological information: a formal account.

Schneider, T. D. 1999: Measuring molecular information. *Journal of Theoretical Biology*, 201, 87–92.

Shannon, C. E. 1948: A mathematical theory of communication. *Bell Systems Technical Journal*, 27, 379–423, 623–56.

Sterelny, K. 2000: The "genetic program" program: a commentary on Maynard Smith on information in biology. *Philosophy of Science*, 67, 195–202.

— and Griffiths, P. E. 1999: *Sex and Death: An Introduction to Philosophy of Biology*. Chicago: The University of Chicago Press.

Watson, J. D., Tooze, J., and Kurtz, D. T. 1983: *Recombinant DNA: A Short Course*. New York: W. H. Freeman.

Williams, G. C. 1966: *Adaptation and Natural Selection*. Princeton, NJ: Princeton University Press.

Further reading

Kay, L. 2000: *Who Wrote the Book of Life?: A History of the Genetic Code*. Stanford, CA: Stanford University Press.

Schrödinger, E. 1944: *What is Life? The Physical Aspect of the Living Cell.* Cambridge: Cambridge University Press.

Thiéffry, D. and Sarkar, S. 1998: Forty years under the central dogma. *Trends in Biochemical Sciences,* 32, 312–16.

Yockey, H. P. 1992: *Information Theory and Molecular Biology.* Cambridge: Cambridge University Press.

CHAPTER FOURTEEN

Genes do not Encode Information for Phenotypic Traits

Peter Godfrey-Smith

14.1 Introduction

How can a philosopher sit here and deny that genes encode information for phenotypic traits? Surely this is an empirical, factual, scientific matter, and one that biology has investigated and answered? Surely I must realize that biology textbooks now routinely talk about genes encoding information? Isn't the discovery of the "genetic code" one of the triumphs of twentieth-century science? Indeed, doesn't much of the research done by biologists now focus on working out *which* genes encode information for *which* phenotypic traits? So is this one of those unfortunate cases of philosophers, in their comfortable armchairs, trying to dictate science to the scientists?

I can understand if that is your initial reaction to my claim. I hope to persuade you, however, that it is indeed a mistake to say that genes encode information for phenotypic traits. My aim is not to overturn empirical results or deny core elements of the textbook description of how genes work. Instead, I will be trying to argue that if we look closely at the details of what biology has learned, it becomes clear that it is a mistake to talk about genes encoding phenotypic traits. This common way of talking is not, as it seems to be, a straightforward and concise summary of how genes work. Instead, it involves a distortion of what biologists have actually learned.

My denial that genes encode information for phenotypic traits is based, as you might expect, on a particular way of interpreting some key terms. My argument is that *if* we understand "encode information" and "phenotypic traits" in a particular way, then genes do not encode information for phenotypic traits. I do think that my interpretations of these terms are reasonable ones, but I will also discuss other interpretations of them.

More specifically, my claims in this chapter rest on two distinctions. One is a distinction between two senses of "information." This distinction will take some time to outline but, roughly speaking, there are two ways of using the language of "information" to describe events and processes in the world. One way is relatively harmless and uncontroversial. In this sense, any process at all in which there is a reliable correlation between two states can be described in terms of information. This is the sense in which dark clouds carry information about bad weather, and tree rings carry information about the age of the tree. The second sense of "information" is more controversial and problematic. Here, when we speak of information we mean that one thing functions in some way as a representation of another, and has a definite semantic content. Something more than a mere correlation or ordinary causal connection is involved. The rings in a tree are not functioning as representations or signals in the life of the tree; they are just indicators that happen to be useful to observers like us. The words on this page, in contrast, make use of a conventional symbol system to convey particular meanings.

The second crucial distinction that I will make is less complicated and less controversial. What counts as a "phenotypic trait" for the purposes of this discussion? Clearly, the properties of whole organisms, such as size, shape, and behavioral patterns, count as part of the phenotype. But might we also include the more immediate causal products of particular genes, which are mRNA molecules and protein molecules? In one very broad sense, the phenotype might be seen as including all of the physical properties of an organism other than its genetic composition. In this sense, even individual proteins are part of the phenotype. But that is *not* the sense that I will use in this chapter. Instead, when I say "phenotypic traits" I mean the familiar properties of whole organisms studied by biologists – size, shape, color, structure, behavior patterns, and so on.

Now I can say something about the sense in which I deny that genes encode information for phenotypic traits. It is certainly possible to describe the connection between genes and phenotypic traits in informational terms, in the first of the two senses of "information" above. This applies however one interprets the term "phenotype." But now consider the second sense of "information," the semantic sense. Do genes have a role in the production of the phenotype that can be described in terms of messages, representations, meaning, or codes?

Clearly, genes are not sending deliberate signals, and they do not use a representation system maintained by social convention, like the words on this page. But we should not get too hung up on these issues. We can ask a different, and more difficult question: Do genes operate in the production of phenotypes in a way that involves some kind of special process that is at least *strongly analogous* to familiar cases of representation? Might there be a useful and general sense of "representation" or "instruction" in which both genes and human sentences play a similar and special kind of role? In answering this question, a lot hangs on how we understand "phenotype." If we were to understand individual protein molecules as part of the "phenotype" of an organism, then I think a good case can be made for the genes having a special informational role. But that is not the ordinary sense of "phenotype." If we ask whether genes have a special informational role in the production of the phenotype in the ordinary sense – the structure and behavior of a whole organism – then the answer is "No."

14.2 More on the Two Senses of Information

Let us look more closely at the concept (or concepts) of information.

The first sense of information outlined above is basically the sense used by Shannon in the original formulations of information theory (Shannon, 1948). I can be fairly brief here, because this topic is covered in more detail in Sarkar's companion chapter 13. Information, in Shannon's sense, exists just about everywhere. A "source" of information is anything that has a range of possible states (such as the weather, or the age of a tree). And whenever the reduction of uncertainty at one place is correlated to some extent with the reduction of uncertainty at another, we have a "signal" that is carrying information. The sky, for example, can have a range of different states (cloudy, clear, and so on). The weather two hours from now also has a range of different states. If the state of the sky now is correlated (perhaps not perfectly, but at least to some extent) with the future state of the weather, then the sky carries information about the future weather.

In the literature on genes and information, this is sometimes called the "causal" sense of information. Or people say that one thing carries "causal information" about another (Griffiths, 2001). This terminology is fine, although causation *per se* is not what is required here. What is required is a correlation (Dretske, 1981). So my preference is to talk of this first sense of information as information in the *Shannon* sense, after the main inventor of information theory, Claude Shannon.

Information, in this sense, is everywhere. It does connect genes with phenotypic traits, but it connects them in both directions and connects both of these with environmental conditions as well. Here it is important to remember that all phenotypic traits are causally influenced by both genes and by environmental conditions. In some cases, a trait might be much *more* sensitive to one or the other of these – to slight differences in the genes, or in environmental circumstances. But that does not change the fact that all traits arise from an interaction, of some kind, between genetic and environmental conditions.

So let us look at how we might use informational language of this kind to describe genes. Whenever we describe any process using this sort of informational language, we treat some factors as background conditions or "channel conditions." If we want to treat genes as carrying information about phenotypic traits, we will regard the role of the environment as a mere background condition. Then if the presence of gene G correlates reliably with the appearance of phenotype X, we can say that gene G carries information about the phenotype, and carries the information that X will be present. But we can also turn the situation around; we can treat genes as background, and look at the correlations between different environmental conditions and phenotypic traits. From this point of view, it is environmental conditions that carry information, not the genes (Oyama, 1985; Griffiths and Gray, 1994). And informational relationships run "backward" as well as "forward" in these cases. The phenotype carries information about the genes, just as the genes carry information about the phenotype.

As the reader will see, I am trying to suggest that when someone claims that genes carry information about phenotypes in the Shannon sense, there is very little being claimed. In this sense of information, it is not a novel and important part of biological theory that genes and phenotypes have an informational connection between them. But I do not want to suggest that this sense of information is useless or unimportant in biology. On the contrary, Shannon's concept of information provides a very useful and concise way of describing masses of physical associations and connections, especially because information in the Shannon sense can be described mathematically. A lot of discussion in contemporary biology is facilitated by this conceptual framework. For example, biologists often talk about the information that genetic variation within and between populations carries about evolutionary relationships. It turns out that "junk" DNA carries more of this information than does DNA that is used to make proteins. So informational description of this kind has definite utility. But, obviously, in cases such as these, the information in the genes is something that only *we*, the observers and describers of genetic systems, use. This information is not part of any explanation of the biological role that the *genes* play within organisms. Here, genes are being used by us just as we might use tree rings.

Some of the talk by biologists about how genes carry information about phenotypes can be understood as using the Shannon sense of information only. But not all of it can be interpreted in this way. For example, biologists are often (though not always) reluctant to say that the environment contains information that specifies phenotypes, in the same sense as genes. And the word "code" is important here; there is not an environmental code in the way there is a genetic code. Biologists often treat genes as having an informational role in a stronger sense than the Shannon sense.

So let us look at the other sense of "information." I should say immediately here that the word "sense" is a bit misleading, because it suggests there is a *single* additional sense of the word "information." Some might think that this is true, but I do not. When we leave the precise Shannon sense of information, we encounter an unruly collection of different concepts. We encounter the large and controversial domain of *semantic* properties – properties that involve meaning, representation, reference, truth, coding, and so on. Despite a massive effort by philosophers and others over many years (especially the past 100 years), I think we do not have a very good handle on this set of phenomena.

The two domains where the idea of meaning and representation seem most familiar to us are the domains of *language* and *thought*. Philosophy of language and philosophy of mind have both seen attempts to formulate general theories of meaning and representation, that might be used to help us describe semantic phenomena in other areas. (For theories in the philosophy of mind, see Stich and Warfield (1994); for philosophy of language, see Devitt and Sterelny (1999).) People who are trying to distinguish genuine semantic properties from information in the Shannon sense often point to the capacity for error, and the ability to represent nonactual situations, as marks of semantic phenomena. In the simple clouds-and-rain case, there is no sense in which the clouds could *misrepresent* the weather. The correlation between clouds and rain might fail to hold in some particular case, but that does not imply that the clouds said something *false*. If I tell you a lie in this chapter, however, my words have indeed been used to say something false. I can also use these words to describe a situation that I know does not obtain, such as my having won the lottery. These are not features of information in the mere Shannon sense. Despite much effort, philosophers have not made much headway in giving a general theory of semantic properties. (Some would disagree, but that's how it seems to me.)

Biologists began to use the terminology of information and coding back in the middle of the twentieth century (for the history of this trend, see Kay, 2000). Biologists do not all have the same thing in mind when they talk of genes carrying information. Some see it as a mere metaphor, or as loose and picturesque talk. Others mean no more than information in the Shannon sense. But in recent years, the idea of genetic information has been taken more and more seriously by some biologists (Williams, 1992; Maynard Smith, 2000), and these biologists have clearly had more than the simple Shannon sense of information in mind.

So do genes encode information for phenotypic traits in something more than the Shannon sense? Answering this question is made awkward by the absence of a good philosophical theory of semantic properties, the absence of a good *test* that we could apply to genes. Most philosophers will agree that we do not have a clear diagnostic question to ask. Here is the method that I will apply in this chapter. My approach will be to see whether there are properties of genes that have an important analogy with familiar and central cases of languages and symbol systems. The test is not so much whether these analogies *look* striking, but whether the analogies are *important within biological theory*. To what extent does describing genes in informational terms help us to understand how genetic mechanisms work? That will be how I approach the question addressed by my and Sarkar's chapters.

14.3 What Genes Do and How They Do It

My argument about genes and information will be based on some detailed facts about how genes exert their influence on the phenotype. My argument is not directed against the idea that genes have an important *causal* role in producing the phenotypes of organisms; no one can deny that. My argument concerns *how* they exercise that causal role. I claim that this causal role does not involve the interpretation of an encoded message in which genes specify phenotypic traits. So let us look in some detail at what genes do and how they do it.

What genes do, fundamentally, is make protein molecules, make a few other molecules, and contribute to the regulation of genes in these same activities. And it is not really very accurate to say that genes "make" proteins. Genes act as templates that specify a particular crucial feature of each protein molecule – the linear order of the protein's building blocks, which are individual amino acids.

Genes are made of DNA, which is, of course, a long two-stranded molecule arranged as a double helix. Each strand of DNA is a chain of four building blocks (the "bases" C, A, G, and T), held together by a backbone. Two main steps are distinguished in the causal chain between DNA and a finished protein molecule. "Transcription" is the process in which DNA gives rise to mRNA ("messenger RNA"). Then "translation" generates the protein itself, using the mRNA.

mRNA is a similar molecule to DNA – another chain of four building blocks, but single-stranded rather than double-stranded. The molecule of mRNA produced during transcription is formed using a stretch of DNA directly as a template, and the mRNA contains a sequence of bases that corresponds, by a standard rule, to the sequence of bases in the DNA from which it was derived. In organisms other than bacteria, the mRNA is usually processed before it is used in translation. Then, elsewhere in the cell (at the ribosomes), the mRNA is used to direct the formation of another long-chained molecule. This time, the chained molecule is a chain of amino acids – a protein.

In this process of translation, a crucial role is played by another kind of RNA molecule, tRNA (or "transfer RNA"). These are shorter chains of the RNA building blocks, which also bind to single amino acids. Different tRNAs bind to different amino acids (of which there are 20 kinds used in protein synthesis). At the ribosomes, tRNA molecules carrying amino acids with them bind temporarily to specific three-base sequences in the mRNA molecule. So each triplet of bases in the mRNA is associated, via the special chemical properties of tRNA molecules, with a particular amino acid.

When a protein is assembled at a ribosome, a chain of amino acids is produced whose sequence corresponds, by a near-universal rule, to the sequence of bases in the mRNA. "The genetic code" is, strictly speaking, this rule connecting RNA base triplets with amino acids. This interpretation of the RNA determines the interpretation of the DNA from which the mRNA is derived. As there are four bases in the mRNA (almost the same four as in DNA), there are 64 possible triplets. Of these, 61 specify particular amino acids; some amino acids are specified by as many as six different triplets. The three remaining triplets are "stop" signals. The chain of amino acids then folds (and may be processed in other ways) to produce a finished protein.

Protein structure is described at four different levels, of which the *primary* and *tertiary* are most important for our purposes. The primary structure of a protein is its linear sequence of amino acids. The tertiary structure is the three-dimensional folded shape of a single amino-acid chain. The causal role of a protein within an organism depends greatly on its tertiary structure.

There is much more to all these processes, of course. But this outline describes the core of the process by which any genetic message is expressed. I should also add a quick remark about how genes operate in the *regulation* of these same processes. Generalizing greatly, the main way this occurs is by genes producing proteins that physically interact with particular stretches of DNA itself, in a way that either impedes or facilitates the process of transcription. So the process of protein synthesis feeds back on itself in complex ways. There are also other mechanisms by which genes can regulate the activities and products of the genetic system.

In this sketch, I have used standard terminology from genetics, including some standard terminology that seems to be asserting an important role for something like information. We have "transcription," "translation," and a "code" linking DNA/RNA sequences to amino-acid sequences. And I do not object, or at least do not object much, to this collection of terminologies (for a few minor complaints, see Godfrey-Smith, 1999). There are indeed some very special features of the role of genes within protein synthesis, features that seem to me to justify, to some extent, a description in terms of codes, messages, and so on. I will focus on these in the next section. The point to make here is that everything I have said concerns the connection between a gene and the protein molecules that it produces. Nothing has been said yet about the "downstream" effects of the protein itself. And here, a rather different story must be told.

There is not a single, unified biological process by which proteins have their effects on the organism, in the way in which there is a fairly unified process by which genes give rise to proteins. Instead, proteins have a host of different effects, and a host of different ways of having these effects. Some proteins operate as parts of the physical structure of the organism; others act as enzymes controlling key reactions between other chemicals; others act as hormones; and so on. Proteins are the fundamental "working" molecules in all organisms, but their work involves a tremendously complex mass of cascading and interacting processes. The actions of proteins are part of a huge and varied causal network, some parts of which we know a lot about, and other parts of which we know very little about. Being blunt about it, once a gene has produced a protein molecule, the protein generally goes off and enters a kind of causal soup, or causal spaghetti, where it might affect, and be affected by, a great range of other proteins and other factors, including environmental factors.

Given this picture, how should we describe the causal connection between a gene and a particular aspect of the whole-organism phenotype? It depends very much on the particular case. In some cases, despite all the intervening complexity, a trait can be under rather tight causal control by specific genes. In any environment in which the organism can develop at all, if it has gene G (and a reasonably normal set of genes elsewhere in its genome), it will develop with phenotype X. Slight environmental differences have little effect on the phenotypic outcome. (Traits of this kind are sometimes called "canalized," although that term has several related uses.) In other cases, phenotypic traits have much more sensitivity to particular environmental conditions. Many writers like to describe this second set of cases by saying that the genes here determine a *range* of possible values for a trait, and the environment fills in the *specific* value. I have always been a bit suspicious about this formulation, although it does seem to be helpful sometimes. In general, I think that we have not really developed much of the causal language that we need for describing the relations between genes and phenotypic traits. The most everyday causal language, of the "this causes that" type, is often very misleading, especially because it suggests an overly simple picture in most cases. There are various more specialized forms of language used in genetics, including statistical talk about the "percentage of phenotypic variation explained by genetic variation," and causal talk about the "norms of reaction" of particular genotypes (Lewontin, 1974). Finding the right framework for the causal description of genes is a difficult and interesting business, but it is not the task that we confront here. The task here is the assessment of a particular way of describing what genes do, a way using the language of information and coding.

14.4 The Real but Restricted Role of "Genetic Coding"

I said earlier that philosophers have not succeeded in providing a good analysis of meaning, representation, and other semantic properties. Biologists, however, have become enthusiastic about the description of genes and gene action in term of coding and information. How should we respond to this situation? Should we just accept the biological description uncritically, especially given the shortcomings of philosophical analysis in this area? Or should we suspect that biologists have become careless (or worse) in their enthusiasm for "the informational gene"? Here is the approach that I will take to the problem. I ask: Which attributions of coding and/or informational properties to genes have a *useful theoretical role* within biology? I will argue in this section that there is a real, but restricted, domain in which the attribution of coding properties to genes does help us to solve problems and understand how organisms work. This domain is the explanation of protein synthesis itself.

The picture sketched in the previous section was roughly like this: genes make proteins, and proteins go off and enter a kind of causal spaghetti in which they interact with all sorts of things and may have a variety of roles (including feedback on the operation of the genetic system itself). To fully understand how organisms work, we need to understand all parts of this process. And to some extent, different concepts might be useful in understanding different parts of the whole. Let us focus again on protein synthesis. What is being explained here is how the cell makes proteins, where proteins are long chains of elements that must be put into exactly the right order. How to cells do this? They use the linear order of bases in their DNA to specify the order of amino acids in a protein, and this is done via a two-step *templating* operation. In a way, the concept of a template, or "template surface," is *the* crucial one for understanding how cells use genes to specify the primary structure of proteins (Watson et al., 1987, ch. 3). But the nucleic acids (DNA and RNA) act as a very specific *kind* of template in protein synthesis. Three properties make the nucleic acid templates special.

First, the templates used in making proteins are not proteins themselves, but another kind of molecule. DNA templates for mRNA which, after processing, templates for the protein. Given that the building blocks of a nucleic acid sequence (DNA and RNA bases) are not identical to the building blocks of the protein (amino acids), there must be a rule of specificity linking the two. In actual organisms, this rule is largely (though not completely) fixed and invariant. So a great variety of proteins are constructed from different DNA templates, in the same and in different organisms, via the same compact and general rule.

Secondly, the specification of proteins by these templates is *combinatorially* structured; the rule has a definite part-to-whole organization, involving the free rearrangement of the same fixed components. This structure exists at two levels. Most obviously, the elements of the templates that are specific for particular amino acids are triplets of nucleic acid bases. And, in addition, during translation a given triplet specifies the same amino acid regardless of its neighboring triplets. (There are some exceptions to this principle, but it holds in general.) The interpretation of a long sequence of bases is (again, in general) a simple and fixed function of the interpretation of its component triplets.

Thirdly, the rule linking base triplets with amino acids is believed to be largely "arbitrary." Some controversy surrounds this point, and by "arbitrary" I mean something specific. I mean that nothing about the chemistry of a particular amino acid is responsible for it corresponding to a particular base triplet. Contingent features of the tRNA molecules, and the enzymes that attach the amino acids to tRNAs, determine which triplets go with which amino acids. If the tRNA molecules had different sequences, or if the enzymes that work with them operated differently, a given base triplet could have an entirely different interpretation in protein synthesis. This need not have been true during the earliest evolution of the genetic code, but given the way protein synthesis works now, there is a kind of arbitrariness in the system.

So, my first main claim in this section is that the peculiar way in which proteins are made does justify talk of genes as "coding." This way of describing genes is, indeed, picking out some real and distinctive features of genetic mechanisms. Genes help make proteins by acting as templates, and this involves a combinatorial and arbitrary rule, largely static and universal, which connects nucleic acid sequences with amino-acid sequences. The analogy with human symbol systems such as languages is quite strong. So it is reasonable to say that genes encode information specifying amino-acid sequences in proteins, where this involves more than just the minimal, Shannon sense of information.

These features were not the ones picked out by the first discussions of "coding" in genetics. Indeed, the physicist Erwin Schrödinger, who was the first to use this terminology here, had nothing more than a *predictability* relation in mind when he initially introduced the idea of a "code-script" in his book in *What is Life?* (1992, p. 21). Tellingly though, Schrödinger made it clear in a later passage in *What is Life?* that he was thinking of systems with combinatorial properties, such as Morse code (p. 61). In any case, I claim that the positive, problem-solving content in the idea of "genetic coding" is its picking out these special features of the way nucleic acids act as templates for amino-acid sequences.

So, proteins are made by having their amino-acid sequence coded for. What does this conclusion tell us about the more general issues concerning the link between genes and phenotypes? It tells us that genetic causation has peculiarities, because at one specific place *within* the causal chains linking genes and phenotypic traits, we have some very special processes and mechanisms. But these unusual features of genetic causation concern *how* genes manage to have some of their *immediate* effects. These features of genetic causation do not extend beyond the local process in which the protein in question is being made.

In recent years, however, enthusiasm for the semantic characterization of genes has been unstoppable. This has led to the extension of "coding" talk, to the point at which the concept of genetic coding is now used to describe and distinguish the *entire causal paths* in which genes are involved. Such descriptions *seem* to bear real theoretical weight; the suggestion is that among all the causal paths leading to the development of an adult organism, some of these causal paths are distinctive because they involve the expression of a genetically encoded message specifying the process. But I suggest that once we leave the context of explaining protein synthesis, semantic descriptions of genes have a very different status. These further kinds of semantic description of genes have no empirical basis and make no contribution to our understanding.

Consequently, it is a mistake to use the idea of genetic coding or genetic information to pick out a distinction between the characteristics of whole organisms: a distinction between traits that are "coded for" and traits that are not. All that can be coded for is the primary structure of a protein molecule; not even higher-level protein structure is coded for, strictly speaking. To say this is not to deny the long causal reach of genes. A protein is made by being coded for, and the protein might have a key causal role in some far more complex and "distal" phenotypic trait. But I deny that positing a *coding* relation, as opposed to an ordinary causal relation, helps us at all in understanding the downstream consequences of the production of a protein.

In a recent paper (Godfrey-Smith, 2000), I used an elaborate thought experiment to try to make this point persuasive. I will sketch the central ideas here (though if the reader is already persuaded, it is fine to skip to the start of the next section). The thought experiment is based on the history of twentieth-century biology. The point of the thought experiment is to ask: If there was no genetic coding, which parts of biology would have to be different?

For some time in the years before Watson and Crick's work, it was thought that proteins themselves were the natural candidates for being the molecules that made up genes. One way for this to work is for genes to be little *samples* of the protein molecules needed by the cell, stored in the chromosomes. A gene would make other proteins, and would be replicated, by acting as a template *for itself.* There seemed no feasible way for amino-acid chains to template for themselves directly. So there might be a set of "connector" molecules that could bind on each side to the same amino acid. There would have to be 20 types of these connectors, for the 20 amino acids. They would enable an indirect like-with-like template mechanism.

Imagine an alternative possible biology in which this is what genes were like; genes would be little samples of protein. In such a situation, there would be no such thing as genetic coding. Instead of being a sort of *representation* of a protein, the gene would be a *sample* of the protein. So there would be no messenger, no translation, no expression, and no interpretation. Coding talk makes sense in genetics because the ordering of amino acids in proteins is done by means of *another* kind of molecule, the nucleic acids, which contain components that can be mapped with a special combinatorial rule onto the order of the amino acids in proteins.

Despite the absence of genetic coding in this situation, rather little else need be different. In the causal processes found in the development and metabolism of organisms, *one part* of every causal process would differ from our actual situation. The way in which genes place amino acids in the right order in protein synthesis would be different in each case. But the causal stories *from* the point at which each individual protein molecule has been produced could be exactly, or nearly, the same. Once a protein is produced, it makes no difference whether it was coded for or run off from a sample.

In my recent paper (Godfrey-Smith, 2000), I discuss the consequences of this thought experiment more fully. The main point is that the *overall* patterns found in the biological world need not be much different in the scenario we are envisioning, although many of the details would have to be different. The claim that I am making is that once the amino acids are placed in order, the role for genetic coding is over. After that, it is up to the proteins to do whatever they can do.

14.5 Two Objections and Replies

The previous section argued that the semantic description of genes is justified in one specific domain, but only in that domain. Basically, I broke the causal chain between a gene and its phenotypic effects into two parts. The first part connects a gene with

284

its protein products, and the second part connects the proteins with their myriad "downstream" effects on the organism's phenotype. The term "downstream" here does involve a simplification, because of the role of some proteins in regulating gene action, but I do not think this simplification affects my main points. I argued that genes do code for proteins, but this does not imply that they code for the proteins' further effects on the phenotype.

But what reason is there to insist that the "coding for . . ." relation only extends as far as I say it does? This is the first objection that I will discuss. In the case of *causal* description, at least of some kinds, we find something different. If A was a cause of B, and B was a cause of C, then A was also a cause of C. This principle is not completely uncontroversial, and it only applies to some kind of causal description. But it does seem to apply for much of the time. The usual term for this is "transitivity." Any relation R is transitive if: given that A has R to B, and B has R to C, then A must also have R to C. Causation is often said to be a transitive relation.

The situation we have in the case of genes, as I have described it so far, is that a gene codes for a protein and the protein goes off and affects various phenotypic traits. The protein might have its most obvious effect just on one trait, but it will have some causal involvement (if only as background) with many others. Let us look at a simplified case, though. Suppose that gene G makes protein P, which then reliably causes the organism to have trait X. This, I should emphasize, is a *highly* simplified scenario. Much of the point of the preceding sections was to argue that although a gene can specify a protein's structure in a fairly direct and straightforward way (at least in many cases), the downstream causal role of a protein is a complex and interactive affair, which usually cannot be described simply in terms of the protein "causing trait X." But consider the simplified case for now. What is to stop us from saying that because gene G codes for protein P, and protein P causes trait X, gene G codes for trait X? This is not exactly the same as a question about transitivity, because we do not have the same relation between G and P, and between P and X. But it is related, obviously.

In a sense, nothing stops us from saying that G codes for X, in a case like this. This will be a new, specialized sense of the word "code." It is possible to concoct all sorts of novel ways of describing the roles that genes have, by mixing causal and semantic language together and giving explicit definitions of what one means to say. But are there any good reasons to describe things in this way?

One thing to note is that this treatment of the idea of coding would not follow the general pattern that we find in the case of other kinds of semantic properties. Here, I have in mind the semantic properties of more familiar examples of representation, such as thoughts and sentences within public languages. For example, suppose you know that if you order the extra-large pizza from your favorite pizza place, your action will have the consequence that the delivery arrives late. (Smaller pizzas tend to arrive on time.) This fact does not imply that when you order the extra-large pizza you are also ordering them to make the delivery late. The likely or inevitable *effects* of a message are not all part of the *content* of the message. The semantic content of your message is just: *bring me an extra-large pizza*. Your order is complied with if and only if the order leads to the delivery person bringing you an extra-large pizza.

Aside from being late, your ordering of the extra-large pizza might have all sorts of further downstream effects, both systematic and accidental. Who knows what might ultimately result from the lateness of a pizza, or (to use an older example) the loss of a horseshoe nail? Still, although the causal chain starting from your pizza order can extend indefinitely, you only ordered the extra-large pizza. The semantic content of the message is specific to a certain portion of the resulting causal chain. If you order P and P causes X, that does not mean that you ordered X. You might, in some cases, *intend* X to occur as well. But the fact that X is caused by P does not imply that it was part of the message itself.

So, in the everyday cases, we do not use semantic concepts in the way outlined in my statement of the objection to my position. That does not mean that we *couldn't* use the language of genetic coding in this way if we wanted to. All sorts of novel forms of description can be introduced, by modifying existing terms. But is there any good reason to describe genes in this way? The idea of a genetic code linking nucleic acid bases and amino acids is part of a theoretical framework that solved a real problem in biology. That problem is restricted, however, to the domain of protein synthesis. We do not gain any further understanding – and we make distortion and over-simplification quite likely – when we extend the semantic description of genes beyond this domain. Genes can have a causal role that extends far beyond the production of proteins, but proteins are all that a gene can code for.

I will discuss one other possible view here, which I will also express in the form of an objection to my position. My argument has been that there is no reason to see genes as coding for anything beyond the primary structure of proteins. Once the protein has been made, the downstream effects of the protein involve a mass of ordinary causal relations. But perhaps, it might be argued, the situation looks different once we look at the relation between gene and phenotype in an *evolutionary* context.

Particular genes are favored by natural selection because of their overall effects on the organism's phenotype, even if these effects are indirect and are dependent on many other factors. For some philosophers and biologists, the key to understanding the semantic properties of genes is to approach the question from an evolutionary point of view. They have argued that if a gene has been favored by natural selection because it tends to bring about a particular phenotypic effect, X, then the gene *represents* X, or carries the information that X is to appear in the phenotype. This suggestion has been made by Sterelny, Smith and Dickison (1996) and also by Maynard Smith (2000). When natural selection favors a gene because of some particular phenotypic effect, that effect becomes something that the gene *represents*, as well as merely *causes*.

One way in which to understand this proposal is to look at the concept of *biological function*. There are many senses of the word "function," and many even within biology. But in one sense of the term, the function of a structure is the effect it has that has been responsible for its being selected for. When someone claims that the function of a bird's long tail is to attract mates, or the function of a spider's web is to catch insects, it is not just being claimed that the tail or the web tends to have this effect. Rather, these are the effects that the tail and the web are *supposed* to have. Catching insects is what the web is *for*, in an evolutionary sense (Wright, 1973; for a collection on these issues, see also Buller, 1999). A number of philosophers have thought that there is a very close relation between functional properties, in this special evolutionary sense, and semantic properties (Millikan, 1984). For example, this sense of "function" is one that permits claims about *malfunction*, or error (see section 14.2 above). When the bird's tail attracts a hunter rather than a mate, it is not performing its function.

So, the advocates of an evolutionary approach to genetic information argue that when a gene has been favored by selection because it causes X, it acquires the *function* of causing X. This, in turn, justifies the claim that the gene instructs, encodes, represents, or carries information specifying that the phenotype is to include X. This proposal would only assign informational properties to genes that have been favored by natural selection, but that might not be a bad problem for the view.

Might this be a good way to defend the idea that genes code for more than proteins? In reply to this proposal, I want to first emphasize the fact that lots of biological structures have functions, in the evolutionary sense, without carrying information or representing anything. The spider's web does not represent prey, although it has the function of catching prey. Legs are for walking, but they do not represent walking. Something can have the function of producing a particular effect in biological processes, without representing or coding for that effect. The suggestion made by Sterelny, Smith and Dickison is based on *denying* this principle, for a special set of cases. They claim that all structures whose function is to operate within biological development, and bring about a certain effect on the phenotype, *represent* the outcomes of those developmental processes. In these cases, they argue, the connection between function and representation is especially tight. But I don't see any good reason to believe this claim. For example, consider all the other structures that operate in biological development alongside genes, and are part of the machinery by which genes exert their effects. Consider, for instance, the tRNA molecules and the enzymes that operate with them in translation processes. These structures have functions, in the evolutionary sense, and their role involves the same processes that genes are involved in. But they are not usually said to represent anything, or carry information about anything, and there seems no reason at all to say that they do (I discuss this evolutionary proposal about information in more detail in Godfrey-Smith, 1999).

So I think the proposal of Sterelny, Smith, and Dickison and of Maynard Smith is mistaken. Some particular genes are favored by natural selection because of their overall effects on phenotype, and can hence be described using an evolutionary concept of function. Genes *also* have a role in protein synthesis that justifies a kind of informational description, because of the special features of the "genetic code." But I emphasize the "also" in the previous sentence; these are two quite separate facts about genes. It is a mistake to put them together and say that genes encode the phenotypic traits of whole organisms.

14.6 Conclusion

I will end by quickly summarizing the main points of my argument. There are two senses of information, the Shannon sense and a stronger, "semantic" sense. Information in the Shannon sense is everywhere; it does connect genes with phenotypic traits, but it connects them in both directions and it connects environmental conditions with phenotypic traits as well. I have no argument with the idea that genes contain Shannon-type information about the organism's phenotype, but this is a very uncontroversial claim that does not say very much.

No one has a good general theory of semantic properties, like representation and meaning. But the role of genes within protein synthesis is, in a sense, an informational or representational role. Genes specify the primary structure of protein molecules via a compact, code-like rule that is combinatorially structured and (in a sense) arbitrary. These facts provide some justification for describing genes as coding for the primary structure of protein molecules. But these facts provide no justification for the idea that genes encode the phenotypes of whole organisms, or indeed of anything beyond protein primary structure. It is not just loose talk but positively misleading to over-extend the informational description of genes, and see genes as encoding information for phenotypic traits.

Bibliography

Buller, D. (ed.), 1999: *Function, Selection, and Design.* Albany, NY: SUNY Press.

Devitt, M. and Sterelny, K. 1999: *Language and Reality*, 2nd edn. Cambridge, MA: The MIT Press.

Dretske, F. 1981: *Knowledge and the Flow of Information.* Cambridge, MA: The MIT Press.

Godfrey-Smith, P. 1999: Genes and codes: lessons from the philosophy of mind? In V. Hardcastle (ed.), *Where Biology Meets Psychology: Philosophical Essays.* Cambridge, MA: The MIT Press, 305–31.

—— 2000: On the theoretical role of "genetic coding." *Philosophy of Science*, 67, 26–44.

Griffiths, P. E. 2001: Genetic information: a metaphor in search of a theory. *Philosophy of Science*, 68, 394–412.

—— and Gray, R. 1994: Developmental systems and evolutionary explanation. *Journal of Philosophy*, 91, 277–304.

Kay, L. 2000: *Who Wrote the Book of Life? A History of the Genetic Code.* Palo Alto, CA: Stanford University Press.

Lewontin, R. C. 1974: The analysis of variance and the analysis of cause. Reprinted in Levins, R. and Lewontin, R. C. 1985: *The Dialectical Biologist.* Cambridge, MA: Harvard University Press.

Maynard Smith, J. 2000: The concept of information in biology. *Philosophy of Science*, 67, 177–94.

Millikan, R. 1984: *Language, Thought and Other Biological Categories.* Cambridge, MA: The MIT Press.

Oyama, S. 1985: *The Ontogeny of Information.* Cambridge: Cambridge University Press.

Schrödinger, E. 1992: *What is Life?* Cambridge: Cambridge University Press (originally published in 1944).

Shannon, C. 1948: A mathematical theory of communication. *Bell Systems Technical Journal*, 27, 279–423, 623–56.

Sterelny, K., Smith, K., and Dickison, M. 1996: The extended replicator. *Biology and Philosophy*, 11, 377–403.

Stich, S. and Warfield, T. 1994: *Mental Representation: A Reader.* Oxford: Blackwell.

Watson, J., Hopkins, N., Roberts, J., Steitz, J. A., and Weiner, A. 1987: *The Molecular Biology of the Gene*, 4th edn. Menlo Park, CA: Benjamin/Cummins.

Williams, G. C. 1992: *Natural Selection: Levels, Domains, and Challenges*. Oxford: Oxford University Press.

Wright, L. 1973: Functions. *Philosophical Review*, 82, 139–68.

IS THE MIND A SYSTEM OF MODULES SHAPED BY NATURAL SELECTION?

Evolutionary psychology (EP) is a new field that attempts to understand the mind in evolutionary terms. It posits that the mind consists of many "modules," each one performing a very specific kind of task, and operating independently of the others. These modules evolved because they were good solutions to adaptive problems faced by our ancestors. Within this framework, we can gain an understanding of the mind by discovering these modules and the environmental pressures that led to their being naturally selected. Peter Carruthers, in his chapter, presents some of the empirical evidence that bolsters the EP program, and then responds to an important philosophical objection to the modular conception of mental architecture. Fiona Cowie and Jim Woodward concede that the mind has been shaped by natural selection and that the mind consists of a variety of distinct capacities; but surely EP is not merely asserting this banal point. A major problem, they contend, is that evolutionary psychologists are often unclear about what a "module" is supposed to be. They argue that there is no unambiguous sense of the word "module" that is faithful to the usage of evolutionary psychologists and that would also render the central claim of EP true. In particular, the assumption that individual modules can evolve in isolation from one another is at odds with much of what we know about the evolution of brains. This exchange shows how empirical and conceptual issues can merge, blurring the line between philosophy of science and science itself.

The Mind is a System of Modules Shaped by Natural Selection

Peter Carruthers

This chapter defends the positive thesis that constitutes its title. It argues, first, that the mind has been shaped by natural selection; and, secondly, that the result of that shaping process is a modular mental architecture. The arguments presented are all broadly empirical in character, drawing on evidence provided by biologists, neuro-scientists, and psychologists (evolutionary, cognitive, and developmental), as well as by researchers in artificial intelligence. Yet the conclusion is at odds with the manifest image of ourselves provided both by introspection and by common-sense psychology. The chapter concludes by sketching how a modular architecture might be developed to account for the patently unconstrained character of human thought, which has served as an assumption in a number of recent philosophical attacks on mental modularity.

15.1 Introduction: Evolutionary Psychology and Modularity

If we are to address the topic picked out by the title of this chapter, then there are two main questions for us to answer:

1 Is the human mind, as well as the human body and brain, a product of natural selection? Is the human mind an adaptation/collection of adaptations?
2 Supposing that the answer to question 1 is positive, has the effect of natural selection on the human mind been to impose on it a modular architecture?

The reasons why a positive answer should be returned to question 1 are quite straight-forward, and are not disputed within the present debate. Everyone should agree that the human mind is an adaptation, provided that one thinks (i) that the mind is causally and explanatorily relevant in the production of behavior, and supposing that one agrees (ii) that the mind is significantly structured. For assumption (i) will make the mind a major determinant of evolutionary fitness, and so a prime *candidate* to be shaped by natural selection. And since natural selection is the only serious contender for explaining the appearance of functional complexity in the biological world (Williams, 1966; Dawkins, 1976, 1986), assumption (ii) means that we shall have good reason to think that the mind has *actually* been shaped by evolution. These points will be taken for granted in what follows.

The main burden of this chapter will be to motivate a positive answer to question 2, then. This actually breaks down into two sub-questions: (a) *Does* the mind have a modular architecture? (b) Supposing that it does, is that architecture a product of natural selection or, rather, does it get created within each individual through general learning and some sort of process of gradual modularization? Most of my attention will be devoted to question (2a), in sections 15.2 and 15.3 below. In section 15.4, I shall turn briefly to question (2b).

The thesis to be defended here – that the mind consists largely of evolved modules – is the claim of "massive modularity," which has been proposed and argued for in recent decades by evolutionary psychologists (Gallistel, 1990; Tooby and Cosmides, 1992; Sperber, 1996; Pinker, 1999). It is important to realize at the outset that evolutionary psychology is a broad church (somewhat like utilitarianism), embracing a variety of different theoretical claims and approaches. So one should be wary of arguments that take the form: "Many evolutionary psychologists claim that *P. P* is implausible. So evolutionary psychology is implausible." (No one would ever dream of arguing like this against utilitarianism. One would, of course, have to explore other avenues and options within a utilitarian perspective before concluding with a negative verdict.) In what follows, I shall try to present the thesis of massive modularity in what seems to me its strongest light. But the reader should be aware that there are other ways in which the approach might be developed, and that some of these might eventually turn out to be better. Evolutionary psychology is best seen as a *research program* (in the sense of Lakatos, 1970), not a fixed body of theory.

15.1.1 What are modules?

A natural first thought would be that the answer to our main question might depend crucially upon what one means by "module." To the question, "Does the mind consist of evolved modules?" one might naturally reply, "That largely depends on what you understand by a 'module'." And in a way, this is obviously quite correct. However, it would be a mistake to clog up our inquiry by attempting an analysis of what we should mean by "module" in advance. As generally happens in science, a proper understanding of our terms should be the *outcome* of empirical and theoretical enquiry, not something which is stipulated at the start.

According to Fodor's now-classic (1983) account, for example, modules are stipulated to be domain-specific innately specified processing systems, with their own proprietary transducers, and delivering "shallow" (nonconceptual) outputs; they are held to be mandatory in their operation, swift in their processing, encapsulated from and inaccessible to the rest of cognition, associated with particular neural structures, liable to specific and characteristic patterns of breakdown, and to develop according to a paced and distinctively arranged sequence of growth. But such an account was designed to apply specifically to *input* systems, such as vision or touch. Those in the cognitive sciences who have wished to extend the notion of "modularity" to apply to the central systems that generate new beliefs from old (such as our "theory of mind," or the faculty of common-sense psychology), or which generate new desires, would obviously have to modify that account a great deal. Yet many in philosophy remain strangely fixated on Fodor's original analysis, using it as a stick with which to beat some of the more ambitious claims for mental modularity (see, e.g., Currie and Sterelny, 1999).

I propose, in contrast, that we should begin our work with "module" understood in its weakest and loosest everyday sense, to mean something like, "isolable functional sub-component." Thus a company organized in modular fashion has separate units that operate independently and perform distinct functions. And a hi-fi system that can be purchased on a modular basis is one in which the separate component parts can be bought independently, in which some at least of those components can operate independently of the others – you can have the tape-deck without the CD player – and in which different versions of the same part can be substituted for others without altering the remainder. Note that even this weak notion of modularity isn't anodyne in relation to the mind. Specifically, functionalism about mental properties might be correct without it being the case that personal memory, say, forms an isolable type that can be damaged or lost while leaving the rest of the mind intact. And when this notion is conjoined with the claim that different modules will have at least partially distinct selective histories, the result is anything but anodyne. This weak everyday notion should be enough to get us started. In the end, a module will be whatever is warranted by the various *arguments for* mental modularity.

These arguments are of various different types, and have somewhat different strengths. And actually they warrant rather different notions of modularity too; so our final answer to the question concerning the meaning of "modularity" might have to be a multiple one. I will not have space to develop this point fully in the present chapter, but let me just assert the following. We should expect modules to be relatively *encapsulated*, in the sense of processing their inputs independently of most of the information stored elsewhere in the mind. We should expect many (but not necessarily all) central modules to be *domain specific*, in the sense of having been designed to operate on particular sorts of conceptual inputs or conceptual structures. We should expect that some (but only some) modules will deploy algorithms that are unique to them, not being redeployed elsewhere in the mind to subserve other functions. And, of course (if we are to believe in *evolved* modules), we should expect that the development of modules is innately channeled to some significant degree, and that many of them, at least, will be liable to local impairment.

Notice that we need *not* be committed to saying that modules are "elegantly engineered" processing systems, which have simple and streamlined internal structures, and which exist independently of other such systems. On the contrary, we should expect that many modules are *kludges* (Clark, 1987), recruiting and cobbling together in quite inelegant ways resources that existed antecedently. (This is routine in biology, where the evolution of any new structure has to begin from what was already present. Thus the penguin's flippers, used for swimming, evolved from the wings that it once used for flight.) We should also expect many modules to consist of arrangements of sub-modules, which are linked together in complex ways to fulfill the function in question. (A glance at any of the now-familiar wiring diagrams for visual cortex will give the flavor of what I have in mind.) And, for similar reasons, we should expect many modules and sub-modules to have multiple functions, passing on their outputs to a variety of other systems for different purposes. In that case, distinct modules may share parts in common with one another. (This need not mean that they cannot be selectively damaged, since it may be possible to damage some of the links in a system without destroying any of the shared components.)

I shall begin our discussion of the arguments for mental modularity in section 15.2, with a number of broadly programmatic/methodological lines of reasoning for the conclusion that we should *expect* the mind to be modular in structure. In section 15.3 I shall look briefly at some of the evidence – developmental, pathological, and experimental – for thinking that the mind really does have such an architecture. Then in section 15.4 I shall consider whether modular architectures are innately structured adaptations, or result from some sort of process of modularization-through-learning.

Finally, in section 15.5 I shall take up what seems to many people to be the most powerful objection to a modular conception of the human mind, deriving from the nondomain-specific and unconstrained character of human thinking. (We can combine together highly disparate concepts in our thoughts, and we can apparently do so at will.) In that section, I shall sketch a language-involving cognitive architecture according to which a modular natural language faculty serves to integrate the outputs of other modular systems, and where that language faculty can be used to generate new contents in imagination. But there won't be space, here, to explain these ideas fully; a mere sketch will have to suffice (for further development, see Carruthers, 2003b, 2003c, forthcoming).

15.1.2 Belief/desire psychology and learning

There is one final point that I want to emphasize and develop in this introduction, before we get down to the arguments. This is that, contrary to many people's first impressions, evolved modules aren't at all inconsistent with learning. Opponents of evolutionary psychology (e.g., Dupré, 2001) are apt to suppose that what is at issue are a suite of evolved *behavioral dispositions*. Hence they think that such an approach will minimize the role of learning in both human development and in mature behaviors. And they think, in consequence, that evolutionary psychology is incapable of explaining the distinctive *flexibility* of human behavior. But nothing could be further from the truth. There are at least two distinct confusions here, which need to be sorted out. The first is that evolutionary psychology is concerned not with behavioral dispositions *per se* but, rather, with the cognitive systems that operate and interact to produce those dispositions. (In fact, this is the crucial difference between evolutionary psychology and the earlier scientific research program of sociobiology.) And the second is that most of these postulated systems are actually systems of *learning*. I shall establish these points in turn. In this case, rapid assimilation of new information leading to flexibility in behavior is exactly what we should predict.

Evolutionary psychology takes quite seriously a belief/desire (or an information/goal) organization of psychological systems. This is true even in the case of insects, where it turns out that the desert ant has states representing that a food source is 44.64 meters northeast of its nest on a bearing of 16.5 degrees, say, which it can deploy either in the service of the goal of carrying a piece of food in a straight line back to the nest, or in returning directly to the source once again – or, in the case of bees, when the goal is to inform other bees of the location of the food source (Gallistel, 1990, 2000). What has been selected for in the first instance, on this view, are systems for generating beliefs and desires; the behaviors that result from those beliefs and desires can be many and various. (This isn't to deny that insects and other animals will *also* have a suite of innate behaviors and fixed action patterns, of course, in cases in which flexibility isn't required.)

Seen like this, it becomes obvious that the systems in question are learning systems. (At least, this is so in connection with belief-generation; I shall return to the case of desires below.) The system in the ant which uses dead reckoning to figure out the exact distance and direction of a food source in relation to the nest (given the time of day and position of the sun) is a system for *learning* that relationship, or for acquiring a *belief* concerning that relationship. Similarly, the language system in humans is, in early childhood, designed for *learning* the syntax and vocabulary of the surrounding language (albeit using a unique information-rich set of learning algorithms); and in older children and adults it is used to *learn* what someone has just said on a given occasion, extracting this information from patterns in ambient sound.

As already noted, many of these learning systems are hypothesized to deploy learning algorithms that are unique to them. The computations necessary to extract a directional bearing from information about the sun's position in the sky and the time of day and year are obviously very different from those needed to generate the syntax of English from samples of English discourse. But some systems might operate using learning algorithms that are used many times over in the brain. Thus the system that is used in acquiring vocabulary in childhood, and the system in the visual cortex that extracts object shape, may both be designed to do Bayesian inference (Lila Gleitman, personal communication). This is still consistent with the weak sense of "modularity" that we have adopted, since we can claim that the systems in question are highly restricted in their input and output conditions, and are relatively encapsulated in their processing. Each has been designed to take specific kinds of data as input and to generate a certain sort of output. Whether the algorithms used are unique or replicated many times over in other learning systems is much less significant.

15.1.3 Practical reasoning

If it is being taken for granted that belief/desire psychology applies throughout the animal kingdom, however, then a system for practical reasoning is being taken for granted too. This gives rise to a natural objection. This is that practical reasoning can't be modular, because if an organism possesses just a single practical reasoning system (as opposed to distinct systems for distinct domains), then such a system obviously couldn't be domain specific in its input conditions. However, such a system could, nevertheless, be highly restricted in terms of its processing database, and this is all that is needed to secure its modular status in the sense that matters.

For example, we can imagine the following very simple practical reasoning module. It takes as input whatever is the currently strongest desire, P. It then initiates a search for beliefs of the form "$Q \to P$," cueing a search of memory for beliefs of this form and/or keying into action a suite of belief-forming modules to attempt to generate beliefs of this form. When it receives one, it checks its database to see if Q is something for which an existing motor program exists. And it initiates a search of the contents of current perception to see if the circumstances required to bring about Q are actual (i.e., to see not only whether Q is something doable, but doable here and now). If so, it goes ahead and does it. If not, it initiates a further search for beliefs of the form "$R \to Q$," or of the form "Q" (for if Q is something that is already happening or is about to happen, the animal just has to wait in order to get what it wants – it doesn't need to do anything more), and so on. Perhaps the system also has a simple stopping rule: if you have to go more than n number of conditionals deep, stop and move on to the next strongest desire.[1]

Note that the sort of module described above would be input unrestricted. Since, in principle, almost anything can be the antecedent of a conditional whose consequent is something desired (or whose consequent is the antecedent of a further conditional whose consequent . . . , and so on), any belief can in principle be taken as input by the module. But what the module can *do with* such inputs is, I am supposing, extremely limited. All it can do is the practical reasoning equivalent of *modus ponens* (I want P; if Q then P; Q is something I can do here-and-now; so I'll do Q), as well as collapsing conditionals ($R \to Q$, $Q \to P$, so $R \to P$), and initiating searches for information of a certain sort by other systems. It can't even do conjunction of inputs, I am supposing, let alone anything fancier. Would such a system deserve to be called a "module," despite its lack of input-encapsulation? It seems to me plain that it would. For it could be a dissociable system of the mind, selected for in evolution to fulfill a specific function, genetically channeled in development, and with a distinct neural realization. And because of its process-encapsulation, its implementation ought to be computationally tractable.

However, it is plain that human practical reasoning is not at all like this. There seem to be no specific limits on the kinds of reasoning in which we can engage while thinking about what to do. We can reason conjunctively, disjunctively, to and from universal or existential claims, and so forth. And contents from all the various allegedly modular domains can be combined together in the course of such reasoning. This makes the practical reasoning system look like an archetypal holistic, a-modular (and hence computationally *in*tractable) central system, of just the sort that Fodor thinks make the prospects for a worked-out computational psychology exceedingly dim (Fodor, 1983, 2000). I shall return to this problem briefly in section 15.5 below, and again at greater length in future work (Carruthers, forthcoming).

---

1 Although I have described this practical reasoning module as "very simple" (in relation to the sorts of reasoning of which humans are actually capable), its algorithms would not, by any means, be computationally trivial ones. Each of the component tasks should be computationally *tractable*, however – at least, so far as I can see (although intuitions of computational tractability are notoriously unreliable).

15.1.4 Acquiring desires

Desires aren't learned in any normal sense of the term "learning," of course. Yet much of evolutionary psychology is concerned with the genesis of human motivational states. This is an area in which we need to construct a new concept, in fact – the desiderative equivalent of learning. Learning is a process that issues in true beliefs, or beliefs that are close enough to the truth to support (or at least not to hinder) individual fitness.[2] But desires, too, need to be formed in ways that will support (or not hinder) individual fitness. Some desires are instrumental ones, of course, being derived from ultimate goals together with beliefs about the means that would be sufficient for realizing those goals. But it is hardly very plausible that all acquired desires are formed in this way.

Anti-modular theorists such as Dupré (2001) are apt to talk vaguely about the influence of surrounding culture, at this point. Somehow, goals such as a woman's desire to purchase a wrinkle-removing skin cream, or an older man's desire to be seen in the company of a beautiful young girl, are supposed to be caused by cultural influences of one sort or another – prevailing attitudes to women, perceived power structures, media images, and so forth. But it is left entirely unclear what the mechanism of such influences is supposed to be. How do facts about culture generate new desires? We are not told, beyond vague (and obviously inadequate) appeals to imitation.

In contrast, evolutionary psychology postulates a rich network of systems for generating new desires in the light of input from the environment and background beliefs. Many of these desires will be "ultimate," in the sense that they haven't been produced by reasoning backward from the means sufficient to fulfill some other desire. But they will still have been produced by inferences taking place in systems dedicated to creating desires of that sort. A desire to have sex with a specific person in a particular context, for example, won't (of course) have been produced by reasoning that such an act is likely to fulfill some sort of evolutionary goal of producing many healthy descendants. Rather, it will have been generated by some system (a module) that has evolved for the purpose, which takes as input a variety of kinds of perceptual and nonperceptual information, and then generates, when appropriate, a desire of some given strength. (Whether that desire is then acted upon will, of course, depend upon the other desires that the subject possesses at the time, and on his or her relevant beliefs.)

The issue is not, then, the extent to which learning is involved in the causation of forms of human behavior. Both evolutionary psychologists and their opponents can agree that learning is ubiquitous. Nor is the issue even whether the algorithms used in learning are domain general (being suitable for extracting many different kinds of information) or, rather, are specific to a particular domain (although the domain speci-

---

2   This isn't meant to be a definition, of course. If there are innate beliefs, then evolution might also be a process that issues in true beliefs, but evolving isn't learning. What is distinctive of learning is that it should involve some method (not necessarily a *general* one, let alone one which we already have a name for, such as "enumerative induction") for extracting information from the environment within at least the lifetime of the individual organism. And what distinguishes learning from mere triggering is that it is a process that admits of a correct cognitive description – learning is a cognitive as opposed to a brute-biological process.

ficity of learning algorithms is certainly an interesting question). Rather, it is whether the mechanisms that engage in learning are multiple, and have been specifically and separately designed by evolution to extract information (or to generate fitness-enhancing goals) concerning a given domain.[3]

15.2 Why we should Expect the Mind to be Modular

In this section, I shall review a number of general arguments for the thesis that we should expect the evolved structure of the mind to be modular. Considerations of space mean that our discussion of these arguments will have to be quite brisk. For more detailed elaboration, see Tooby and Cosmides (1992).

15.2.1 The argument from biology

One argument for massive modularity appeals to considerations deriving from evolutionary biology in general. The way in which evolution of new systems or structures characteristically operates is by "bolting on" new special-purpose items to the existing repertoire. First, there will be a specific evolutionary pressure – some task or problem that recurs regularly enough and that, if a system can be developed which can solve it and solve it quickly and reliably, will confer fitness advantages on those who possess that system. Then, secondly, some system that is targeted specifically on that task or problem will emerge and become universal in the population. Often, admittedly, these domain-specific systems may emerge by utilizing, coopting, and linking together resources that were antecedently available; and hence they may appear quite inelegant when seen in engineering terms. But they will still have been designed for a specific purpose.

Another way of putting the point is that in biology generally, distinct functions predict distinct (if often overlapping) mechanisms to fulfill those functions (Gallistel, 2000). No one supposes that there could be a general-purpose sensory organ, which could fulfill all of the functions of sight, hearing, taste, touch and smell. On the contrary, what we expect to find – and what we do find – are distinct organs, specialized for the distinctive structure of each domain, and which have been shaped by natural selection to fulfill the function in question. Similarly, no one expects to find that there is a general-purpose organ fulfilling the functions of both a heart and a liver, or fulfilling the functions of digestion and respiration. Likewise, then, in the case of the mind: one should expect that distinct mental functions – estimating numerosity, predicting the effects of a collision, reasoning about the mental states of another person, and so on – are likely to be realized in distinct cognitive learning mechanisms, which have been selected and honed for that very purpose.

---

3 Note that once the evolutionary psychologist's thesis is seen to be restricted to the genetically channeled character of a suite of learning mechanisms, rather than the innateness of most of the contents of the mind or anything of that sort, then the force of the argument from the relative paucity of genes in relation the large number of neurons in the brain (Elman et al., 1996) is much reduced. And see Marcus (forthcoming) for a very nice demonstration of how a small number of genes can be used to build banks of distinctively organized neurons.

15.2.2 Could general learning evolve?

A different – though closely related – consideration is negative, arguing that a general-purpose problem-solver *couldn't evolve*, partly because it would always be outcompeted by a suite of special-purpose conceptual modules. A general-purpose learning system would, inevitably, have to be very slow and unwieldy in relation to any set of domain-specific competitors. One point here is that such a system would face the problem of combinatorial explosion as it tried to search through the maze of information and options available to it (see section 15.2.3 below). Another point is that it would either have to process many different learning tasks at once using the same learning apparatus (and how would *that* be organized, except on a modular basis?), or it would have to tackle those tasks sequentially, leading to significant delays and bottlenecks.

Yet another point, however, is that many learning tasks simply cannot be solved without substantial innate assumptions about the domain being learned; which argues for the existence of a number of distinct learning mechanisms within which these assumptions can be embedded. The most famous such domain is language, where so-called "poverty of stimulus" arguments show that there must be an innately structured language acquisition device, in order for language learning to be possible (Chomsky, 1988; Crain and Pietroski, 2001; Laurence and Margolis, 2001). But similar arguments can be constructed for other domains: such as common-sense psychology, where a rich causal story has to be extracted somehow (and by the age of four, by all normal children) from the behavioral and introspective data (Carruthers, 1992; Botterill and Carruthers, 1999); and also for normative reasoning, where children again manifest an early grasp of a highly abstract set of concepts and principles, this time lacking any straightforward empirical basis (Cummins, 1996; Núñez and Harris, 1998; Dwyer, 1999).

Further arguments relate more specifically to the mechanisms charged with generating desires. For many of the factors which promote long-term fitness are too subtle to be noticed or learned within the lifetime of an individual; in which case there couldn't be a general-purpose problem-solver with the overall goal "promote fitness" or anything of the kind (Tooby and Cosmides, forthcoming). On the contrary, a whole suite of fitness-promoting goals will have to be provided for, which will then require a corresponding set of desire-generating computational systems.

Consider, for example, the surprising prediction that in certain social species where the reproductive success of males can vary a great deal as a function of fitness and status (such as deer and humans), females should vary their reproductive strategies (Trivers and Willard, 1973). Low-fitness, low-status, females should invest in female offspring, since these offer their best chance of passing on their genes to future generations; high-fitness, high-status, females, by contrast, should invest in male offspring. In deer, it seems that the mechanism by which this is effected is noncognitive, somehow altering the birth ratios, since does that are in poor physical condition are more likely to *give birth to* female offspring. In humans, on the other hand, it would appear that the mechanism is a cognitive one, operating via the mother's desire (or absence of desire) to have another child quickly, and/or via her degree of investment in the child that she has. (But see Grant (1998) for evidence that suggests that human mothers may also be capable of controlling the sex of an unborn infant, to some small degree.) Thus, low-status women in the USA (measured by income and by the presence or absence of an investing male partner) whose first child is a daughter are likely to wait longer before giving birth to a second child; they are more likely to breast-feed a female child; and they will also breast-feed for a significantly greater time; with high-status women displaying the converse pattern (Gaulin and Robbins, 1991).

Roughly speaking, the prediction is that low-status women should want daughters, whereas high-status women should want sons. And this prediction does seem to be supported, both by the results mentioned above and by extensive data from other measures of parental investment around the globe, including rates of male and female infanticide (Hrdy, 1999). But, of course, no one thinks that the women are reasoning backwards, from a desire to be as reproductively successful as possible to the means most likely to realize that goal. For it requires sophisticated evolutionary–biological reasoning to figure the thing out. Rather, the suggestion should be that evolution has favored a desire-generating mechanism in human women that is sensitive to a variety of indications of expected fitness, and which has been selected for because of its long-term effects on reproductive success.

15.2.3 The argument from computational psychology

Perhaps the most important argument in support of mental modularity for our purposes, however, simply reverses the direction of Fodor's (1983, 2000) argument for pessimism concerning the prospects for computational psychology. It goes like this:

(1) The mind is computationally realized.
(2) A-modular, or holistic, processes are computationally intractable.
(3) So the mind must consist wholly or largely of modular systems.

Now, in a way Fodor doesn't deny either of the premises in this argument; and nor does he deny that the conclusion follows. Rather, he believes that we have independent reasons to think that the conclusion is false; and he believes that we cannot even *begin* to see how a-modular processes could be computationally realized. So he thinks that we had better give up attempting to do computational psychology (in respect of central cognition) for the foreseeable future. Fortunately, however, his reasons for thinking that central cognition is holistic in character are poor ones. For they depend upon the assumption that scientific inquiry (social and public though it is) forms a useful model for the processes of individual cognition; and this supposition turns out to be incorrect (Carruthers, 2003a; but see also section 15.5 below for brief discussion of a related argument against mental modularity).

Premise (1) is the guiding assumption that lies behind all work in computational psychology, hence gaining a degree of inductive support from the successes of the computationalist research program. Just about the only people who reject premise (1) are those who endorse an extreme form of distributed connectionism, believing that the brain (or significant portions of it, dedicated to central processes) forms one vast connectionist network, in which there are no local representations. The successes of the distributed connectionist program have been limited, however, mostly being confined to various forms of pattern recognition; and there are principled reasons for thinking that such models cannot explain the kinds of one-shot learning of which humans and other animals are manifestly capable (Horgan and Tienson, 1996; Marcus, 2001). Indeed, even the alleged neurological plausibility of connectionist models is now pretty thoroughly undermined, as more and more discoveries are made concerning localist representation in the brain.

Premise (2), too, is almost universally accepted, and has been since the early days of computational modeling of vision. You only have to *begin* thinking in engineering terms about how to realize cognitive processes in computational ones to see that the tasks will need to be broken down into separate problems that can be processed separately (and preferably in parallel). And this is, indeed, exactly what we find in the organization of the visual system. What this premise then does, is to impose on proposed modular systems quite a tight *encapsulation* constraint. For any processor that had to access the full set of the agent's background beliefs (or even a significant subset thereof) would be faced with an unmanageable combinatorial explosion. We should therefore expect the mind to consist of a set of processing systems which are not only modular in the sense of being distinct isolable components, but which operate in isolation from most of the information that is available elsewhere in the mind. (I should emphasize that this point concerns not the *inputs* to modular systems but, rather, the *processing databases* which are accessed by those systems in executing their algorithms; see Carruthers (2003a) and Sperber (2003).)

Modularism is now routinely assumed by just about everyone working in artificial intelligence, in fact (McDermott, 2001). So anyone wishing to *deny* the thesis of massive modularity is forced to take on a heavy burden. It must be claimed either that minds *aren't* computationally realized, or that we haven't the faintest idea how they can be. And, either way, it becomes quite mysterious how mind can exist in a physical universe. (This isn't to say that modularism doesn't have it's own problems, of course. The main ones will be discussed in section 15.5 below.) Modularism in psychology is now warranted in the same sense and to the same degree as the assumption of *mechanism* in biology was warranted prior to the discovery of the double-helix structure of DNA. Biologists in the middle part of the twentieth century were surely justified in believing that the laws of inheritance must be realized *somehow* in biochemical mechanisms, although they couldn't yet say how. In the same way, we are now warranted in believing that the mind must be realized in the operations of a set of modular computational processes, even though there is much that we cannot yet explain.

15.3 Evidence that the Mind is Modular

There are powerful arguments of a general sort, then, for the conclusion that we should *expect* the mind to have a modular organization. In this section, we will review some of the evidence that this expectation is actually fulfilled. Once again, our exposition will have to be extremely brisk.

15.3.1 Developmental evidence

According to the modularist hypothesis, the human mind is made up of isolable components. And the expectations are that such a modular architecture is innate or innately channeled, and also that many of the modular components will operate in accordance with algorithms that are innate, or will make innate assumptions about the domains that they concern.

A variety of kinds of developmental evidence bears on, and supports, these proposals. One point is that developmental psychologists now mostly agree (in marked contrast with the earlier views of Piaget and his supporters) that cognitive development is a domain-specific process. It proceeds at different speeds in different domains (naïve physics, naïve psychology, naïve biology, mathematical understanding, and so on), and the cognitive structures that extract information concerning these domains would seem to be very different from one another. Instead of advancing on a broad front through some sort of general-learning process, it seems that different aspects of our cognition are acquired according to their own separate timetables and trajectories.

Another point is that some degree of competence in at least some of these domains is demonstrable at a very early stage in infancy – in some cases as young as four months. There is now robust evidence of early competence in a simple form of contact-mechanics, as well as in the rudiments of social understanding (Spelke, 1994; Baillargeon, 1995; Woodward, 1998; Phillips et al., 2002). And in other domains too, children acquire competence remarkably early considering the abstract non-observational character of the concepts involved. Thus, children appear to have a good understanding of normative notions such as "should," "must," "permissible," and so on by the age of three or four (Cummins, 1996; Núñez and Harris, 1998).

Under this general heading also fall the various "poverty of stimulus arguments," which have been run so decisively in the case of linguistic knowledge, but which can also be mounted with a good deal of plausibility in other domains too, such as "theory of mind" and moral belief (Carruthers, 1992; Dwyer, 1999). To the extent that it is very hard to see how children could acquire their competencies by the ages at which they do – using only general-learning systems to do it, and given the sorts of evidence available to them – it will be plausible to postulate an innately channeled domain-specific learning module to do the job.

15.3.2 Pathological evidence

The moral of the evidence from human pathology (whether developmental or resulting from later brain injury) is, roughly speaking, that *everything dissociates from everything else* (Shallice, 1988; Tager-Flusberg, 1999). (The data here must inevitably be messy and complicated, of course. For it is known that the same genes can be involved in a number of distinct functions, and any brain damage can be more or less extensive, also having effects on multiple functions at once.) In development, language can be damaged while everything else remains normal ("specific language impairment," or SLI); theory of mind can be damaged while language and physical/spatial thinking are normal (autism); and both theory of mind and language can be normal while physical/spatial thought are severely damaged (Williams syndrome). Moreover, so-called "general intelligence" can be very impaired while other systems are relatively spared (Downs syndrome).

Similarly amongst adults, aphasias can involve severe loss of linguistic function while much else remains undamaged. Thus aphasic people can often still run many aspects of their own lives, and interact successfully with other people and with the physical world. One severely a-grammatic aphasic man, for example, has been shown to have intact theory of mind abilities, and also to be quite adept at reasoning about physical causes, such as identifying the locus of breakdown in a complex machine (Varley, 1998, 2002). There is also evidence that brain-damaged individuals can lose their capacity to reason normally about biological kinds (folk-biology) while all else is left intact (Job and Surian, 1998), and that a capacity to reason about social contracts can be lost independently of the ability to reason about risks and dangers (Stone et al., 2002); and so on and so forth.

15.3.3 Experimental evidence

There is not as much experimental evidence bearing on the question of modularity of mind as there could be, since most investigators have not thought to go looking for such evidence. This is because experimental psychology remains just as dominated by the empiricist and general-learning-theory assumptions that continue to exert such a hold over philosophy and much of the social sciences. What evidence there is, however, is highly suggestive.

One piece of evidence concerns the existence of a geometric module in both rats and humans (and presumably in many other mammalian species, at least). Rats shown a food source in a rectangular enclosure, who are then disoriented and replaced in that space, will search equally often in the two geometrically equivalent corners (e.g., those having a long wall on the left and a short wall on the right). And they do this despite the fact that there can be highly salient cues, which rats can perfectly well recognize and use in other circumstances, such as heavy scenting or patterning of one of the walls. It appears that, in conditions of disorientation at least, rats cannot integrate geometric information with information of other kinds (Cheng, 1986).

In these same circumstances, human children before the age of six or seven behave just like rats. When disoriented they rely only on geometric information even if, for example, one wall of the room is brightly colored while the others are neutral (Hermer and Spelke, 1994, 1996). And it turns out that human adults, too, are subject to just these effects when they are required to shadow speech (tying up the resources of the language faculty), but not when they have to shadow a complex rhythm (Hermer-Vazquez et al., 1999). In fact, these results provide one of the main sources of evidence for the thesis to be sketched in section 15.5 below, that it is the language faculty that enables information from a variety of otherwise-isolated modular systems to be integrated into a single (natural language) representation.

Another set of experimental evidence concerns the existence of distinct systems for reasoning about social contracts, on the one hand, and about risks and threats, on the other. (Each of these systems was originally predicted on the basis of evolutionary considerations.) When presented with problems that are structurally exactly similar, except that one concerns a social contract of some sort (often involving the possibility of cheating) and the other concerns a significant risk to self or other, people will adopt very different reasoning strategies. Indeed, presented with just one social contract problem, or one risk-problem, people will reason differently depending upon which role in the contract or situation they are cued to identify with (Cosmides and Tooby, 1992; Gigerenzer and Hug, 1992; Fiddick et al., 2000). (Of course, these data do not *per se* entail the *modularity* of the systems in question. But standard practice in science is that surprising predictions that are made by a theory and turn out to be correct then serve to support that theory even though they do not entail it.)

Finally, there is evidence showing that even the very heartland of empiricist learning-theory – namely classical or associationist conditioning – is actually subserved by a special-purpose computational system that is designed to predict varying temporal contingencies, and which was selected for because of the role that it plays in successful foraging (Gallistel, 2000). There are a number of well-established findings that are important here. One is that when intervals between training trials are kept proportional to the delay between stimulus and reinforcement, then increased delays in reinforcement have no effect on the rate of acquisition. Another is that the number of reinforcements required for acquisition of a new form of behavior is left unaffected by inserting significant numbers of unreinforced trials for every reinforced one. These and other facts are extremely puzzling from a perspective which sees learning as matter of building associative strengths; whereas they can readily be explained within a computational model which assumes that what the animals are really doing is estimating likelihoods and calculating rates of return (Gallistel and Gibson, 2001).

15.4 Adaptation versus Learned Modularization

There is good reason to think, then, that the human mind has a modular organization, to some significant degree. But what of the suggestion that modularization may actually be the outcome of some sort of general-learning process, a product of *over-learning*? On this model, all that would be given at the outset of development are some general-learning abilities and a suite of domain-specific attention biases, which together serve to build a set of organized and automatically operating bodies of knowledge and skill. Then theory of mind, for example, would be a module in the same sense and to the same extent that chess-playing ability in an experienced Grand Master is a module (Karmiloff-Smith, 1992).

Such accounts face a great many difficulties, however. One is that they fail to address the full range of arguments sketched in section 15.2 above for the conclusion that we should expect the mind to have a given (innate) mental architecture. Another is that, in postulating that development is driven by general-learning abilities, the approach has severe difficulty in accounting for the fact that developmental profiles are similar across individuals and across cultures, despite wide variations in the richness and quantity of stimuli that subjects have experienced, and despite equally wide variations in general intelligence (which one would think should correlate well with general-learning ability).

In fact, the thesis of gradual modularization is a product of confusion concerning the relevant alternatives. The contrasting evolutionary psychology view is *not* that modules are there, fully formed, at birth and are realized in specific lumps of neural tissue, in the way that Elman et al. (1996) suppose. So the evidence of neural plasticity in the developing cortex, for example, is simply irrelevant (Samuels, 1998). Rather, the view is that evolved modules develop according to genetically channeled time-scales and profiles, emerging as learning-mechanisms that will serve to build increasingly elaborate bodies of knowledge throughout the lifetime of the individual. When sent up against this, its real opponent, the thesis of modularization through general learning, is not, in my view, a serious contender.

15.5 A Problem for Modularity: The Unconstrained Character of Thought

The arguments that the mind will contain at least *a great many* evolved modules are compelling, then. But what is the full extent of the mind's modularity? It seems obvious to many people that our minds must at least contain a very large and significant central arena, in which thoughts are formed and inferences drawn, which is *non*modular in character. Now this isn't the argument for the holistic character of belief-formation (Fodor, 1983, 2000), which we considered briefly and dismissed in section 15.2.3 above. Rather, it is the fact that we are, manifestly, unconstrained in the way that we can combine together concepts in our thoughts, crossing any putatively modular boundaries. I can be thinking about beliefs one moment and horses the next, and then wonder why I am thinking about both horses and beliefs, in effect then combining concepts from these disparate domains into a single thought.

If, in contrast, the human mind were wholly constructed out of modules, then one might expect that there would be severe limits on the structure and complexity of the kinds of thought that we can think. For some, at least, of these modular systems would be domain specific in character, only handling a given range of proprietary concepts. And for sure there should be limits on information-flow through a modular architecture, since one would expect that while some modules would provide their outputs as input to some others, not every module would be linked with every other one. In that case, there should be some combinations of content that we would find it difficult or impossible to entertain.

Now, one response to this difficulty would be to concede that there is, indeed, a nonmodular arena in which thoughts can be formed; but to continue to insist that this is embedded within an otherwise modular architecture (Cosmides and Tooby, 2001). A better-motivated response, however, is to claim that integration of thought-contents across modular domains is actually subserved by an existing module, namely the natural-language faculty (Carruthers, 1998, 2003b; Hermer-Vazquez et al., 1999). For almost everyone agrees that the language faculty is a distinct module of the mind; yet it is manifest that it would need to take inputs from any other conceptual modules, so that the outputs of those modules should be reportable in speech. The language system will be ideally *positioned*, then, to facilitate the integration of modular contents; and the abstract and recursive nature of natural language syntax would serve to make such an integration possible in reality.

Three major difficulties remain. One is that not only can we *form* cross-modular thoughts, but we can do new things *with* them – we can use them as premises in reasoning, derive new information from them, and so on. Would this require a suite of nonmodular consumer systems, positioned downstream of the natural language representations (a nonmodular central arena once again)? Arguably not. A case can be made that the further use of cross-modular linguistic representations can be explained in terms of the deployment of a variety of existing modular processes, with perhaps some minor additions and adjustments (Carruthers, 2003c).

The second problem doesn't so much concern the *use* of cross-modular thoughts but, rather, their *creation* or mode of generation. We seem to be able to frame thoughts with arbitrarily novel contents in fantasy and imagination, for example. (So I can easily suppose – and perhaps for the first time in all of human history – that there is a red dragon on the roof, dreaming of diamonds.) How is this to be explained? Would *this* require some sort of radically a-modular thought-forming arena? Again, arguably not. Admittedly, we do have the capacity to *suppose*, and we can put together concepts in novel and creative ways in our suppositions. But this capacity might be a relatively simple addition, built onto the back of the language faculty; and it may be that the function of pretend play in childhood is precisely to construct and develop it (Carruthers, 2002).

Finally, we need to explain how distinctively human forms of practical reasoning are possible, built on the back of the simple processing-encapsulated practical reasoning module which we inherited from our ancestors, together with interactions involving the language system as envisaged above. Here too, it may be possible to construct an account that remains faithful to the spirit of massive modularity (Carruthers, forthcoming).

It should be acknowledged that these are hard problems. But then *everyone* here faces hard problems. As noted earlier, if we gave up on massive modularity then we might lose one set of problems, but we would, instead, face the task of explaining how holistic, a-modular, processes can be computationally realized. Since no one currently believes that *this* problem can be solved, it seems better to continue exploring the resources available to a massive modularist.

15.6 Conclusion

I have argued that there is a powerful case to be made in support of the thesis that forms the title of this chapter, although there has not been the space here to do more than sketch the outlines of that case. There is good reason to think that natural selection has imposed on the human mind a modular architecture. Moreover, there are no overwhelming reasons for thinking otherwise. I should emphasize, however, that the case I have been making is a broadly empirical one, and is therefore subject to empirical falsification. And in the end, of course, our question won't be settled by philosophers, but by scientists.

Acknowledgments

I am grateful to the participants at a Georgetown University philosophy colloquium for discussion of this material.

Bibliography

Baillargeon, R. 1995: Physical reasoning in infancy. In Gazzaniga, op. cit., pp. 181–204.

Barkow, J., Cosmides, L., and Tooby, J. (eds.) 1992: *The Adapted Mind*. New York: Oxford University Press.

Botterill, G. and Carruthers, P. 1999: *The Philosophy of Psychology*. Cambridge: Cambridge University Press.

Carruthers, P. 1992: *Human Knowledge and Human Nature*. Oxford: Oxford University Press.

— 1998: Thinking in language?. In P. Carruthers and J. Boucher (eds.), *Language and Thought*. Cambridge: Cambridge University Press.

— 2002: Human creativity. *The British Journal for the Philosophy of Science*, 53, 225–49.

— 2003a: Moderately massive modularity. In A. O'Hear (ed.), *Mind and Persons*. Cambridge: Cambridge University Press.

— 2003b: The cognitive functions of language. *Behavioral and Brain Sciences*, 25 (in press).

— 2003c: On Fodor's problem. *Mind and Language*, 18, 502–23.

— forthcoming: Practical reasoning in a modular mind.

Cheng, K. 1986: A purely geometric module in the rat's spatial representation. *Cognition*, 23, 149–78.

Chomsky, N. 1988: *Language and Problems of Knowledge*. Cambridge, MA: The MIT Press.

Clark, A. 1987: The kludge in the machine. *Mind and Language*, 2, 277–300.

Cosmides, L. and Tooby, J. 1992: Cognitive adaptations for social exchange. In Barkow et al., op. cit., pp. 163–228.

— and— 2001: Unraveling the enigma of human intelligence. In R. Sternberg and J. Kaufman (eds.), *The Evolution of Intelligence*. Mahwah, NJ: Lawrence Erlbaum Associates.

Crain, S. and Pietroski, P. 2001: Nature, nurture and universal grammar. *Linguistics and Philosophy*, 24, 139–86.

Cummins, D. 1996: Evidence for the innateness of deontic reasoning. *Mind and Language*, 11, 160–90.

Currie, G. and Sterelny, K. 1999: How to think about the modularity of mind-reading. *Philosophical Quarterly*, 50, 145–60.

Dawkins, R. 1976: *The Selfish Gene*. Oxford: Oxford University Press.

— 1986: *The Blind Watchmaker*. New York: Norton.

Dupré, J. 2001: *Human Nature and the Limits of Science*. Oxford: Oxford University Press.

Dwyer, S. 1999: Moral competence. In K. Murasugi and R. Stainton (eds.), *Philosophy and Linguistics*. Boulder, CO: Westview Press.

Elman, J., Bates, E., Johnson, M., Karmiloff-Smith, A., Parisi, D., and Plunkett, K. 1996: *Rethinking Innateness: A Connectionist Perspective on Development*. Cambridge, MA: The MIT Press.

Fiddick, L., Cosmides, L., and Tooby, J. 2000: No interpretation without representation: the role of domain-specific representations and inferences in the Wason selection task. *Cognition*, 77, 1–79.

Fodor, J. 1983: *The Modularity of Mind*. Cambridge, MA: The MIT Press.

— 2000: *The Mind Doesn't Work That Way*. Cambridge, MA: The MIT Press.

Gallistel, R. 1990: *The Organization of Learning*. Cambridge, MA: The MIT Press.

— 2000: The replacement of general-purpose learning models with adaptively specialized learning modules. In Gazzaniga, op. cit., pp. 1,179–91.

— and Gibson, J. 2001: Time, rate and conditioning. *Psychological Review*, 108, 289–344.

Gaulin, S. and Robbins, C. 1991: Trivers–Willard effect in contemporary North American society. *American Journal of Physical Anthropology*, 85, 61–9.

Gazzaniga, M. (ed.) 2000: *The New Cognitive Neurosciences*, 2nd edn. Cambridge, MA: The MIT Press.

Gigerenzer, G. and Hug, K. 1992: Domain-specific reasoning: social contracts, cheating and perspective change. *Cognition*, 43, 127–71.

Grant, V. 1998: *Maternal Personality, Evolution and the Sex Ratio*. London: Routledge.

Hermer, L. and Spelke, E. 1994: A geometric process for spatial reorientation in young children. *Nature*, 370, 57–9.

— and — 1996: Modularity and development: the case of spatial reorientation. *Cognition*, 61, 195–232.

Hermer-Vazquez, L., Spelke, E., and Katsnelson, A. 1999: Sources of flexibility in human cognition. *Cognitive Psychology*, 39, 3–36.

Horgan, T. and Tienson, J. 1996: *Connectionism and Philosophy of Psychology*. Cambridge, MA: The MIT Press.

Hrdy, S. 1999: *Mother Nature: A History of Mothers, Infants and Natural Selection*. New York: Pantheon.

Job, R. and Surian, L. 1998: A neurocognitive mechanism for folk biology? *Behavioral and Brain Sciences*, 21, 577.

Karmiloff-Smith, A. 1992: *Beyond Modularity*. Cambridge, MA: The MIT Press.

Lakatos, I. 1970: The methodology of scientific research programmes. In I. Lakatos and A. Musgrave (eds.), *Criticism and the Growth of Knowledge*. Cambridge: Cambridge University Press, 91–196.

Laurence, S. and Margolis, E. 2001: The poverty of the stimulus argument. *The British Journal for the Philosophy of Science*, 52, 217–76.

Marcus, G. 2001: *The Algebraic Mind*. Cambridge, MA: The MIT Press.

— forthcoming: What developmental biology can tell us about innateness. In P. Carruthers, S. Laurence, and S. Stich (eds.), *The Innate Mind: Structure and Contents*.

McDermott, D. 2001: *Mind and Mechanism*. Cambridge, MA: The MIT Press.

Núñez, M. and Harris, P. 1998: Psychological and deontic concepts. *Mind and Language*, 13, 153–70.

Phillips, A., Wellman, H., and Spelke, E. 2002: Infants' ability to connect gaze and emotional expression to intentional action. *Cognition*, 85, 53–78.

Pinker, S. 1999: *How the Mind Works*. New York: Norton.

Samuels, R. 1998: What brains won't tell us about the mind: a critique of the neurobiological argument against representational nativism. *Mind and Language*, 13, 548–70.

Shallice, T. 1988: *From Neuropsychology to Mental Structure*. Cambridge: Cambridge University Press.

Spelke, E. 1994: Initial knowledge: six suggestions. *Cognition*, 50, 433–47.

Sperber, D. 1996: *Explaining Culture*. Oxford: Blackwell.

— 2003: In defense of massive modularity. In I. Dupoux (ed.), *Language, Brain and Cognitive Development. Festschrift for Jacques Mehler*. Cambridge, MA: The MIT Press.

Stone, V., Cosmides, L., Tooby, J., Kroll, N., and Wright, R. 2002: Selective impairment of reasoning about social exchange in a patient with bilateral limbic system damage. *Proceedings of the National Academy of Science*, 99, 11,531–6.

Tager-Flusberg, H. (ed.) 1999: *Neurodevelopmental Disorders*. Cambridge, MA: The MIT Press.

Tooby, J. and Cosmides, L. 1992: The psychological foundations of culture. In Barkow et al., op. cit., pp. 19–136.

— and — forthcoming: Motivation and the debate on representational and non-representational innateness. In P. Carruthers, S. Laurence, and S. Stich (eds.), *The Innate Mind: Structure and Contents*.

Trivers, R. and Willard, D. 1973: Natural selection of parental ability to vary the sex ratio in offspring. *Science*, 179, 90–2.

Varley, R. 1998: Aphasic language, aphasic thought. In P. Carruthers and J. Boucher (eds.), *Language and Thought*. Cambridge: Cambridge University Press, 128–45.

— 2002: Science without grammar: scientific reasoning in severe agrammatic aphasia. In P. Carruthers, S. Stich, and M. Siegal (eds.), *The Cognitive Basis of Science*. Cambridge: Cambridge University Press, 99–116.

Williams, G. 1966: *Adaptation and Natural Selection*. Princeton, NJ: Princeton University Press.

Woodward, A. 1998: Infants selectively encode the goal object of an actor's reach. *Cognition*, 69, 1–34.

CHAPTER SIXTEEN

The Mind is not (just) a System of Modules Shaped (just) by Natural Selection

James Woodward and Fiona Cowie

16.1 Did the Mind Evolve by Natural Selection?

Of course our minds and brains evolved by natural selection! They aren't the result of divine intervention or fabrication by space aliens. Nor are they solely products of drift or any other naturalistic alternative to selection. That natural selection profoundly "shaped" the mind and brain is accepted by both evolutionary psychologists and by virtually all of their most vigorous critics.

What, then, is at issue in the debate surrounding evolutionary psychology (hereafter, "EP")? First, there are disagreements about the likely intellectual payoffs of EP's characteristic research strategy. EP employs a "reverse engineering" methodology: the researcher (i) notes some competence or behavior, (ii) conjectures that it is a solution to some "adaptive problem" faced by our tree- or savanna-dwelling ancestors, and then (iii) proposes that natural selection engineered a specialized psychological mechanism or "module" to produce that competence or behavior. Some EP researchers also offer (iv) behavioral or psychological evidence for the proposed module. But, as we shall see, this evidence is rarely compelling, and other relevant evidence (from, for example, neurobiology, genetics, or developmental biology) is often not cited. Critics of EP, like us, think that this methodology is unlikely to yield much insight.

We also dispute EP's views about the structure of the human mind, the way in which it develops, and the relation between evolution and mental architecture. Evolutionary psychologists claim that the mind is "massively modular." It is composed of a variety of more or less independent "organs," each of which is devoted to the performance of a particular kind of task, and each of which develops in a largely genetically determined manner. EP's hypothesis of massive mental modularity is not just the uncontroversial idea that the mind/brain consists, at some level of analysis, of components that operate according to distinctive principles. For, as we argue in section 16.4, EP endows its modules with a number of additional properties such as informational encapsulation (section 16.4.3) and independent evolvability (section 16.4.4). In addition, EP also makes specific claims about which modules we have. Thus, the modules at issue in EP are not, for example, small groups of neurons but are instead the complex processing structures that underlie high-level cognitive tasks such as "cheater detection."

EP's views about mental structure and development are motivated by two very general evolutionary considerations. First, EP holds that evolution is likely to have favored strongly modular mental architectures. Secondly, and relatedly, EP holds that mental modules are the fairly direct products of natural selection. This picture requires that the different modules must be independently evolvable: they must have independent genetic bases so that natural selection can act to change one module independently of the others. It also means that while EP theorists are careful to say in their "official" pronouncements that they allow a role for learning and other environmental influences, their more detailed arguments typically assume that the development of modules is tightly genetically constrained.

There are problems with all of these assumptions. First, there is no reason to think that evolution "must" produce modular minds. Evolutionary psychologists (e.g., Cosmides and Tooby, 1994; Tooby and Cosmides, 1990; Carruthers, this volume) argue that general-purpose psychological mechanisms would not have evolved because they are too slow, and require too much background knowledge and computational space for the making of life-or-death judgments. Specialized modules, on the other hand, deliver fast and economical decisions on matters that crucially affect an organism's fitness, and so would have been preferred by natural selection. However, it is simply wrong to suppose that modules are invariably (or even usually) superior to general-purpose devices. What sorts of mental organization will be favored by selection depends entirely on the details of the selection pressures an organism is subject to and its genetic structure. As Sober (1994) shows, such factors as how variable the environment is, the costs of making various sorts of mistakes, the costs of building various sorts of discriminative abilities into the organism, and so on, can have large effects on the relative fitnesses of general-purpose versus more specialized strategies. In addition, the ability to adapt quickly (i.e., within the individual's lifetime) to changing circumstances is vital for organisms that inhabit unstable environments (Maynard Smith et al., 1985; Sterelny 2003). Indeed, there is evidence that both the physical (in particular, climatic) and social environments inhabited by early hominids were highly unstable (Potts, 1996, 1998; Allman, 1999). There thus would have been considerable selective pressure favoring the evolution of cognitive mechanisms allowing the rapid assimilation of new information and behavioral flexibility, rather than innately specified modules (for more on this issue, see section 16.3 below, and Woodward and Cowie, forthcoming).

Secondly, EP's view that the modules existing in the adult mind are largely genetically specified (or are the products of learning mechanisms that are themselves genetically constrained to produce a particular module as output) is inconsistent with what is known about the role of experience-dependent learning and development in shaping the mature mind. As we argue in section 16.2.3, whatever modular processing mechanisms the adult mind contains emerge from a complex developmental process. Less modular structures and capacities that are present in infants interact with both the environment and genetic mechanisms to generate new competences that were not directly selected for.

Thirdly, the notion of a module is itself quite unclear. As we show in section 16.4, there are several different (and noncoextensive) criteria for modularity employed in the EP literature. Researchers move back and forth among different notions of modularity, illicitly taking evidence for modularity in one sense to bear on modularity in other senses. They also tend to conflate issues about the modularity of processing in the adult mind with quite separate issues to do with the role of modules in development and learning. These unclarities make EP's claim that the mind is "a system of modules" somewhat difficult to assess. Our view, defended in section 16.3, is that the mind is not just a collection of specialized modules. Although our minds probably do contain modules in some sense(s) of that term, these structures are unlikely to correspond to the modules (for cheater detection, mate selection, predator avoidance, and so on) postulated in EP.

16.2 Reverse Engineering: A Backward Step for Psychology

EP is premised on the idea that modern human mental organization is a more or less direct reflection of the ways in which hominids evolved to solve the problems posed by their physical and social environments. Thus, by reflecting on the tasks our ancestors must have been able to solve, and by supposing that whatever psychological abilities enabled our ancestors to solve those tasks would have been selected for, evolutionary psychologists seek to map our current psychological organization. Because they also assume that selection engineers a proprietary solution for each of these "adaptive tasks," evolutionary psychologists see the modern mind as "massively modular": it contains numerous specific mechanisms (or "modules") that evolved for specific tasks, and it houses few (if any) general-purpose psychological mechanisms.

One problem with this strategy has already been mentioned: it ignores the possibility that *flexibility* might well have been at a selective premium for hominids inhabiting rapidly changing environments. In this section, we discuss three further problems with EP's adaptationist or "reverse engineering" approach to generating psychological hypotheses.

16.2.1 Reading structure from function

EP believes that since "form follows function" (Tooby and Cosmides, 1997, p. 13), one can infer how the mind is structured from a consideration of what psychological competences would have been selected for in the "environment of evolutionary adaptation" ("EEA")). One reason why EP's reverse engineering strategy is misguided is that you *can't* infer structure from function alone. Instead, formulating and confirming functional and structural hypotheses are highly interrelated endeavors, with information about structure informing hypotheses about function and vice versa.

As an illustration, consider how our thinking about human declarative memory has evolved over the past half century (cf., LeDoux, 1996, ch. 7). By the 1940s, neurophysiologists had concluded that memory is distributed over the whole brain, not localized in a particular region (a structural hypothesis). But then came the patient H.M., who had had much of both temporal lobes removed to treat severe epilepsy. Post-operatively, H.M. remembered much of what had happened to him prior to the surgery and could form new short-term memories lasting a few seconds. However, he was unable to form new long-term memories. H.M. thus indicated that short-term memory and long-term memory are distinct (a functional hypothesis), that they are supported by different brain systems (a structural hypothesis), and that the areas responsible for the formation of new long-term memories are different from those allowing storage of the old ones (structural). Also prior to this, the limbic system – including the hippocampus and amygdala – had been thought to comprise the emotional circuitry of the brain (functional). But the hippocampus was one of the areas that was so badly compromised in H.M., and in other patients with severe memory deficits (structural), which indicated that the limbic system was also involved in cognitive functioning (functional) and which suggested that the hippocampus was the seat of memory (structural). As the workings of the hippocampus were further investigated (structural), it was found to be especially implicated in learning and memory of spatial information (functional). Furthermore, since all of the patients on whom the early hippocampal memory story had been based had also had damage to the amygdala (structural), this was an indication that the amygdala was also involved in memory (functional). [This latter claim is still controversial (functional), given that later studies have shown that hippocampal lesions alone will produce amnesia (structural).]

This vignette illustrates how views about functions are (or should be) highly sensitive to structural information. It thus underscores the naïveté of the assumption (endemic in EP) that one can accurately individuate psychological functions by enumerating the tasks that the mind can perform. Evolutionary psychologists try to avoid this difficulty by inferring functions not (or not just) from behavioral data about what our minds can do at present but, rather, from their ideas about which psychological capacities were selected for back in the EEA. In effect, then, evolutionary psychologists think of psychological functions as *biological* functions (in the sense of Wright, 1973): capacities that the mind had in the past that are still present because they were selected for, rather than as functions in the sense of what the mind does at present, regardless of whether they were selected for (*causal role* functions in the sense of Cummins, 1975).

Prima facie, however, this move compounds, rather than solving, the problem just discussed. After all, if it's hard to delineate the functional anatomy of our own minds on the basis of merely behavioral evidence, it's even harder to limn the minds of our ancestors by speculating about what they did and the selection pressures that they faced: biological functions are typically *tougher* to figure out than causal role functions. For one thing, as Lewontin has repeatedly pointed out (e.g., 1990), cognitive functions leave no unambiguous marks on the hominid fossil record and humans have no close living relatives whose homologous psychological capacities might allow inferences about ancestral functioning. In addition, as Stolz and Griffiths (2002) argue, the evolutionary or "adaptive" problems faced by an organism cannot be specified independently of the organism's capacities (and/or the structures that underlie those capacities). If you didn't know, to take their example, that a given fossil bird had a reinforced beak and skull (like a modern woodpecker), you would be unable to reconstruct its niche (living in a forest), its habits (eating insects living under the bark of trees), or the adaptive problems (getting at the insects) and selection pressures that it faced. In the absence of detailed knowledge of what the mind is actually like, speculating about the adaptive problems faced by hominids in the EEA is like speculating about the niche and feeding habits of a headless fossil bird. Thus, EP's strategy of inferring the mind's functional architecture from speculations about its biological function(s) is seriously off track.

16.2.2 The one-to-one assumption

The epistemological problems just outlined are quite endemic to adaptationist reasoning about the mind. However, there is a second problem with EP's view of the relation between structure and function: EP assumes that once a psychological function is somehow identified, it is legitimate to postulate a *single* mechanism – a "module" – that performs that function. As Carruthers puts it:

> ... in biology generally, distinct functions predict distinct ... mechanisms to fulfill those functions ... [Hence] one should expect that distinct mental functions – estimating numerosity, predicting the effects of a collision, reasoning about the mental states of another person, and so on – are likely to be realized in distinct cognitive learning mechanisms ..." (this volume, p. 300)

This "one-to-one" assumption is not a dispensable part of EP methodology. If a single mechanism could subserve many different functions, or if a single function required the cooperation of a number of different mechanisms, then the characteristic EP procedure of inferring mechanisms from functions would be undermined. For in that case, there would be many different alternative hypotheses about the mechanisms involved in the performance a given function, and the identification of the function itself would provide no evidence about which of these alternatives was correct. The one-to-one hypothesis avoids this difficulty by assuming that the only possibility is that a distinct mechanism performs each function.

Given the central role played in the EP methodology by the "one-to-one" assumption, it is then a real problem for EP that this assumption embodies a serious misapprehension about how natural selection works. Far from "characteristically [operating] by 'bolting on' new special-purpose items to the existing repertoire" (Carruthers, this volume, p. 300), natural selection usually operates by jury-rigging what is already there to perform new tasks instead of (or in addition to) the old ones. Feathers originally evolved for thermal regulation, and subsequently were exapted for flight and mating displays as well. Vertebrate limbs originally evolved for swimming, and subsequently were fitted for walking, climbing, flying, and manipulation. At the genetic level too, exaptation and multifunctionality are common, both within organisms and across species. The Hox genes that control the development of a chicken's legs and feet, for instance, also control development of its wings. Moreover, the self-same genes are responsible (with only very minor changes in sequencing) for limb development in all tetrapods – wings, claws, paws, flippers, flukes, and hands all have the same genetic origins (Gilbert, 2000, pp. 503–21; Davidson, 2001, pp. 167–76).

Exaptation and multifunctionality are undoubtedly also features of the mind and brain. If a given mechanism M_1 carries out some task, T_1, and in so doing processes information that is relevant to some other task, T_2, then M_1 could well be selected because of its role in performing T_2 in addition to T_1. For example, the processes of object identification may generate information that is relevant to depth perception. If so, those processes may be recruited for both functions and we would have two functions utilizing a single mechanism. On the other hand, what is intuitively a single task may involve multiple mechanisms cobbled together over time: T_2 may involve M_2 and M_3 in addition to M_1. Depth perception looks like this: mechanisms that are at least partly distinct, both anatomically and phylogenetically, are involved in the processing of the various depth "cues," such as binocular disparity, occlusion, texture gradients, and so on.

The reuse of old materials for new purposes, with all the redundancy and *ad hoc* interconnectedness that it implies, is characteristic of selection's "tinkering" mode of operation. Because natural selection typically does not operate by designing new, single-purpose devices to solve new environmental challenges, EP's one-to-one assumption is highly dubious.

Another problem with the one-to-one assumption concerns EP's individuation of functions or tasks. Consider the detection of numerosity. How should we decide whether this is one psychological function subserved by a single module (as Carruthers assumes; this volume, p. 300) or several functions subserved by several modules? The detection of numerosity, after all, is actually a highly complex task. It involves (for example) object detection and individuation, which involve (for example) depth and edge perception, which involve (for example) perception of luminance and color boundaries, and so on. The detection of numerosity is a function carried out by the performance of other, simpler functions: functions are nested. They are also shared. Just as the detection of numerosity itself can play a role in higher-level functions (say, performing a task in a psychology experiment), all of the lower-level functions just discussed play roles in the performance of other tasks: depth perception also subserves motion detection; perception of color boundaries subserves depth perception; object individuation subserves object recognition; and so on. Given that functions are both nested and shared in this manner, it is hard to see how evolutionary psychologists – relying only on the one-to-one assumption and eschewing the sorts of detailed investigations into neural and cognitive mechanisms described in section 16.2.1 – could have any principled reason for saying that a given function (such as the perception of numerosity) is carried out by one module or many – and similarly for face-recognition, cheater detection, and the various other capacities that are the focus of EP theorizing.

The observations in sections 16.2.1 and 16.2.2 clearly undermine EP's assumptions that mechanisms or modules and functions correspond in a neat 1:1 manner and that, as a result, the existence of modules can be inferred from a specification of the tasks that the mind performs. Of course, one could read EP as simply *stipulating* a notion of "module" such that each function is *ipso facto* performed by one and only one module. But such a reading of EP's structural hypotheses trivializes them. In addition, this "thin" interpretation of what a module consists of is inconsistent with the fact that the modules postulated in EP are virtually always assumed to have other properties, such as being independent targets of selection, being independently disruptable, being informationally encapsulated, and so on (see section 16.4).

16.2.3 The role of learning and development

Another crucial limitation of EP's methodology is its misunderstanding of the role of learning and development in shaping the mature mind. It's not that evolutionary psychologists assign no role at all to learning and development. It is rather that they think of these processes as strongly genetically pre-specified in a way that has little empirical support. This "pre-formationist" picture engenders a crucial misspecification in the EP literature of what stands in need of adaptive explanation.

Evolutionary psychologists take some behavior or capacity possessed by mature humans – say, mate preferences, or cheater detection, or the desire to rape – and then proceed to give an adaptive explanation of the postulated mechanism underlying that behavior or capacity (cf., e.g., Thornhill and Thornhill, 1987, 1992 on rape; Wright, 1994 on family relations). But if learning plays an important role in the acquisition of these mechanisms or forms of behavior, then what really needs adaptive explanation is *the processes underlying the development of those mechanisms.*

Admittedly, some evolutionary psychologists do see their task as involving the explanation of development – see Carruthers' emphasis on evolved "learning mechanisms" as giving rise to various modules (this volume, pp. 300, 307). However, the assumption here seems to be that if some competence (and the module, M, underlying it) are adaptations built by natural selection, then either (i) the unfolding of M is directly genetically pre-specified; or (ii) M is produced by a "learning module," L, which is itself built by the genes and tightly constrained to produce M as its output. On this view, the relationship between L and M is very direct: to the extent that experience plays any role at all in the development of M, it merely serves to "trigger" a cascade of effects in L, the outcome of which is tightly genetically constrained.

However, there are a number of serious flaws in this reasoning, even assuming that a given processing module M in the adult mind is indeed an adaptation built by natural selection. First, as a number of psychologists, biologists, and philosophers of biology have emphasized, adaptive traits may be "coded for" in the environment (cf., Oyama, 1985; Sterelny and Griffiths, 1999; Sterelny, 2003). That is, instead of building M into the genes (either directly or indirectly via learning mechanism L), natural selection may have given us dispositions to *construct an environment E* in which M would arise as a result of learning and/or other developmental mechanisms which are *not* genetically determined to produce M. For example, rather than building in a "folk psychology" module, evolution may have given us dispositions to create the kinds of social and familial environments in which children's generalized developmental and learning abilities enable them to acquire knowledge of other minds.

A second problem here concerns the relation of current evidence from neurobiology and genetics to EP's assumption that modules like M or L are "innate or innately channeled" (Carruthers, this volume, p. 304). Several writers (e.g., Bates, 1994; Bates et al., 1998) have advanced a simple counting argument against the notion that numerous cognitive modules (with all their detailed representations and complex algorithms) are genetically specified. Human beings have approximately 30,000–70,000 genes (Venter et al., 2001; Shouse, 2002). By contrast, there are an estimated 10^{14} synaptic connections in the brain. Thus, it is argued, there are too few genes by many orders of magnitude to code for or specify even a small portion of these connections.

We find this argument suggestive but not decisive. The role of regulatory genes and networks in governing the expression of structural genes probably generates many more combinatorial possibilities than the figure of 30,000 genes suggests. Still, the counting argument does draw attention to the need for evolutionary psychologists to explain, consistently with what is known about brain development, how cognitive modules could be genetically specified. This, we think, is a nontrivial task, especially *vis-à-vis* the cerebral cortex, which is known to play a central role in the sorts of high-level cognitive tasks (such as language acquisition, cheater detection, theory of mind, and so on) that figure in EP theorizing. For while the gross architectural features of the cortex do appear to be genetically specified, there is considerable evidence that the cortex is in other respects initially relatively undifferentiated and equipotent. In particular, the patterns of synaptic and dendritic connections that develop in different cortical areas – and presumably correspond to the representations (of syntax, folk psychology, and so on) which EP's modules contain – are very heavily influenced by sensory inputs, and influenced in a way that the evolutionary psychologist's "triggering" metaphor seems ill equipped to capture. Indeed, many areas of cortex have the capacity to acquire fundamentally different sorts of representations depending on experience. For instance, the cortical areas normally devoted to visual processing in sighted subjects are used for tactile tasks, such as Braille reading, in congenitally blind subjects, and auditory cortex is recruited for processing sign language in deaf subjects (e.g., Büchel, Price, and Friston, 1998; Nishimura et al., 1999). This phenomenon of "cross-modal plasticity" makes it very hard to see how the cortex could contain innate representations specialized for specific cognitive or learning tasks, and undermines EP's notion that the development of cognitive modules such as M or L is genetically driven. We think that until we hear more about the ways in which the genetic and regulatory mechanisms needed to build the mental modules postulated in EP actually work, we are entitled to view EP's developmental story – or, really, its lack of such a story – with suspicion.

16.2.4 Non-Darwinian traits

Such suspicions are reinforced by consideration of a final shortcoming of EP's reverse engineering strategy; namely, its blindness to the fact that many psychological traits may not be susceptible of direct Darwinian explanation at all. First, while we agree with Carruthers (this volume, p. 294) that the entire mind is unlikely to be the product of drift or some other nonselective process, it's by no means impossible that particular psychological mechanisms might be the results of such processes. Developmental, allometric, and physicochemical factors are all known to play significant roles in neural functioning and organization, and may well turn out to be responsible for some psychological traits as well.

Alternatively, some psychological mechanisms might be "spandrels" in the sense of Gould and Lewontin (1979). That is, they might be lucky byproducts of traits that evolved for other purposes. There's evidence, for instance, that our capacity to organize continuous acoustical signals into linguistically relevant segments (phonemes) is a byproduct of the way in which mammalian brains happen to have evolved to process auditory information. Of course, byproducts that happen to be advantageous may themselves be subject to positive selection pressure – they may become "secondary adaptations." But the possibilities that psychological mechanisms are spandrels or mere secondary adaptations undermine, in different ways, EP's assumption that each psychological mechanism is built to order to solve a distinct adaptive problem. The spandrels possibility puts into doubt EP's assumption that modules are *optimal* or near-optimal solutions to adaptive problems: a turtle's fins may be optimized for propelling a heavy body through water, but they are far from an optimal means of crossing the sand at nesting time. And the possibility that some mental mechanisms are exaptations further undermines EP's one-to-one assumption, discussed in section 16.2.2: complex exaptations (such as, arguably, the human capacity for language or cheater detection) are often cobbled together from multiple mechanisms that are designed (and still used) for other purposes. While one can certainly *call* such complex secondary adaptations *single* mechanisms or modules, it's unclear that they can be attributed the other features commonly ascribed to modules, such as informational encapsulation or independent disruptability (see below, sections 16.4.3 and 16.4.4).

16.3 The Mind as a System of Modules

EP claims not just that the mind contains various mental modules, but that it is a *system* of modules. In this section, we examine the arguments for this claim. (We assume here, for the sake of argument, that the notion of a "module" is relatively clear. This assumption will be criticized in section 16.4.)

The main argument for the claim that the mind is a system of modules is originally due to John Tooby and Leda Cosmides. They claim that domain-specific modules would inevitably be selected for, because relatively content-independent (or general-purpose) architectures are in principle not viable objects of selection (e.g., Tooby and Cosmides, 1990, 1992; Cosmides and Tooby, 1994, 1995; and for a forceful statement of EP's "massive modularity" hypothesis, see also Samuels, 1998). There are two arguments given for this claim. First, general learning mechanisms face the "frame problem." Unless the factors relevant to a problem are delineated in advance, general-purpose inference mechanisms face a massive combinatorial explosion – and their owners get eaten before they can reproduce (see section 16.3.1). Secondly, Chomsky's poverty of the stimulus argument for the existence of a language-learning module is generalized to show that general-purpose inference is ineffective in the face of *any* learning problem. For one thing, there will always be more hypotheses compatible with the available data than the learner can effectively test. For another thing, testing is itself problematic. There are no domain-neutral criteria for success: evaluating foraging strategies involves different measures from those used to test hypotheses about cheaters. Worse, there are some hypotheses and strategies that an individual cannot evaluate at all – mate selection strategies would be an example, assuming, of course, that the appropriate measure here is inclusive fitness (see section 16.3.2). The upshot is that hominids equipped only with general-purpose inference or learning mechanisms wouldn't have survived in the EEA. Additional constraints on learning mechanisms are clearly needed, and those are what modular architectures supply.

16.3.1 Combinatorial explosion and the "frame problem"

Fodor (1983) maintained that many or most cognitive (or "central") processes are nonmodular, since reasoning, deliberation, planning, and so on must potentially have access to everything an agent knows. He recognized that this meant that such nonmodular processes are subject to the so-called "frame problem" – the problem of specifying what information is relevant to which problem – and for this reason, speculated that they would prove unamenable to cognitive-scientific investigation. The pessimism of evolutionary psychologists is deeper even than Fodor's: they view the frame problem not just as an obstacle to *theorizing about* central processors but, rather, to their very existence! Carruthers (this volume, p. 303), for instance, argues that "any processor that had to access the full set of the agent's background beliefs . . . would be faced with an unmanageable combinatorial explosion" and hence concludes that "the mind . . . consist[s] of a set of processing systems which . . . operate in isolation from most of the information that is available elsewhere in the mind." EP thus (dis)solves the frame problem by assuming that the processes underlying decision-making and behavior are modular: they neither have nor need access to the bulk of the agent's beliefs and desires.

Whether this is a satisfactory solution to the frame problem depends on what one takes that problem to be. *If* human reasoning, deliberation, and planning processes can generate satisfactory decisions and behavior without access to large numbers of the agent's beliefs and desires, then this will indeed be an important point in favor of the modularist picture. However, it seems plain that in many cases, reasoning and so on *cannot* issue in even minimally satisfactory decisions and forms of behavior without such access – consider, for instance, the range of factors bearing on a decision to cooperate with a conspecific. If this is so, EP's claim to have solved the frame problem is undermined, and the modularist must confront the question of how our processes of reasoning, deliberation, and planning *could* have access to so many and so varied of our background beliefs and desires. Presumably, evolutionary psychologists cannot invoke a single, hard-wired "decision-making module" here, for natural selection clearly cannot anticipate all the decisions that we potentially face in a lifetime; moreover, the beliefs and desires that are relevant to these decisions vary with context and hence cannot be pre-specified. Suppose that it is instead suggested that a *group* of encapsulated modules collaborate in the planning and execution of complicated actions. In that case, we must ask how their operations are coordinated. There seem to be two options. One is that there is a fixed hierarchy of modules, such that each module sends its outputs to the next one up in the hierarchy, and so on, until a behavioral command is outputted. Alternatively, there is some kind of "module integration module" (what Samuels, Stich, and Tremoulet (1999) unironically call a "Resource Allocation Module"), which takes the outputs of various lower-level modules, evaluates them, and issues in the same behavioral instruction – Carruthers (this volume, section 15.6) proposes that "an existing module . . . the natural-language faculty" (p. 307) performs this integrative task.

But neither of these alternatives is plausible. An evolved, hard-wired hierarchy of modules is vulnerable to the same objections as the decision-making module: our behaviors are simply too complex, and the mental processes giving rise to them too varied, for the frame problem to be solved by a pre-specified hierarchy. This leaves us with the idea of a module integration module, which takes in the deliverances of all the other modules whose computations are potentially relevant to a given problem and decides what to do with them. But a "module" that can (i) assess which of the plethora of modular outputs are important in a given context and (ii) decide what outcome is desirable and then (iii) figure out which behaviors (and in what order) will result in that outcome *isn't a module (in the EP sense) at all*! Instead, it's function-ally equivalent to Fodor's Central Processor and, assuming that the frame and com-binatorial explosion problems are real problems, it raises them all over again. As soon as one looks in detail at how a massively modular mind is supposed to work, one sees that the frame problem is not an argument *for* the theory that the mind is mas-sively modular; instead, it's an argument *against* that thesis![1]

16.3.2 The argument from the poverty of the stimulus

Suppose that a poverty of the stimulus argument has convinced us that some hypoth-esis or skill which people acquire could not have been learned just from the evidence available. This shows us that additional constraints, not present in the evidence, are required for successful learning. Evolutionary psychologists (like other proponents of poverty of the stimulus arguments) are quick to assume that the constraints in question must be (i) representational, (ii) cognitively sophisticated, and (iii) specific to various common-sense domains or subject matters. Thus, for instance, we are told that the necessary constraints are "theories" of various sorts (e.g., universal grammar, theory of mind). And because the content of these theories so far outruns the avail-able data, this view suggests in turn (iv) that the needed constraints on learning are embodied in innately specified modules (Language Acquisition Devices, Theory of Mind modules, and so on).

However, this picture itself outruns what is warranted by the poverty of the stim-ulus argument. For that argument indicates only that some constraints are needed, not what *kinds* of constraints those are. Thus, learning may be subserved by other types of constraint in addition to (or instead of) the sophisticated representational constraints postulated in EP. There might, for instance, be perceptual biases of various sorts, or dispositions to direct our attention to certain kinds of stimuli, or facts about our reward structures that encourage certain sorts of behavior rather than others. For

---

1   We concede that our discussion does not even begin to explain how human beings manage to take account of a wide range of background information and act flexibly and reasonably. But, as we have shown in this section, modular theories are in far worse shape. They not only fail to provide a positive account of how the problem is solved, but also make assumptions that are inconsistent with the fact that we do (somehow) solve the problem. Alternatively, and to the extent that they do attempt to accommodate this fact, they are forced to abandon basic commitments of the modular account.

example, there is evidence that subcortical mechanisms preferentially direct infants' visual attention to objects that fit a loosely face-like template, and that reward mechanisms release chemicals that make infants feel good when attending to such stimuli (Johnson, 1997). By themselves, these mechanisms are incapable of generating the full range of adult face-recognition behavior. However, they do help in reducing the underdetermination problem faced by the child (why focus on faces rather than elbows? why focus on eyes rather than chins?), and the preferential looking and attending that they produce may result in the gradual construction of cortical circuits that behave like a "face-recognition module."

Other possible constraints are developmental or architectural. Chronotopic factors governing the timing of different aspects of development can reduce underdetermination by guiding the *sequencing* of various learning tasks: learning the grammar of a language is easier if you already have a representation of its phonemes, for example. In addition, although the detailed pattern of synaptic connections that develops in the cortex is experience-dependent, the gross architecture of the cortex (e.g., different areas' characteristic laminar structures and basic circuitry types) may well be genetically specified (cf., section 16.2.3). These architectural features do not themselves amount to innate representations or modules, yet they may help the brain to solve learning problems by biasing certain areas to assume some tasks rather than others, or encouraging certain sorts of representations rather than others to develop in response to sensory input. As these examples show, it is a mistake to suppose, as evolutionary psychologists frequently do, that the only two possibilities are either a completely unconstrained, general-purpose learner or a heavily modular learner preequipped with large bodies of domain-specific knowledge.

One final point deserves to be made about EP's claim that the mind is a "system of modules." Both of the arguments discussed in this section are arguments for a very strong version of the modularity hypothesis, namely that the mind contains *nothing but* modules. As already indicated, we don't think that the evidence for this "massively modular" view of the mind is at all compelling. However, there is also a more "modest" modularity hypothesis to the effect that the mind contains *some* modules. (For example, Fodor's (1983) modularity hypothesis was modest: it postulated both modular sensory mechanisms and nonmodular central processing mechanisms. Modesty also embraces the possibility that some cognitive (as opposed to sensory) processing is modular.) Our discussion so far leaves it open that some kind of modest modularity thesis is correct. In the next section, however, we argue that the notion of a module, as deployed in EP, is fundamentally unclear. Thus, while the mind may indeed contain some "modules" (in some sense of that word), we will see in section 16.4 that even advocates of modest modularity need to clarify considerably what their thesis amounts to.

16.4 In Search of Mental Modules

We turn now to the question of what modules are. We argue that the various different criteria used for modularity in the EP literature are far from coextensive and thus lead to quite different notions of a "module." We also emphasize that these different modularity claims require (but often do not get) different sorts of supporting evidence. We conclude that EP's widespread failure to recognize these points both weakens its case for the modularity of mind and undermines the status of the specific cognitive modules it postulates.

16.4.1 Modularity and neural specificity

As Carruthers notes (section 15.1.1, pp. [2–3]) and as we will be lamenting in this section, the meaning of term "module" in EP is highly elastic. However, one negative point about EP's notion of modularity has been foreshadowed in previous sections: it bears little relation to the neuroscientist's notion of *neural specificity*. This is the idea, first, that different brain regions are (relatively) specialized to different tasks. In most people, for instance, the left hemisphere is dominant in language processing, with Broca's area, Wernicke's area, and the left thalamus playing a central role in word production (Indefrey and Levelt, 2000, p. 854). Secondly, the idea of neural specificity embraces the fact that the representations and computations that are used in different brain regions and for different tasks may be quite diverse. For example, the perception of an object's color involves the representation of its spectral properties by the three retinal cone types, adjusted so as to compensate for properties of the ambient light (Wandell, 2000), whereas perception of sounds involves the representation of low-level acoustical features as onset time, pitch, and location, followed by the computation of higher-order properties such as timbre, resulting ultimately in the representation of items of speech, music, or other types of noise (Shamma, 2000).

Now, if all that were meant by EP's claims that perceptual and cognitive processing and mechanisms are "domain specific" or "modular" were that such processes and mechanisms are neurally localized and involve different kinds of computations over different kinds of representations, we would readily agree. Not even the most rabid anti-modularist doubts, for example, that retinal cones are ineffective at extracting acoustical information. However, as we have already suggested, adherents of EP generally have something much stronger in mind than this.

As evidence for this, consider first the fact that the neural specialization described above is typically *relative,* rather than absolute. Cells in a certain area may respond especially strongly to certain kinds of inputs or may be particularly active in the execution of a certain task. But, as neural imaging data are increasingly making clear, they will typically also respond, though less vigorously, to many other inputs and task demands. Andersen et al. (2000), for example, give evidence that the posterior parietal cortex, classically thought to be specialized for attention and spatial awareness, is also involved in the planning of goal-directed behavior. Similarly, DeAngelis, Cumming, and Newsome (2000) argue that cortical area MT, normally held to be highly specialized for motion detection, is also implicated in the perception of stereoscopic depth.

Just as the same brain areas may subserve different tasks, many tasks that common sense might count as unitary can involve activation of numerous different brain regions. Face recognition, for example, involves not only the areas in the fusiform gyrus that are cited in lesion and dissociation studies, but also the parahippocampal gyrus, the hippocampus, the superior temporal sulcus, the amygdala, and the insula (McCarthy, 2000). Likewise, the production of verbs involves areas in the left frontal cortex, anterior cingulate, posterior temporal lobe, and right cerebellum (Posner and Raichle, 1994/1997, p. 120). At the neural level, then, tasks such as recognizing a face or producing a spoken word are performed by a "single mechanism" only in a very attenuated and task-relative sense.

This sharing of tasks by the same neural areas and the distribution of tasks over numerous different areas contrasts strongly with EP's talk of distinct modules devoted to distinct cognitive and perceptual tasks. Hence, evolutionary psychologists' claims about "domain-specific" or "dedicated" modules should not be confused with the facts about neural specificity just described. But if that's the case, what *does* EP's talk of "dedicated" or "domain-specific" processing amount to?

Evolutionary psychologists answer that one needs to distinguish between what Marr (1982) called "implementational level" details on the one hand, and theories at the "computational level" on the other (cf., Griffiths, forthcoming). Since their theories are at the psychological or computational level, we should not expect the modules that they postulate to be reflected in the nitty gritty of neural organization. As Cosmides and Tooby put it, EP "is more closely allied with the cognitive level of explanation than with any other level of proximate causation" (1987, p. 284).

But while the urge to theorize at one level of description while ignoring constraints from other levels is endemic to cognitive science, we think that it is a mistake. No psychologist should ignore the neurosciences, because psychological theories must be implementable in brains and, as is increasingly becoming apparent, this constraint is an extremely strong one. It is doubly a mistake for evolutionary psychologists to neglect facts about how psychological tasks are performed by the brain. First, as section 16.2.1 made clear, the individuation of psychological functions must be constrained by implementational information. Secondly, as section 16.3.1 urged, one cannot usefully theorize about how natural selection operates on the mind and brain while neglecting implementational issues. Thirdly, a sharp psychological/implementational divide undermines one of EP's central sources of evidence: if EP's modules have nothing to do with the brain, it is hard to see the relevance of the sorts of neuroscientific data (about localization, dissociations, and so on) that are frequently cited in the EP literature (cf., Carruthers, this volume, sections 15.4.2 and 15.4.3; and see Pinker, 1999). Most importantly, though, neglecting implementational constraints threatens to leach EP's notions of modularity and task-specificity of any real content. If the notion of a module is not tied to claims about neural specificity, what does it amount to? In what follows, we review several features that have been ascribed to modules and examine their interrelations.

16.4.2 Modularity and dissociability

One feature that is often ascribed to modules in EP is dissociability, or independent disruptability, the idea being that if two modules are distinct, then it should be possible (at least in principle) to interfere with the operation of each one without affecting the operation of other.[2] As we have already observed (section 16.2.1), EP lacks an intrinsic characterization of modules that would allow one to determine directly whether one independently identified mechanism has dissociated from another. Instead, modules are characterized functionally, in terms of the tasks that they are assumed to perform, and the dissociations that are actually observed are dissociations between *tasks* (e.g., between production of words and comprehension of grammatical sentences). It is these dissociations among tasks that are taken to be evidence for the existence of independent modules. Thus, Carruthers (this volume, section 15.4.2) and Pinker (1994, pp. 49ff), for instance, argue that the double dissociation between general cognitive tasks and language production and comprehension tasks seen in subjects with Specific Language Impairment and Williams syndrome is strong evidence that there is a task-specific mental module underlying language.

While the evidential significance of dissociations is a complicated subject to which we cannot do justice here, such inferences are far more problematic than is generally appreciated.[3] First, there are a number of intuitively nonmodular architectures that can give rise to double dissociations among tasks (cf., Shallice, 1988, pp. 245ff). Secondly, it is crucial to distinguish between dissociations arising from developmental disorders and dissociations resulting from injuries to (or experimental manipulations of) adult brains. The former bear on mechanisms of learning or development, and the latter on mature psychological competences. Thirdly, inferences from a double dissociation of capacities to the distinctness of modules generally require additional empirical assumptions, such as: (i) a "universality" assumption to the effect that both normal and abnormal subjects share a cognitive architecture (excluding the damaged modules in abnormals); (ii) a "subtraction" assumption to the effect that brain damage only removes modules or the connections between them, and it does not engender any significant neural reorganization; and (iii) various "gating" assumptions about whether the destruction of one or all connections between modules involved in a task is necessary for disruption of the task (cf., Shallice, 1988, pp. 218ff; Glymour, 2001, pp. 135–6, 143–4).

These assumptions are empirically questionable, especially when the dissociations in question are developmental or genetic in origin. First, subjects with genetic abnormalities (or childhood brain injuries) differ from normal subjects in many different ways. Secondly, incapacities that appear early in childhood are known to call forth compensatory psychological strategies and substantial neural reorganization. Hence, and contrary to what Carruthers and Pinker imply, it is extremely unlikely that subjects with Specific Language Impairment differ from normal subjects only in having impaired language function. Instead, as many empirical studies attest, such subjects have numerous other cognitive and perceptual deficits as well.[4] Thus, the cleanness of the dissociation between language and general cognitive abilities is undercut – as

2   Dissociations are often thought to be particularly compelling evidence of independent modules when there is a "double dissociation" of tasks; that is, when a pair of individuals is observed, one of whom can perform task A but not task B, and the other of whom can perform B but not A.
3   This issue is the subject of considerable discussion. For surveys, see Shallice (1988) and Glymour (2001).
4   See, for example, Vargha-Khadem and Passingham (1990), Anderson, Brown, and Tallal (1993), and Merzenich et al. (1996).

is EP's inference from that dissociation to the existence of distinct modules underlying those abilities.

We conclude this section by again acknowledging that there is a very "thin" notion of module such that, given certain other assumptions, a double dissociation entails modularity (in that sense). For example, if we simply assume that a distinct module underlies each distinct capacity (cf., the "one-to-one"' assumption discussed in section 16.2.2), with all normal subjects sharing the same architecture, and if we count a dissociation of capacities in any two people as indicating that those capacities are distinct (across all subjects), then we have an unproblematic inference from dissociability to distinctness of modules. However, this pretty inference is bought at the cost of a not-very-interesting notion of "module." As soon as we begin to invest modules with other, "thicker" properties – such as informational encapsulation or independent evolvability (sections 16.4.3 and 16.4.4) – the inference becomes far less compelling, as these properties do not necessarily apply to modules as distinguished by the dissociability criterion.

16.4.3 Modules and encapsulation

Modules are also often said to be *informationally encapsulated* in the sense that other psychological systems have access only to the information that is the output of the module; the processing that goes on within it is not accessible to, or influenced by, information or processes in other parts of the mind (Fodor, 1983). However, it is not clear how useful this feature is in picking out distinct cognitive mechanisms. First, informational encapsulation is often a relative, rather than an all-or-nothing, matter. It's plausible that some brain or psychological mechanisms may be completely informationally isolated from *some* other mechanisms (in the sense that there are no circumstances in which mechanism A is internally influenced by mechanism B). But many, if not virtually all, mechanisms are influenced in their internal operations by *some* other mechanisms – or at least this is true if we don't trivialize the notion of an "internal operation" (see below). Relatedly, informational encapsulation often seems to be task-relative. Whether mechanism or brain region A is influenced in its internal processing by information or processing in mechanism or region B may vary depending on the tasks that A and B are engaged in.

As an illustration of these points, consider the role of attention in many psychological processes. There is evidence that although low-level visual processing, such as occurs in the primary visual area V1, is often relatively encapsulated, it can be modified by higher-level processes involving visual attention, which occur in other neural regions (Luck and Hillyard, 2000). This kind of result raises familiar issues about EP's individuation of tasks: Are the processes in V1 performing *different* tasks or functions depending on whether attention is involved? It also undermines the usefulness of the encapsulation criterion for modularity. Does the fact that the processing of a visual stimulus by V1 is altered depending on whether subjects pay attention to that stimulus show that V1 is unencapsulated with respect to tasks that involve attention but encapsulated with respect to other tasks not involving attention? If so, are there *two* modules associated with V1, one operative when attention is involved and the other when it's not? Peter Carruthers (private communication) suggests that

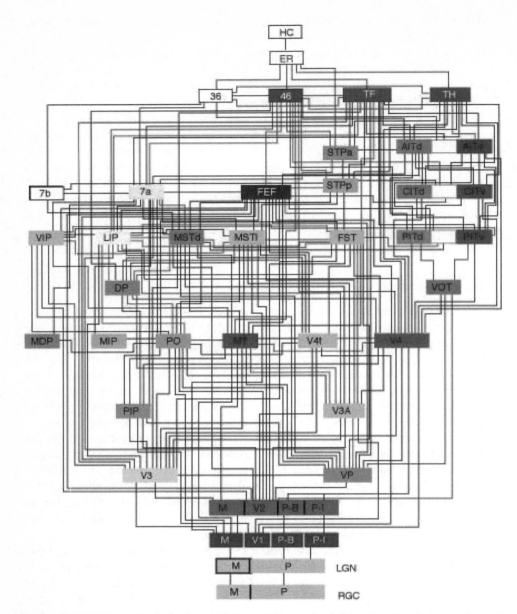

Figure 16.1 The hierarchy of visual areas in the macaque brain, showing 32 areas of visual cortex and their linkages, together with some subcortical and nonvisual connections (from Felleman and Van Essen, 1991; reproduced by courtesy of the authors).

if attention sometimes influences processing in V1, then attention should count as an *input* to V1, not an influence on its internal processing. Hence, he argues, processing in V1 *is* encapsulated after all. Our response is that the notion of informational encapsulation only makes sense if there is some basis for distinguishing between inputs to a module and processes that influence the internal operation of the module, for it's the latter kind of influence that claims of encapsulation deny. If *any* informational influence on the internal processing of a mechanism can be reconceptualized as an input to that mechanism, and if influence *via* input is consistent with encapsulation, then the notion of informational encapsulation is vacuous.

We have already argued (section 16.3.1) that dissociability is a dubious criterion for modularity. It's also of little help in the present connection, for contrary to what is often assumed, encapsulation bears no simple connection to dissociability. Consider the well-known diagram of the macaque visual system due to Felleman and Van Essen (1991), shown in figure 16.1.

We see here some 32 cortical "areas," as well as some subcortical areas. These areas are differentially sensitive to different sorts of stimuli and/or specialized for different sorts of processing (although typically not in an all-or-nothing fashion). Assuming that they are also susceptible of at least some degree of dissociation, we would appear to have (by the 1:1 and dissociability criteria) as many as 32 distinct modules depicted in this diagram. However, these cortical areas are also highly interconnected: Van Essen traced 197 linkages (equal to roughly 40 percent of the $(32 \times 31)/2 \approx 500$ linkages that are in principle possible). Most of these linkages appear to be reciprocal, which indicates that there is no simple sequential or hierarchical direction of information flow among the postulated modules; instead, each module talks to (and is talked to by) numerous others at numerous different stages of visual processing. This raises serious questions about how the dissociability criterion is supposed to line up with the encapsulation criterion for modularity and how the latter criterion is to be interpreted. Is the sort of interconnectedness found in figure 16.1 consistent with these areas being distinct modules? If so, it looks as though encapsulation (and perhaps modularity as well) come in degrees, rather than being all-or-nothing matters, in which case we need (i) some measure of degree of encapsulation and (ii) a theory about how this bears on judgments of modularity. If, on the other hand, modularists prefer to say that this degree of interconnectedness is *in*consistent with the idea that the areas form distinct modules, then it follows that distinctness of function and dissociability are not reliable criteria for individuating modules.

16.4.4 Modules and independent evolvability

Still another criterion for modularity is that modules are *independent targets of natural selection.* That is, selection must be able to change each of them independently of the others. This feature of modules is presupposed by EP's characteristic view of organisms as confronting a large collection of separate adaptive problems, each of which gets an independent evolved module by way of solution.

The independent evolvability criterion, however, is again problematic. For if a trait is to be an independent target of selection, it must be what Sterelny and Griffiths (1999) call a "mosaic" rather than a "connected" trait.[5] To use one of their examples, skin color is a plausible candidate for a mosaic trait because "it can evolve with relatively little change in the rest of the organism" (1999, p. 320). By contrast, having

---

5   Gilbert (2000, pp. 693ff) calls this a requirement of "modularity" – not to be confused with the cognitive modularity that concerns EP.

two lungs is a connected trait: you can't change this trait without changing a great deal else in the organism, because lung number is influenced by the genes and developmental mechanisms that govern the bilateral symmetry of the organism. Hence, natural selection can only influence lung number by influencing these genes and developmental mechanisms, and this in turn would affect many other phenotypical features. Since lung number is not an independent target of selection – since it is part of the "bigger package" (Sterelny and Griffiths, 1999, p. 320) – it would be a mistake to try to give an adaptive explanation of our having two lungs *simpliciter*. Instead, what needs to be explained is the evolution of bilateral symmetry.

Evolutionary psychologists assume that modules are independently evolvable; that is, that they are mosaic traits (such as skin color) rather than connected traits (such as having two lungs). However, there is evidence that many human cognitive abilities may be connected rather than mosaic traits. For example, Finlay and Darlington (1995) show that brain structures change in size across species in a highly coordinated and predictable manner: homologous structures enlarge at different but stable rates when compared to overall brain size. It is thought that these regularities reflect deeply entrenched developmental constraints on neurogenesis, suggesting that while natural selection can increase (or decrease) the size of the brain as a whole, the sizes of particular cortical regions cannot be changed independently, even in response to specific and pressing selective problems. Thus, natural selection may *not* be able to "fine tune" the cortical regions responsible for (say) cheater detection or the perception of numerosity independently of the (allegedly) distinct cognitive modules that underlie other cognitive capacities such as face recognition or language.

A further question concerns the relationship between the independent evolvability criterion and the other features of modules discussed above. We submit that there is no connection between these properties: independent evolvability does not entail, and is not entailed by, either independent disruptability or informational encapsulation. Indeed, it is a consequence of the arguments presented in this chapter that there is *no connection whatsoever* between *any* of the properties – independent disruptability, informational encapsulation, innateness, and independent evolvability – that are commonly ascribed to modules.

This is important, because it undermines a pattern of argument that is highly prevalent in the EP literature. Evolutionary psychologists provide evidence for the existence of a module in *some* sense (e.g., in the sense that performance on two tasks dissociates) and then go on to assume (without argument) that the module in question satisfies the other criteria discussed above as well. Thus, they slide from hypotheses of modularity in one of the various "thin" senses that we have discussed in this chapter to claims about the existence of modules in a much "thicker" and more substantive sense.

This slide is wholly unjustified. As an illustration, consider Cosmides and Tooby's (1992) well-known experiments on the Wason selection task and their subsequent hypothesis of a "cheater-detection" module. *Prima facie*, what their experimental results show is that people behave differently (and in some respects more reliably) when dealing with conditionals framed as rules governing social exchange than they do when dealing with conditionals with other contents. Even if we accept that these results establish differential performance on cheater-detection tasks *tout court* (and not just those that involve conditionals – itself a big jump), they do *not* constitute evidence for the existence of a distinct cheater-detection module in any more robust sense. That is, they do not even remotely suggest that cheater detection is subserved by an independently disruptable, informationally encapsulated psychological mechanism that has been subject to distinct selection pressures and that, as a consequence, is genetically specified, "innate," and so on. It is of course conceivable, although (we think) unlikely, that a cheater-detection module possessing all these features exists; our point is that Tooby and Cosmides' experiment provides no evidence that it does.

Our overall argument in section 16.4 can be put as follows. Interpreted one way (as involving a sufficiently "thin" conception of a module), EP's claims about modularity amount to little more than redescriptions of certain experimental results or evolutionary psychologists' functional speculations. So construed, claims about the existence of "mental modules" are uncontroversial – but also uninteresting. Modularity claims become more contentful and more interesting as the "thin" notion of a module is extended to include the other properties described above. However, not only is the evidence that would support such extensions rarely provided, but what we know about the brain makes it unlikely that there could be "thick" mental modules for the sorts of high-level cognitive capacities that are EP's main theoretical focus.

16.5 Conclusion

Much of the appeal of EP derives from the fact that it appears to provide a way of "biologizing" cognitive science, with evolutionary considerations supposedly providing powerful additional constraints on psychological theorizing. We think that this appearance is misleading. Evolutionary psychologists largely ignore the biological evidence that has the strongest scientific credentials and is most directly relevant to their claims about psychological mechanisms. This includes not only evidence from neurobiology, genetics, and developmental biology, but also any evidence from evolutionary biology, ethology, and population genetics that threatens to undermine their armchair adaptationism. Their methods assume, wrongly, that one can usefully speculate about biological and psychological functions in ignorance of information about structure, genes, and development. Their central theoretical concept – modularity – is left fundamentally unclear. And their picture of the mind as "massively modular" fails to do justice to many of its most important features, such as its capacity to engage in long-range planning and its remarkable cognitive and behavioral flexibility.

Bibliography

Allman, J. M. 1998: *Evolving Brains*. New York: Scientific American Library/W. H. Freeman.

Andersen, R. A., Batista, A. P., Snyder, L. H., Buneo, C. A., and Cohen, Y. E. 2000: Programming to look and reach in the posterior parietal cortex. In Gazzaniga, op. cit., pp. 515–24.

Anderson, K., Brown, C., and Tallal, P. 1993: Developmental language disorders: evidence for a basic processing deficit. *Current Opinion in Neurology and Neurosurgery*, 6, 98–106.

Barkow, J., Cosmides, L., and Tooby, J. (eds.) 1992: *The Adapted Mind: Evolutionary Psychology and the Generation of Culture*. New York: Oxford University Press.

Bates, E. 1994: Modularity, domain-specificity and the development of language. In D. C. Gajdusek, G. M. McKhann, and C. L. Bolis (eds.), *Evolution and Neurology of Language. Discussions in Neuroscience*, 10(1–2), 136–49.

——Ellman, J., Johnson, M., Karmiloff-Smith, A., Parisi, D., and Plunkett, K. 1998: Innateness and emergentism. In W. Bechtel and G. Graham (eds.), *A Companion to Cognitive Science*. Oxford: Blackwell, 590–601.

Büchel, C., Price, C., and Friston, K. 1998: A multimodal language region in the ventral visual pathway. *Nature, 394*, 274–7.

Cosmides, L. and Tooby, J. 1987: From evolution to behavior: evolutionary psychology as the missing link. In J. Dupré (ed.), *The Latest on the Best: Essays on Optimality and Evolution*. Cambridge, MA: The MIT Press, 277–307.

——and —— 1992: Cognitive adaptations for social exchange. In Barkow et al., op. cit., pp. 163–243.

——and —— 1995: From evolution to adaptations to behavior: toward an integrated evolutionary psychology. In R. Wong (ed.), *Biological Perspectives on Motivated and Cognitive Activities*. Norwood, NJ: Ablex, 11–74.

——and —— 1994: Origins of domain specificity: the evolution of functional organization. In L. Hirschfeld and S. Gelman (eds.), *Mapping the Mind: Domain Specificity in Cognition and Culture*. Cambridge: Cambridge University Press, 85–116.

Cummins, R. 1975: Functional analysis. *Journal of Philosophy*, 72, 741–64.

Davidson, E. H. 2001: *Genomic Regulatory Systems: Development and Evolution*. San Diego: Academic Press.

DeAngelis, G. C., Cumming, B. G., and Newsome, W. T. 2000: A new role for cortical area MT: the perception of stereoscopic depth. In Gazzaniga, op. cit., pp. 305–14.

Felleman, D. J. and Van Essen, D. C. 1991: *Distributed hierarchical processing in primate cerebral cortex. Cerebral Cortex*, 1, 1–47.

Finlay, B. L. and Darlington, R. B. 1995: Linked regularities in the development and evolution of mammalian brains. *Science, 268*, 1,578–84.

Fodor, J. A. 1983: *The Modularity of Mind: An Essay on Faculty Psychology*. Cambridge, MA: Bradford Books/The MIT Press.

Gazzaniga, M. S. (ed.), *The New Cognitive Neurosciences*, 2nd edn. Cambridge, MA: Bradford Books/The MIT Press.

Gilbert, S. F. 2000: *Developmental Biology*, 6th edn. Sunderland, MA: Sinauer Associates.

Glymour, C. 2001: *The Mind's Arrows: Bayes' Nets and Graphical Causal Models in Psychology*. Cambridge, MA: Bradford Books/The MIT Press.

Gould, S. J. and Lewontin, R. C. 1979: The spandrels of San Marco and the Panglossian paradigm: a critique of the adaptationist program. *Proceedings of the Royal Society of London*, B205, 581–98. Reprinted in Sober, E. (ed.) 1984: *Conceptual Issues in Evolutionary Biology*. Cambridge, MA: The MIT Press.

Griffiths, P. E. forthcoming: Evolutionary psychology: history and current status. In S. Sarkar and J. Pfeiffer (eds.), *The Philosophy of Science: An Encyclopedia*. New York: Routledge. Preprint available at http://philsci-archive.pitt.edu/archive/00000393

Indefrey, P. and Levelt, W. J. M. 2000: The neural correlates of language production. In Gazzaniga, op. cit., pp. 845–65.

Johnson, M. 1997: *Developmental Cognitive Neuroscience*. Oxford: Blackwell.

LeDoux, J. 1996: *The Emotional Brain: The Mysterious Underpinnings of Emotional Life*. New York: Simon and Schuster.

Lewontin, R. C. 1990: The evolution of cognition: questions we will never answer. In D.

Scarborough and S. Sternberg (eds.), *An Invitation to Cognitive Science*, 2nd edn, Vol. 4: *Methods Models and Conceptual Issues*. Cambridge, MA: The MIT Press, 107–32.

Luck, S. J. and Hillyard, S. A. 2000: The operation of selective attention and multiple stages of processing: evidence from human and monkey physiology. In Gazzaniga, op. cit., pp. 687–700.

Marr, D. 1982: *Vision*. New York: W. H Freeman.

Maynard Smith, J., Burian, R., Kauffman, S., Alberch, P., Campbell, J., Goodwin, B., Lande, R., Raup, D., and Wolpert, L. 1985: Developmental constraints and evolution. *Quarterly Review of Biology*, 60, 285–7.

McCarthy, G. 2000: Physiological studies of face processing in humans. In Gazzaniga, op. cit., pp. 393–409.

Merzenich, M., Jenkins, W., Johnston, P. S., Schreiner, C., Miller, S. L., and Tallal, P. 1996: Temporal processing deficits of language-learning impaired children ameliorated by training. *Science*, 271, 77–81.

Nishimura, H., Hashikawa, K., Doi, K., Iwaki, T., Watanabe, Y., Kusuoka, H., Nishimura, T., and Kubo, T. 1999: Sign language "heard" in the auditory cortex. *Nature*, 397, 116.

Oyama, S. 1985: *The Ontogeny of Information*. Cambridge: Cambridge University Press.

Pinker, S. 1994: *The Language Instinct*. New York: HarperCollins.

—— 1999: *How the Mind Works*. New York: Norton.

Posner, M. I. and Raichle, M. E. 1994/1997: *Images of Mind*. New York: Scientific American Library/HPHLP.

Potts, R. 1996: *Humanity's Descent: The Consequences of Ecological Instability*. New York: Avon.

—— 1998: Variability selection in hominid evolution. *Evolutionary Anthropology*, 7(3), 81–96.

Samuels, R. 1998: Evolutionary psychology and the massive modularity hypothesis. *The British Journal for the Philosophy of Science*, 49, 575–602.

—— Stich, S. P., and Tremoulet, P. D. 1999: Rethinking rationality: from bleak implications to Darwinian modules. In E. Lepore and Z. Pylyshyn (eds.), *What is Cognitive Science?* Malden, MA: Blackwell, 74–120.

Shallice, T. 1988: *From Neuropsychology to Mental Structure*. Cambridge: Cambridge University Press.

Shamma, S. A. 2000: Physiological basis of timbre perception. In Gazzaniga, op. cit., pp. 411–23.

Shouse, B. 2002: Human gene count on the rise. *Science*, 295, 1457.

Sober, E. 1994: The adaptive advantage of learning and *a priori* prejudice. In *From a Biological Point of View: Essays in Evolutionary Philosophy*. Cambridge: Cambridge University Press, 50–70.

Sterelny, K. 2003: *Thought in a Hostile World*. Oxford: Blackwell.

—— and Griffiths, P. 1999: *Sex and Death*. Chicago: The University of Chicago Press.

Stolz, K. C. and Griffiths, P. E. 2002: Dancing in the dark: evolutionary psychology and the argument from design. In S. Scher and M. Rauscher (eds.), *Evolutionary Psychology: Alternative Approaches*. Dordrecht: Kluwer, 135–60.

Thornhill, R. and Thornhill, N. 1987: Human rape: the strengths of the evolutionary perspective. In C. Crawford, M. Smith, and D. Krebs (eds.), *Sociobiology and Psychology*. Hillsdale, NJ: Lawrence Erlbaum Associates, 269–92.

—— and —— 1992: The evolutionary psychology of men's coercive sexuality. *Behavioral and Brain Sciences*, 15, 363–421.

Tooby, J. and Cosmides, L. 1990: On the universality of human nature and the uniqueness of the individual: the role of genetics and adaptation. *Journal of Personality*, 58, 17–67.

—— and —— 1992: The psychological foundations of culture. In Barkow et al., op. cit., pp. 19–136.

—— and —— 1997: Evolutionary psychology: a primer. Available online at: *http://www.psych.ucsb.edu/research/cep/primer.html*

Vargha-Khadem, F. and Passingham, R. 1990: Speech and language defects. *Nature*, 346, 226.

Venter, J. C. et al. 2001: The sequence of the human genome. *Science*, 291, 1,304–51.

Wandell, B. A. 2000: Computational neuroimaging: color representations and processing. In Gazzaniga, op. cit., pp. 291–303.

Woodward, J. and Cowie, F. forthcoming: *Naturalizing Human Nature*. New York: Oxford University Press.

Wright, L. 1973: Functions. *Philosophical Review*, 82, 139–68.

Wright, R. 1994: *The Moral Animal: Evolutionary Psychology and Everyday Life*. New York: Pantheon.

Further reading

Barrett, L., Dunbar, R., and Lycett, J. 2002: *Human Evolutionary Psychology*. Princeton, NJ: Princeton University Press.

Fodor, J. A. 2000: *The Mind Doesn't Work that Way: The Scope and Limits of Computational Psychology*. Cambridge, MA: Bradford Books/The MIT Press.

Karniloff-Smith, A. 1995: *Beyond Modularity: A Developmental Perspective on Cognitive Science*. Cambridge, MA: Bradford Books/The MIT Press.

Index

Index